BOUNDAR

BOUNDARY LAYERS

A.D. Young
FRS, FEng

Emeritus Professor of Aeronautical Engineering
Queen Mary College, University of London

BSP PROFESSIONAL BOOKS
OXFORD LONDON EDINBURGH
BOSTON MELBOURNE

First published 1989

British Library
Cataloguing in Publication Data

Young, A.D.
Boundary layers.
1. Fluids. Boundary layer flow
I. Title
532'.052

ISBN 0-632-02122-5

BSP Professional Books
A division of Blackwell Scientific Publications Ltd
Editorial Offices:
Osney Mead, Oxford OX2 0EL
(Orders: Tel. 0865 240201)
8 John Street, London WC1N 2ES
23 Ainslie Place, Edinburgh EH3 6AJ
3 Cambridge Center, Suite 208, Cambridge, MA 02142, USA
107 Barry Street, Carlton, Victoria 3053, Australia

Set by Setrite Typesetters Limited

Printed and bound in Great Britain by Mackays of Chatham PLC, Chatham, Kent

Contents

Preface

It is of interest to note that Prandtl's historic paper in 1904, presenting for the first time the concept and theory of boundary layers, appeared within a year of the momentous first flight of a powered aircraft by the Wright brothers. If aviation as we know it can rightly be said to have started with that flight, its dramatic progress since owes much to the understanding and scientific intuition revealed by Prandtl's paper. The latter not only provided a basic ingredient for the subsequent rapidly developing science of aerodynamics and its applications but it also became a great stimulus to related developments in other branches of engineering involving the flow of fluids, e.g. turbo-machinery design and hydraulics.

Boundary layers have therefore featured for many years in university courses in aeronautical, mechanical, naval and civil engineering, often as part of the general field of fluid mechanics; and departments of applied mathematics and meteorology have also found the subject relevant to their interests. Nevertheless, there are surprisingly few text books devoted to boundary layers. In recent years there have been important developments in the subject, mainly as a result of the rapidly growing use of increasingly powerful computers which have made it possible to tackle complex problems and open up new areas of study. Competitive pressures in aeronautics, as well as in other branches of fluids engineering, have led to a growing demand for improvements in performance and reductions in fuel consumption. Interest has therefore focussed on the accuracy of pressure distribution and drag predictions as well as on means for drag reduction, such as the development of extensive regions of laminar boundary layers; and methods of boundary layer control and manipulation are being vigorously investigated.

These considerations have pointed to the need for a text book devoted to boundary layers and written with engineering undergraduates approaching their final year, as well as postgraduates and young workers in industry and research establishments very much in mind. Such readers imply that the book must be of modest size and price to be within their financial reach. Nevertheless, the scope of such a book should not only cover the traditional material but also the essential features of the more recent developments as well as pointers to future promising growth points. The presentation of the physics of the subject must be paramount and must not be masked by its considerable mathematical content. I have written this book with such a specification as guide.

However, it will be apparent that some compromise must be made between the limitations on the size of the book and the wide ranging coverage desired. Thus, whilst the important use that is being made of computers in helping us to solve the governing equations is fully acknowledged, I have made no attempt to discuss details of the computing techniques used or to present the associated programs and codes. Their inclusion would have doubled the size of the book, and in any case they are to be found in existing specialist text books quoted in the references. There can be very few engineering courses today that do not include the relevant topics of computational analysis. For similar reasons there is little detailed coverage of the experimental techniques that are used in laboratories or flight to explore boundary layers. The subjects of jets, wakes and the flow in pipes and ducts have much in common with boundary layers and some reference is made to them but with little detail. It is felt that an interested reader will find the extension of the material presented to these topics fairly straightforward and again the quoted references should be helpful. On the other hand, the characteristics of boundary layers in compressible flow as well as in incompressible flow have been treated in some detail, since flight at high speeds where compressibility effects cannot be ignored is now commonplace.

The importance of the special features of three dimensional boundary layers is well recognised in the book and essential points are discussed, but the bulk of the material presented refers to two dimensional boundary layers. This is because they are a necessary basis for the subject as a whole and, as is made clear, the subject of three dimensional turbulent boundary layers is not at present as well advanced as its importance would justify and not yet sufficiently developed to provide ready material for a book of this kind.

The book is written on the assumption that the reader has already had an introductory course on the elements of fluid dynamics and compressible flow such as are common in first and second year engineering undergraduate studies. However, on first reading the student might find it advisable to concentrate on the parts of the book dealing with incompressible flow.

I have considered whether some additional detailed comparisons between experimental results and results computed by the various prediction methods for specimen cases of turbulent flow would help the reader more than the general comments that I have made in assessing the relative merits and reliability of the methods. However, it will be clear that many of the methods are still in a state of change and improvement and hence detailed comparisons using 'this year's model' may lose their significance in a relatively short time.

Much of the material presented is based on my notes of lecture courses I have given to undergraduates and postgraduates, mainly at Queen Mary College, University of London, and latterly at the Middle East Technical University, Ankara, Turkey. I have been fortunate in being a member of the Fluid Dynamics Panel of the Advisory Group for Aerospace Research and Development (AGARD) for a number of years, and my involvement in several Symposia sponsored by the Panel has given me much valuable background material on current developments as well as an awareness of future trends which have helped me in writing the book.

The book follows a straightforward and logical scheme of development. The first chapter introduces in non-mathematical terms the basic physical concepts whilst the second presents the theoretical framework from which the subject is developed. The third and fourth chapters are devoted to laminar boundary layers and describe exact and approximate methods of solution of the governing equations for representative cases of practical importance. Chapter 5 deals with the phenomenon of transition from laminar to turbulent flow, a long-standing, fundamental subject of renewed interest to industry, and emphasis is put on the physics of the transition process and its control. Chapter 6 presents current ideas and empiricisms for describing the turbulent boundary layer, whilst the governing equations in time-averaged form are developed in Chapter 7. Chapter 8 is devoted to solutions of forms of the latter when integrated over the boundary layer (integral methods) and their application to the prediction of drag. Chapter 9 deals with turbulence models and their application to the solution of the time-averaged equations (differential methods). Finally, in Chapter 10 a brief review is offered of some complex topics at the frontiers of present industrial and research activity. These include the special features of three dimensional boundary layers, inverse methods of solution, coherent turbulence structures, and the reduction of skin friction in the turbulent boundary layer by modification of these structures e.g. riblets and large eddy break up devices (LEBUs).

In summary, no claim is made that the book is fully comprehensive, and for the more advanced aspects of the subject it provides no more than an introduction. However, I hope that the readers for whom it is written will find it meets their present needs and that it will provide a useful base from which they can explore the more complex developments of the subject that they may wish to master in the future.

I would like to express my appreciation and thanks to Steve Mauldin, of COMIND, Cambridge, for preparing the finished versions of the figures for me at short notice and for his patience and good humour in dealing with the last minute changes that I requested. My wife, Rena, also deserves my thanks and much more for her forbearance during her experience of being a 'book widow' over a period which lasted much longer than I had optimistically forecast.

Alec Young

Acknowledgements

I wish to thank the Advisory Group for Aerospace Research and Development (AGARD) and the authors cited for permission to derive Fig. 5.4 from AG 134, 1969, (H.T. Obremski, M.V. Morkovin and M.T. Landahl, reference 5.8); Fig. 5.5 from Paper 6 (W.G. Saric and A.W. Nayfeh, reference 5.13), and Fig. 5.10 from Paper 1 (L.M. Mack, reference 5.46), of AG CP 224, 1977. Likewise my thanks are due to McGraw–Hill Publishing Company for permission to reproduce Figs 5.3 and 5.7 from 'Boundary Layer Theory' by H. Schlichting, 7th Ed., 1979, reference 2.2. Agreement was kindly granted by P.L. Klebanoff for the use of Fig. 5.12 from reference 5.37 and Figs 6.1, 6.2 and 6.4 from reference 6.1; and agreement was also kindly given by A.R. Wazzan for the reproduction of Fig. 5.2 from reference 5.51 (this figure also appeared in reference 5.8.)

Acknowledgement is gratefully made to the Royal Aeronautical Society and L. Gaudet for permission to derive Figs 6.10 and 8.4 from reference 6.28 and Figs 6.12 and 6.13 from reference 6.34. Thanks are due to the American Institute of Aeronautics and Astronautics for permission to reproduce Fig. 6.18 from R.M. Grabow and C.O. White, AIAA Journal, p. 605ff, 1975, reference 6.46.

It is hoped no material source has gone unacknowledged in the above or in the References. If an omission has occurred I assure those concerned that it was inadvertent and I hereby extend my thanks to them.

Finally, I thank all who have in various ways influenced my ideas in the preparation of this book, in particular, Professor G.M. Lilley who read the book in draft and whose valuable suggestions for improvements were eagerly incorporated wherever possible.

Abbreviations

The following is a list of the main abbreviations used and what they stand for to help the reader who may be unfamiliar with them.

AIAA	American Institute of Aeronautics and Astronautics.
AGARD	Advisory Group for Aerospace Research and Development (NATO).
ARC	Aeronautical Research Council (UK).
ARC R & M	Reports and Memoranda of the ARC.
ARC CP	Current Papers of the ARC.
BAe	British Aerospace.
CUP	Cambridge University Press.
DFVLR	Deutsche Forschungs- und Versuchsanstalt für Luft-und Raumfahrt (German Aerospace Research Institute).
DGLR	Deutsche Gesellschaft für Luft-und Raumfahrt (German Aerospace Society).
ETH	Eidgenössische Technische Hochschule (Federal Institute of Technology) (Switzerland).
JAS	Journal of the Aeronautical Sciences.
Jb	Jahrbuch (Yearbook).
JFM	Journal of Fluid Dynamics.
NACA	The National Advisory Committee for Aeronautics (USA).
NASA	National Aeronautics and Space Administration (USA).
NLR	Nationaal Lucht-en Ruimtevaartlaboratorium (National Aerospace Laboratory) (The Netherlands).
ONERA	Office National d'Etudes et de Recherches Aerospatiales (National Institute for Aerospace Studies and Research) (France).
OUP	Oxford University Press.
RAE	Royal Aircraft (later Aerospace) Establishment (UK).
VKI	Von Karman Institute for Fluid Dynamics (Belgium).
ZAMM	Zeitschrift für angewandte Mathematik und Mechanik (Journal for Applied Mathematics and Mechanics) (Germany).
ZFW	Zeitschrift für Flugwissenschaften und Weltraumforschung (Journal for Aeronautics and Space Research) (Germany).

Chapter 1

Introduction and Some Basic Physical Concepts

1.1 Introduction

At the turn of this century there were two remarkably different and seemingly irreconcilable fields of study concerned with the mechanics of fluids in motion. On the one hand, there was classical hydrodynamics – an elegant, mathematical development of the theory of an inviscid fluid, usually irrotational and incompressible, which slipped freely over containing surfaces and the surfaces of immersed bodies. Some of the world's best scientists in the nineteenth century had contributed to it and it had close parallels with the then rapidly developing field theories of electricity and magnetism. For many problems where the focus was not on the regions of flow close to solid boundaries, its predictions seemed to have relevance to real flows. However, it had some major shortcomings. It predicted no frictional resistance or drag for immersed bodies, contrary to everyday experience, and it could only predict lift on a body if a circulation about it were postulated, but the theory could say nothing about how such circulation could arise. On the other hand, there was hydraulics – a largely empirical subject mainly based on formulae and data sheets developed in the light of experiments and experience by civil and mechanical engineers. It was particularly valuable for dealing with problems arising in the design of fluid machinery, e.g. the flow and losses in pipes, bends and pumps, and it had applications to the design of ships. However, there was little theory to provide a basis whereby the formulae could be justified and confidently generalised.

It was the genius of Prandtl that provided a bridge that linked these two fields of study and so established a logical basis for the subsequent rapid development of modern fluid dynamics, which includes external and internal aerodynamics, gas dynamics and hydrodynamics. That bridge was Prandtl's *boundary layer theory* which he first presented in 1904.[1.1]

The theory rested on certain basic observations. They were:

(1) However small the viscosity of a fluid in motion may be (and for air it is very small) it cannot be ignored. The limit of the flow close to an immersed body or bounding surfaces as the viscosity tends to zero is not the same as for an inviscid fluid.

At the surface of the body the fluid is at rest relative to the body, i.e. there is no slip between the fluid and the surface. (This statement needs qualification for a gas at very low density, where the mean free path of the molecules is not small relative to the body size, then some slip can occur of the mean flow past the body surface. The fluid can then no longer be regarded as a continuum – a basic assumption of boundary layer theory.) Hence near the body the relative fluid velocity increases from zero at the surface to something of the order of the main stream velocity with distance normal to the surface. Therefore, near the body there are significantly large velocity gradients or rates of strain.

(2) Shear stresses due to viscosity are directly related to the rates of strain and can be large where the rates of strain are large. In particular, at the surface these viscous stresses, called frictional stresses there, contribute to the overall drag of the body.

(3) A controlling parameter for flow phenomena where inertia and viscous forces are important is a measure of their ratio, the *Reynolds number* $R = UL/\nu$, where U and L are a typical fluid velocity and body dimension, respectively, and ν is the kinematic viscosity of the fluid $= \mu/\rho$, where μ is the coefficient of viscosity and ρ is the density of the fluid. For Reynolds numbers of interest for most practical applications such as for aircraft, ships, land vehicles, etc., i.e. greater than about 10^4, the regions adjacent a body where the rates of strain are large and the viscous stresses are significant, are thin and become thinner with increase in Reynolds number. Such regions are then graphically referred to as *boundary layers*.

(4) The full equations of motion of a viscous fluid, the so-called *Navier–Stokes equations*, are non-linear equations of formidable difficulty. However, Prandtl observed that the relative thinness of boundary layers at large Reynolds numbers of practical interest permits some welcome simplifications which lead to more readily solvable equations – *the boundary layer equations*. Further, he noted that in the flow outside a boundary layer the viscous terms can generally be neglected and classical inviscid flow theory there applies.

Boundary layers trail off the rear of a body to form a wake (or wakes) downstream of it, and such wakes must likewise be regarded as regions in which the shear stresses can be large and viscosity cannot be ignored. For streamlined shapes the wakes are thin so that again the simplifications of Prandtl's boundary layer theory apply there. They similarly apply to jet flows and to the flow in pipes and ducts.

The problem remains of establishing the boundary conditions at the outer edge of the boundary layer so that the solution of the boundary layer equations merges with that of the external 'inviscid' flow. For many problems concerned with immersed bodies the thinness of the boundary layer leads to the result that these conditions can be equated with acceptable accuracy to those of inviscid flow past the body, the boundary layer then acts in effect like a layer of roller bearings to permit the external flow to 'slip' past the body. This is generally the case when the boundary layer remains close to the body (or *attached*) back to the rearmost point before forming the wake. However, there are cases, usually involving what we shall call *separation* of the boundary layer from the surface, where we can no longer regard the layer as thin and the interaction between it and the external flow is important and must be taken into account in formulating the boundary conditions at the edge of the boundary layer.

The vorticity at a point in a fluid can be defined as twice the instantaneous rate of spin of a small element of fluid centred at the point. It is by definition zero in irrotational flow and its measure is determined by the lateral velocity gradients. Since the boundary layers are regions of large lateral velocity gradients they are also regions in which vorticity is concentrated. Indeed, we can say that in a real isotropic fluid in motion, vorticity is generated at the surfaces of immersed bodies and any bounding surfaces because of the viscous stresses there, and the vorticity is then diffused outwards from the surface by the shear stresses whilst it is convected downstream to form the boundary layer. This process is closely parallel to the diffusion of heat from the surface when its temperature differs from that of the fluid. The thinness of the boundary layer at sufficiently high Reynolds numbers again follows from the fact that the rate of diffusion is then small compared with the convection velocity. Since the circulation round a circuit is equal to the integrated vorticity that threads it (Stokes theorem), we can infer that the source of circulation and vorticity in a real fluid is its viscosity.

In the above we have used some basic terms and concepts with which some readers may not yet be familiar. We shall therefore discuss these concepts and some others in a little more detail in the remainder of this chapter to provide a background for the material in the subsequent chapters. As already noted in the Preface the emphasis in this book will be on unbounded flows past immersed bodies particularly aeronautical applications.

1.2 Stress components

Consider a plane element of surface of area δA, either imaginary in the fluid or on a body immersed in it, then the fluid on one side acts on the surface element with a force F, say, with components N normal to the element and T tangential to it (see Fig. 1.1). Then we refer to the limiting

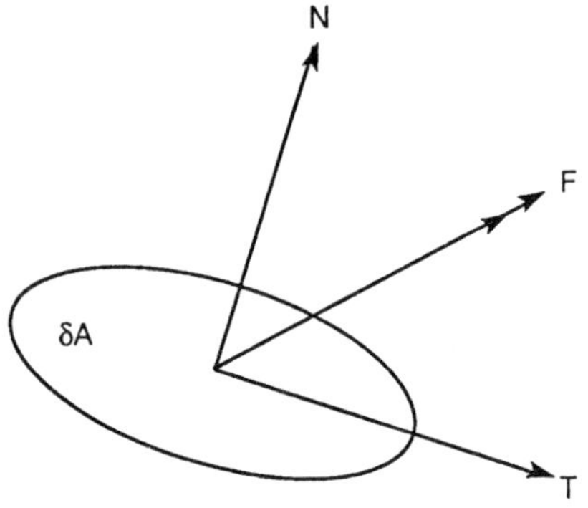

Fig. 1.1

values of $N/\delta A$ and $T/\delta A$ as δA tends to zero as the normal and shear stresses, respectively. These values are dependent on the orientation of the surface element which is kept constant in the process of reducing its area to zero.

It is a fundamental property of a fluid at rest that it cannot sustain a shear stress and remain at rest, but when it is in motion it develops shear stresses related to the rates at which fluid elements change shape, i.e. the rates of strain as determined by the rates of elongation of linear elements and the rates of change of angles between linear elements. The physical property of the fluid which results in these stresses is its *viscosity*. For an inviscid fluid, therefore, $T = 0$ whatever the motion.

If we now consider cartesian axes x, y, z and origin O and we take the area element in the yz plane at O, then N is in the direction of the x axis whilst T can be resolved into two components T_y and T_z along the y and z directions. Thus, we can define three stress components:

$$\tau_{xx} = \text{Lt}\ N/\delta A, \qquad \tau_{xy} = \text{Lt}\ T_y/\delta A, \qquad \tau_{xz} = \text{Lt}\ T_z/\delta A \qquad \text{as } \delta A \to 0$$

We have here adopted the convention that the first suffix in a stress component denotes the positive direction of the normal to the plane element considered, whilst the second suffix denotes the direction along which the component is taken. Similarly, if the area element had been taken normal to the y and z axes in turn a further six stress components would be defined, viz.

$$\tau_{yx}, \tau_{yy}, \tau_{yz} \qquad \text{and} \qquad \tau_{zx}, \tau_{zy}, \tau_{zz}$$

The total array of nine components forms a tensor, the stress tensor, of which the general term can be written $\tau_{\alpha\beta}$. It can easily be shown that the tensor is symmetric, i.e. $\tau_{\alpha\beta} = \tau_{\beta\alpha}$. [The reader is encouraged to do this by taking an elementary cube of sides parallel to the x, y, z axes and equating the moment of the stresses acting on its faces about one of the axes to the corresponding rate of change of angular momentum and then letting the cube shrink to zero.]

The components τ_{xx}, τ_{yy}, τ_{zz} are, in accordance with the above discussion, normal stress components, whilst τ_{yz}, τ_{zx} and τ_{xy} are shear stress components. It is also readily shown that in an inviscid fluid the three normal stress components are equal and invariant of the orientation of the axes, i.e. they are a function only of the position of O. [This can be shown by considering the balance of the forces acting on the fluid and the associated rates of change of momenta in a tetrahedron defined by small elements of the axes and then letting the elements shrink to zero.] If we write this normal stress as $-p$ then we refer to p as the *pressure* at O.

In contrast, in a viscous fluid in motion the normal stress on a surface element at a point O depends on the orientation of the element as well as the position of O. However, in that case it can be shown that the mean of the three cartesian normal stress components is invariant of the orientation of the axes and a function only of the position of O. We then define the pressure at O as

$$p = -(\tau_{xx} + \tau_{yy} + \tau_{zz})/3 = -\tau_{\alpha\alpha}/3 \tag{1.1}$$

We here adopt the usual convention of tensor notation that the repetition of a suffix implies the summation over the three values the suffix can take.

1.3 Viscosity

Consider first a fluid in steady two dimensional shearing motion past a smooth flat plate (see Fig. 1.2). Take the x axis parallel to the plate in the direction of motion and the z axis normal to it and assume the velocity u to be a function of z only with the normal velocity component $w = 0$. Then we know that a viscous shear stress τ_{xz} exists which for a large class of fluids is simply proportional to the velocity gradient $\mathrm{d}u/\mathrm{d}z$. We write

$$\tau_{xz} = \mu \mathrm{d}u/\mathrm{d}z \tag{1.2}$$

where μ is the *coefficient of viscosity* of the fluid. Such a linear relation between the viscous shear stress and the corresponding velocity gradient (or more generally rate of strain) is characteristic of so-called Newtonian fluids, of which air and water are examples. We shall only consider such fluids. A linear relation can be readily shown to follow from the simple kinetic theory of gases – the random motion of the molecules combined with a mean velocity gradient results in a rate of transfer of momentum from the faster moving layers of fluid to the slower and this momentum transfer is manifest as a stress proportional to the velocity gradient and tending to reduce it.

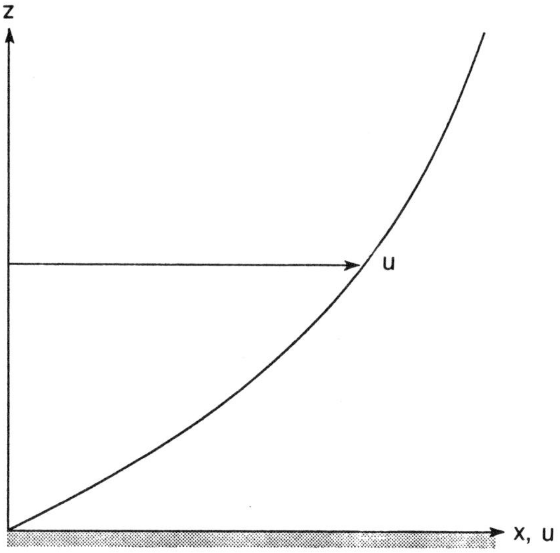

Fig. 1.2 Steady simple shearing motion, $\tau_{xz} = \mu \partial u / \partial z$.

The ratio μ/ρ occurs frequently in fluid dynamic problems where both viscous and inertia forces are present. It is called the *kinematic viscosity* and is denoted by ν. For gases μ increases with increase of temperature, but for liquids it decreases with temperature, whilst the effects of pressure changes are small. Some representative values in SI units for air and water at atmospheric pressure (1.013 N/m^2) are shown in Table 1.1. Data for other materials and temperatures can be found in reference 1.2.

For gases Sutherland's formula is generally accepted:

$$\mu = \text{const. } T^{1.5}/(T + C)$$

where T is the temperature in degrees K, and for air $C = 114$. This can be approximated over the range $100 < T < 300$ by $\mu = \text{const. } T^{8/9}$. For air at

Table 1.1

Temp °C	Air		Water	
	$\mu \times 10^5$ kg/ms	$\nu \times 10^5$ m^2/s	$\mu \times 10^4$ kg/ms	$\nu \times 10^6$ m^2/s
0	1.71	1.32	17.87	1.792
10	1.76	1.41	13.04	1.308
20	1.81	1.50	10.05	1.007
100	2.18	2.30	2.84	0.295

standard temperature and pressure (288.2 K and 1.0132×10^5 N/m^2) $\mu = 1.789 \times 10^{-5}$, $\nu = 1.461 \times 10^{-5}$, and $\rho = 1.225$ kg/m^3.

1.4 Rates of strain

Denote the velocity components at the point (x, y, z) by (u, v, w) and at the neighbouring point $(x + \delta x, y + \delta y, z + \delta z)$ by $(u + \delta u, v + \delta v, w + \delta w)$. Then for small increments we can write:

$$\left.\begin{aligned} \delta u &= \frac{\partial u}{\partial x}\delta x + \frac{\partial u}{\partial y}\delta y + \frac{\partial u}{\partial z}\delta z \\ \delta v &= \frac{\partial v}{\partial x}\delta x + \frac{\partial v}{\partial y}\delta y + \frac{\partial v}{\partial z}\delta z \\ \delta w &= \frac{\partial w}{\partial x}\delta x + \frac{\partial w}{\partial y}\delta y + \frac{\partial w}{\partial z}\delta z \end{aligned}\right\} \quad (1.3)$$

which can be rearranged as:

$$\left.\begin{aligned} \delta u &= a\,\delta x + (h/2)\delta y + (g/2)\delta z + \tfrac{1}{2}(\eta\delta z - \zeta\delta y) \\ \delta v &= (h/2)\delta x + b\,\delta y + (f/2)\delta z + \tfrac{1}{2}(\zeta\delta x - \xi\delta z) \\ \delta w &= (g/2)\delta x + (f/2)\delta y + c\,\delta z + \tfrac{1}{2}(\xi\delta y - \eta\delta x) \end{aligned}\right\} \quad (1.4)$$

where $a = \dfrac{\partial u}{\partial x}$, $b = \dfrac{\partial v}{\partial y}$, $c = \dfrac{\partial w}{\partial z}$

$$f = \frac{\partial w}{\partial y} + \frac{\partial v}{\partial z}, \quad g = \frac{\partial u}{\partial z} + \frac{\partial w}{\partial x}, \quad h = \frac{\partial v}{\partial x} + \frac{\partial u}{\partial y}$$

and (ξ, η, ζ) are the components of the vorticity vector $\boldsymbol{\omega} = \text{curl}\mathbf{U}$, i.e.

$$\xi = \frac{\partial w}{\partial y} - \frac{\partial v}{\partial z}, \quad \eta = \frac{\partial u}{\partial z} - \frac{\partial w}{\partial x}, \quad \zeta = \frac{\partial v}{\partial x} - \frac{\partial u}{\partial y}$$

Now consider the family of quadric surfaces defined by

$$F(X, Y, Z) = aX^2 + bY^2 + cZ^2 + fYZ + gZX + hXY = \text{const.} \quad (1.5)$$

where (X, Y, Z) are cartesian coordinates referred to the point (x, y, z) as origin and with the same axes directions. It can readily be shown that such a system of quadrics has a common set of principal axes. Then at any point (X_0, Y_0, Z_0) the normal to the quadric surface on which it lies has direction cosines proportional to

$$\begin{aligned} \tfrac{1}{2}(\partial F/\partial X)_0 &= aX_0 + (h/2)Y_0 + (g/2)Z_0 \\ \tfrac{1}{2}(\partial F/\partial Y)_0 &= (h/2)X_0 + bY_0 + (f/2)Z_0 \\ \tfrac{1}{2}(\partial F/\partial Z)_0 &= (g/2)X_0 + (f/2)Y_0 + cZ_0 \end{aligned} \quad (1.6)$$

We see that in equation (1.4) the first three terms in each of the expressions for δu, δv and δw represent a motion in which the velocity at a point is in the direction of the normal to the particular quadric of the system defined by equation (1.5) on which the point lies.

If we now choose as axes the principal axes of the system of quadrics and we distinguish them by a prime then the equation of the system takes the form

$$a'X'^2 + b'Y'^2 + c'Z'^2 = \text{const.}$$

so that in each of equations (1.4) the sum of the first three terms transform to

$$a'X',\ b'Y',\ c'Z' \tag{1.7}$$

respectively. Hence, as far as these terms are concerned all lines parallel to the X' axis are elongated at a rate a' times their length, and likewise all lines parallel to the Y' and Z' axes are elongated at rates b' and c' times their lengths, respectively. Such a motion is called a pure rate of strain and the axes are called the principal axes of strain. We see that in such a motion over a short period the axes remain orthogonal but the angles between lines not parallel to the principal axes change continuously.

If we now examine the remaining terms in equation (1.4) we can readily confirm that they represent a pure rate of rotation with angular velocity components about the (x, y, z) axes $(\xi/2, \eta/2, \zeta/2)$. This is consistent with the earlier statement that the vorticity vector is twice the vector representing the angular velocity of a small element at the point O.

Hence, in general the relative motion in the neighbourhood of a point can be resolved into a combination of a pure rate of strain and a pure rate of rotation. Figure 1.3 illustrates this in the case of a simple two dimensional shearing motion. (A) shows an initially square shaped element which after a small time interval distorts into the diamond shape shown as (C). This change is seen to be made up of a pure strain shown in (B) plus a rotation to bring it to (C).

We see that in general the strain rate at a point is completely defined by the quantities a, b, c, f, g, h. We therefore talk of a rate of strain tensor $e_{\alpha\beta}$ where

$$\left.\begin{aligned}
e_{xx} &= 2a = 2\frac{\partial u}{\partial x}, \qquad e_{yy} = 2b = 2\frac{\partial v}{\partial y}\\
e_{zz} &= 2c = 2\frac{\partial w}{\partial z}\\
e_{yz} &= f = \frac{\partial w}{\partial y} + \frac{\partial v}{\partial z}, \qquad e_{zx} = g = \frac{\partial u}{\partial z} + \frac{\partial w}{\partial x}\\
e_{xy} &= h = \frac{\partial v}{\partial x} + \frac{\partial u}{\partial y}\\
\text{or}\quad e_{\alpha\beta} &= \frac{\partial u_\alpha}{\partial x_\beta} + \frac{\partial u_\beta}{\partial x_\alpha}
\end{aligned}\right\} \tag{1.8}$$

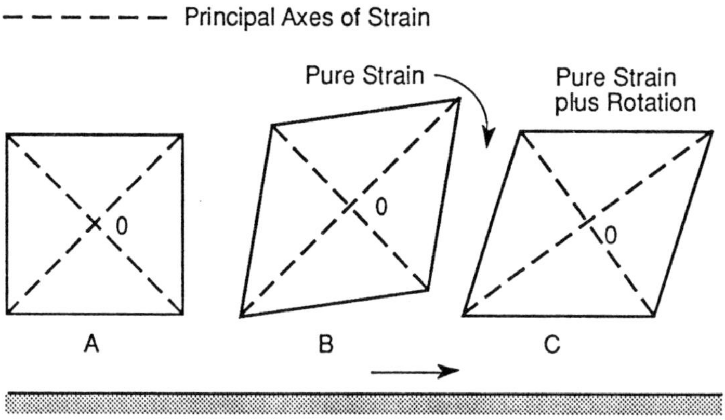

Fig. 1.3

Like the stress tensor the rate of strain tensor is symmetric, i.e. $e_{\alpha\beta} = e_{\beta\alpha}$. The sum

$$e_{\alpha\alpha} = e_{xx} + e_{yy} + e_{zz} = 2\left(\frac{\partial u}{\partial x} + \frac{\partial v}{\partial y} + \frac{\partial w}{\partial z}\right) = 2 \operatorname{div} \mathbf{U} \text{ or } 2\Delta \qquad (1.9)$$

where **U** is the velocity vector (u, v, w). Δ is sometimes called the *dilatation* and is a measure of the rate at which the volume of an element of fluid is changing. Therefore $e_{\alpha\alpha}$ is invariant of the axes chosen. In an incompressible fluid Δ is zero.

1.5 Relation between stress and rate of strain tensors

We recall that in inviscid flow

$$\tau_{yz} = \tau_{zx} = \tau_{xy} = 0$$

and $\quad \tau_{xx} = \tau_{yy} = \tau_{zz} = -p$

Also in viscous flow

$$\tau_{\alpha\alpha}/3 = -p$$

whilst in simple two dimensional viscous shearing flow

$$\tau_{xz} = \mu \frac{\partial u}{\partial z} = \mu\, e_{xz}$$

We now seek the most general form of linear relations between the stress and rate of strain tensor components consistent with these special cases and with the requirement that the form of the relations should not be dependent on the orientation of the cartesian axes chosen. This latter requirement derives from the assumption that the fluid is isotropic, a condition met by all gases and simple homogeneous liquids. It can be shown that this condition also implies that the principal axes of the two tensors are identical. The resulting relations are found to be of the form:

$$\left.\begin{aligned} &\tau_{xx} = -p' + \lambda\Delta + \mu\, e_{xx}, \qquad \tau_{yy} = -p' + \lambda\Delta + \mu\, e_{yy} \\ &\tau_{zz} = -p' + \lambda\Delta + \mu\, e_{zz} \\ &\tau_{yz} = \mu\, e_{yz}, \qquad \tau_{zx} = \mu\, e_{zx}, \qquad \tau_{xy} = \mu\, e_{xy} \\ \text{or} \quad &\tau_{\alpha\beta} = -(p' - \lambda\Delta)\delta_{\alpha\beta} + \mu\, e_{\alpha\beta} \end{aligned}\right\} \tag{1.10}$$

where p', λ and μ are functions of the coordinates of the point considered and independent of the strain rate, also

$$\begin{aligned} \delta_{\alpha\beta} &= 1 \quad \text{if} \quad \alpha = \beta \\ &= 0 \quad \text{if} \quad \alpha \neq \beta \end{aligned}$$

We readily identify μ as the coefficient of viscosity, and λ is sometimes called the second coefficient of viscosity. Since the static pressure

$$p = -\tau_{\alpha\alpha}/3$$

it follows that

$$p = p' - (\lambda + 2\mu/3)\Delta$$

In an incompressible fluid $\Delta = 0$ and then

$$p = p'$$

For a compressible fluid it is usual to assume that the equation of state holds, i.e. p = function of ρ and T, only, whatever the motion. In that case p should be independent of Δ. If this were not so then in any cyclic process with $\Delta \neq 0$ there would be some irreversible dissipation of mechanical energy by the normal stresses and p would differ from the thermodynamic pressure (or the pressure that satisfies the equation of state). In very rapid and intense compression or expansion processes there is some evidence of such dissipation, but for the applications with which we shall be concerned such effects can be safely ignored and p can be identified with the thermodynamic pressure and assumed independent of the dilatation Δ.

It is therefore assumed that the so-called bulk viscosity

$$\beta = \lambda + 2\mu/3 = 0 \tag{1.11}$$

It must be said that there is little experimental evidence to help us determine β, and what there is does not clearly support this relation. Only in the case of a monatomic gas, assumed to consist of spherical molecules, can theory be shown to support it. However, even in cases where $\beta \neq 0$ its contribution to p is very small and, as implied above, it can only play a significant role for problems involving exceptionally large normal rates of strain.

Equations (1.10) and (1.11) then lead to the relations

$$\left.\begin{aligned} \tau_{xx} &= -(p + 2\mu\Delta/3) + \mu\, e_{xx} \\ \tau_{yy} &= -(p + 2\mu\Delta/3) + \mu\, e_{yy} \\ \tau_{zz} &= -(p + 2\mu\Delta/3) + \mu\, e_{zz} \\ \tau_{yz} &= \mu\, e_{yz}, \qquad \tau_{zx} = \mu\, e_{zx}, \qquad \tau_{xy} = \mu\, e_{xy} \\ \text{or} \quad \tau_{\alpha\beta} &= -(p + 2\mu\Delta/3)\delta_{\alpha\beta} + \mu\, e_{\alpha\beta} \end{aligned}\right\} \qquad (1.12)$$

1.6 Laminar flow, transition and turbulent flow

For simplicity we will consider a wing in steady, two dimensional flow (or aerofoil) at a small angle of incidence, as illustrated in Fig. 1.4, but the discussion will be generally applicable to any streamline shape including complete aircraft. It will be assumed that the Reynolds number is high enough for the boundary layers that develop over the wing to be thin relative to a typical dimension of the wing.

We then find that for some distance back from the leading edge the flow in the boundary layer is in smooth streamlines approximately parallel to the wing surface and we refer to the flow as being *laminar*. At some stage, however, depending on the Reynolds number, the pressure distribution over the surface, the smoothness of the surface and the presence or otherwise of external disturbances (e.g. noise) the flow in the boundary becomes unstable and changes in rather a dramatic way from being laminar to become *turbulent*. In the turbulent boundary layer we find irregular fluctuations in velocity magnitude and direction superimposed on the mean flow, but the latter remains roughly parallel to the surface. The process of change from laminar to turbulent flow in the boundary layer is referred to as *transition*. The streamwise extent of the region in which the transition process occurs varies with the Reynolds number. At

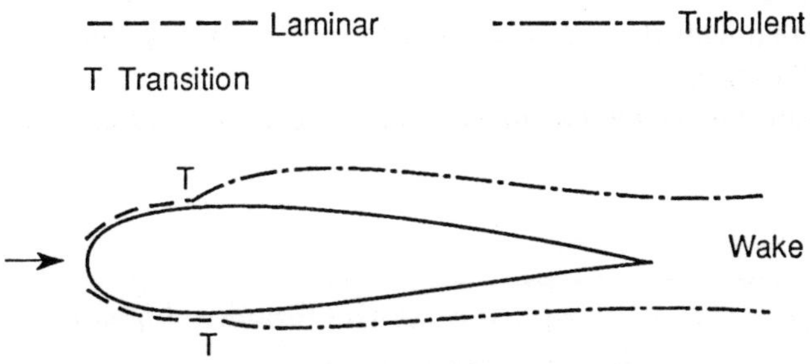

Fig. 1.4 Development of boundary layers on a wing (not to scale).

low Reynolds numbers, i.e. of the order of 5×10^5 or less, in terms of the wing chord, it can be quite a large proportion of the wing chord, but for Reynolds numbers greater than about 2×10^6, the streamwise extent of the transition region is small in relation to the wing chord and for most purposes we can regard the process as almost instantaneous. We then refer to the transition point in two dimensions and to the transition line in three dimensions.

The fluctuations in velocity in the turbulent boundary layer can be of the order of a tenth of the mean velocity. They cause the characteristics of the turbulent boundary layer to differ profoundly from those of the laminar boundary layer. For example, they provide a very powerful mechanism for mixing, and the consequent rates of diffusion of momentum, vorticity and heat are generally very much greater than the analogous rates due to the molecular movements which provide the sole mixing mechanism in the laminar boundary layer. In particular, the turbulent fluctuations in the presence of mean velocity gradients, or rates of strain, give rise to eddy stresses, usually called *Reynolds stresses*. Since the lateral rates of momentum transport due to the relatively large scale eddy movements are much greater than those due to the molecular movements the eddy shear stresses are generally much larger than the viscous shear stresses except in the so-called *viscous sub-layer*. This is a very thin layer adjacent to the surface of the order of a hundredth of the thickness of the turbulent boundary layer, in which the turbulent fluctuations tend to be reduced by the close proximity of the surface, and the dominant shear stresses there are the viscous ones. In particular, the frictional stress at the surface, τ_0, is still given by $\mu(\partial u/\partial z)_0$ there.

Further, the more powerful mixing in the turbulent boundary layer results in the velocity distribution over much of the layer being more nearly constant with distance from the wall, z, but with an associated higher rate of strain very close to the surface. This is illustrated in Fig. 1.5 comparing the velocity distributions in typical laminar and turbulent boundary layers on a flat plate at zero incidence. The much higher rate of strain very close to the surface in the turbulent boundary layer implies that the mean frictional stress at the surface, τ_0, is generally much greater than for a laminar boundary layer at the same Reynolds number. This is illustrated in Fig. 1.6, where we see the local skin friction coefficient $c_{\mathrm{f}} = \tau_0/\frac{1}{2}\rho U^2$ as a function of $R_x = U\,x/\nu$, and the overall skin friction coefficient

$$C_{\mathrm{F}} = \int_0^c c_{\mathrm{f}}\,\mathrm{d}x/c$$

as a function of $R = Uc/\nu$ for laminar and turbulent boundary layers on a flat plate at zero incidence. Here $U = u_{\mathrm{e}}$.

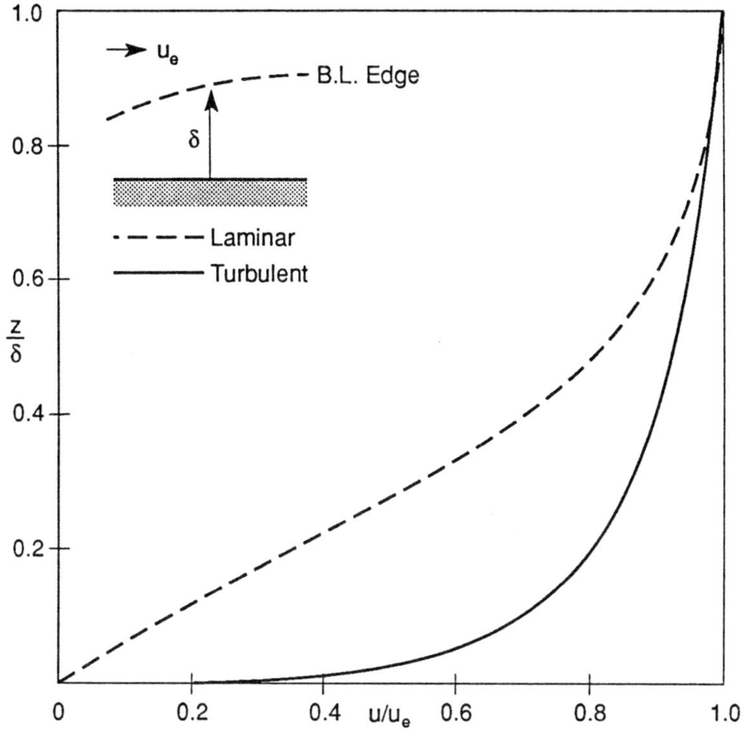

Fig. 1.5 Typical velocity profiles on a flat plate at zero incidence for laminar and turbulent boundary layers.

We must also note that in consequence of the higher mixing rate of the turbulent boundary layer the thickness of the layer grows more rapidly with distance downstream than with the laminar boundary layer. Thus, on a flat plate at zero incidence the turbulent boundary layer grows as $x^{4/5}$, whilst the laminar layer grows as $x^{1/2}$. [In fact, as we shall see, on a flat plate at zero incidence:

$$\delta/x \simeq 0.37/R_x^{\ 1/5} \quad \text{for a turbulent boundary layer}$$
$$\simeq 5/R_x^{\ 1/2} \quad \text{for a laminar boundary layer}$$

Here δ is a measure of the boundary layer thickness.] As is illustrated in Fig. 1.6 the difference in the frictional stresses at the surface between laminar and turbulent flow grows larger as the Reynolds number increases and is so large at the Reynolds numbers of aircraft (10^7 and above) that attention has for many years been focussed on the process of transition and ways of controlling and delaying it. It is a complex process and we

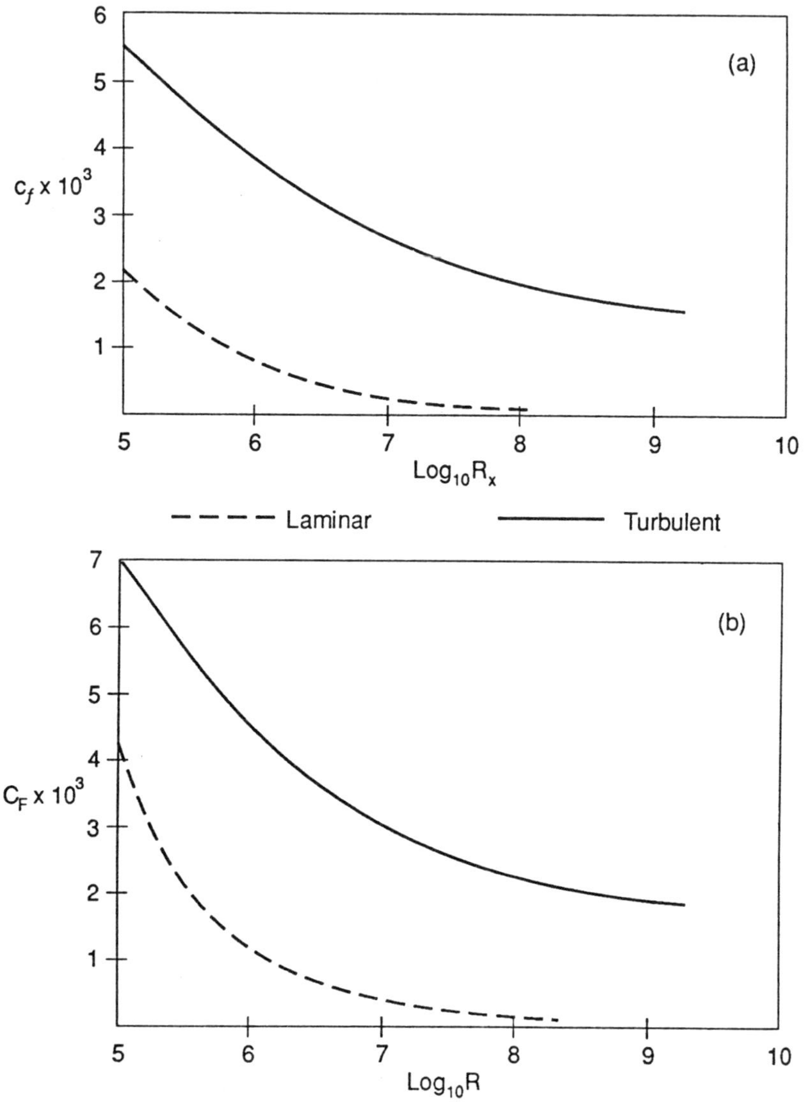

Fig. 1.6 Laminar and turbulent boundary layers on a flat plate at zero incidence. (a) c_f v R_x; (b) C_F v R.

shall discuss it in more detail later (Chapter 5). Factors that are known to hasten it are increase of Reynolds number, rising (adverse) streamwise pressure gradients (i.e. $dp/dx > 0$) imposed by the external flow, surface imperfections (i.e. excrescences, waviness etc.) and disturbances in the external flow (e.g. noise and free stream turbulence).

1.7 Effects of non-uniform pressure distribution; separation

We here continue to confine our attention to steady two dimensional flow.

If the pressure distribution to which the boundary layer on an aerofoil is subjected by the external flow is such that over a range of x it decreases with distance x downstream, i.e. the pressure gradient with respect to x is there negative (or favourable), the velocity in the boundary layer u increases with x along with that of the external flow. However, the relative increase in velocity is more marked in the slower moving strata of the boundary layer close to the surface than in the outer strata, and we speak of the velocity profile of the boundary layer (i.e. the plot of u as a function of the distance normal to the surface, z) as becoming fuller with x. Figure 1.7 illustrates this for a laminar boundary layer, the turbulent boundary layer behaves similarly but for a given pressure change the changes of velocity profile are much less marked.

Likewise, in the presence of an external rising pressure with x, or positive pressure gradient (adverse) the opposite effects occur and the relative decrease in velocity in the strata close to the surface is more marked than for the outer strata. The velocity profile then becomes less full with distance downstream, and indeed it readily becomes strongly inflected (Fig. 1.7). With a large enough pressure rise we can arrive at a stage beyond which the flow close to the surface reverses in direction and moves upstream. That critical stage is marked by the velocity gradient $\partial u/\partial z$ at the surface changing sign from positive to negative with a corresponding change of sign of the frictional stress there. Such a development is illustrated in Fig. 1.8.

It will be seen that there is a locus or line along which the velocity component u is zero (shown dashed) and below it the flow is moving in a direction opposite to the outer flow above the locus. The locus begins on the surface where the gradient $\partial u/\partial z = 0$ and the skin friction is zero, and this point is referred to as the *separation point*. It is so called because the boundary layer can be regarded as separating from the surface there. The reversed flow forms part of a large eddy over which the outer regions of the erstwhile boundary layer ride, and the boundary layer with its vorticity is convected away from the neighbourhood of the surface as a free shear layer as a result of the separation. Depending on a number of factors, such as the subsequent pressure distribution, the section geometry and the Reynolds number, this separated shear layer may subsequently reattach to the surface to form a *separation bubble* (see Fig. 1.9) or it may remain separated to merge into an enlarged wake downstream of the trailing edge or it may separate again subsequent to reattachment. These two latter patterns of flow are characteristic of that of a bluff shape (as distinct from a streamline shape) such as a circular cylinder with its axis

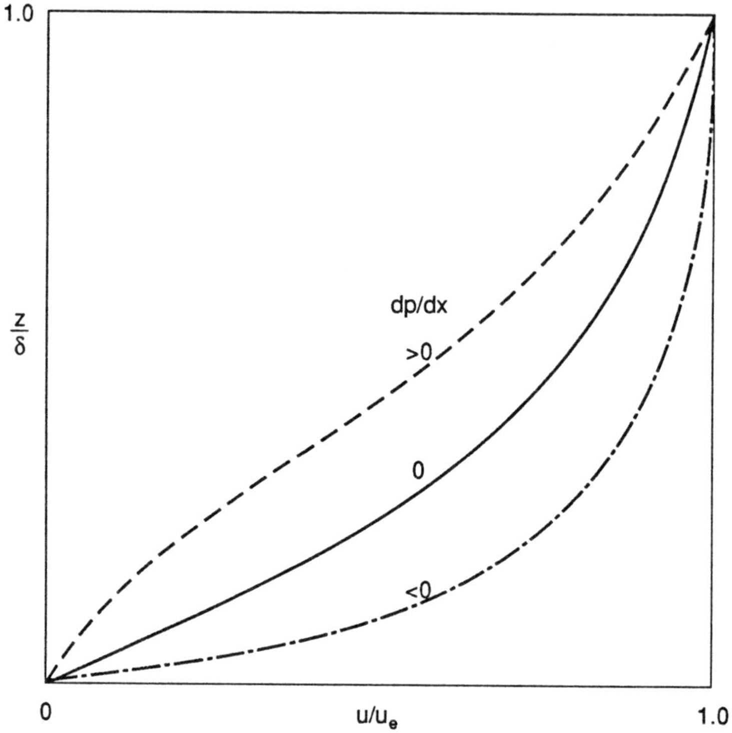

Fig. 1.7 Effect of pressure gradient on the velocity profile.

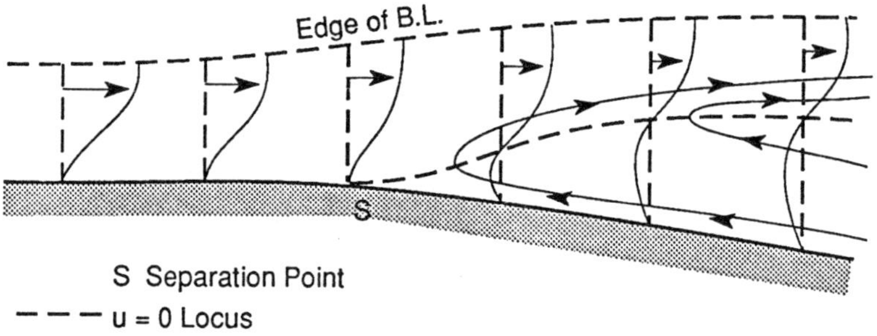

Fig. 1.8 Sketch of the development of boundary layer separation due to adverse pressure gradient.

normal to the main stream direction or a wing section at high incidences above its stalling angle (see Fig. 1.10). In such cases we find that the large eddy or eddies under the separated shear layer are not steady or stable, they move downstream from the wing or body whilst new eddies form, thus the overall flow is oscillatory in character. The wake is then much

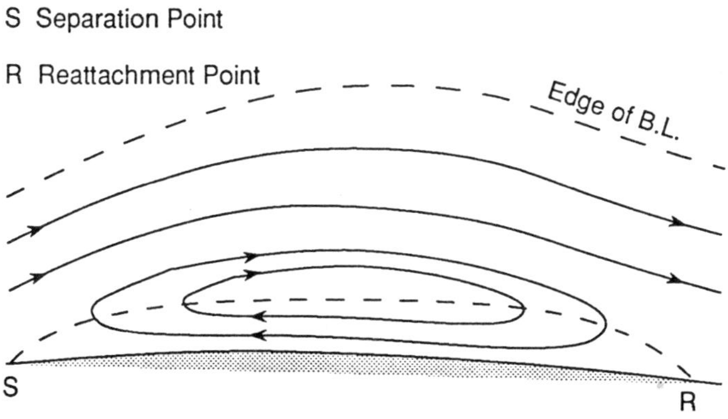

Fig. 1.9 Sketch illustrating a separation bubble.

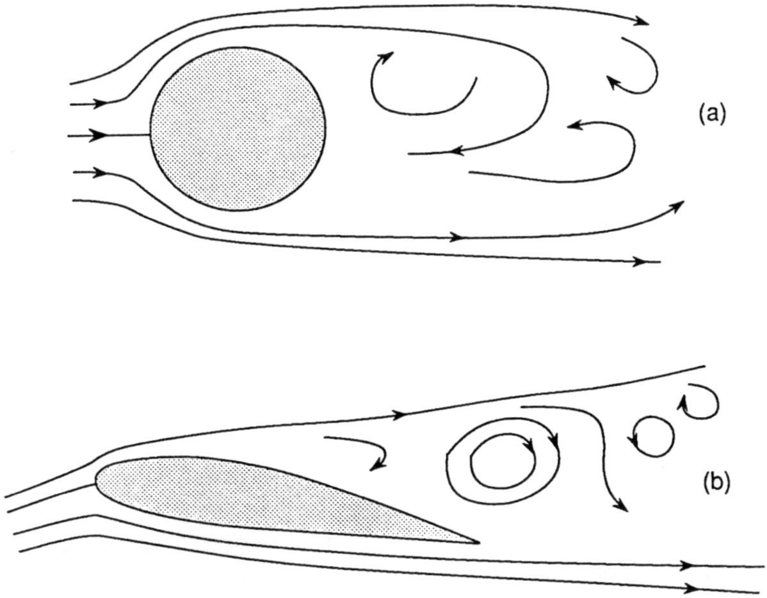

Fig. 1.10 Sketches illustrating separated flow aft of (a) circular cylinder; (b) aerofoil at high angle of incidence.

wider than for a streamline shape with attached boundary layers and considerable mechanical energy can be expended in the continuous generation of the eddies with a consequent reduction of streamwise momentum flux and increase in the drag of the body. The energy of the eddies is convected downstream and ultimately dissipated as heat.

From the foregoing it will be evident that a feature of boundary layer separation in steady two dimensional flow is that the frictional or shear stress at the surface passes through zero and its sign reverses there. Another and perhaps more fundamental feature of separation is the subsequent convection of vorticity away from the surface to lateral distances large compared with those characteristic of the diffusive process in attached boundary layers.

It should be noted that the turbulent boundary layer is more robust than the laminar because of the much stronger mixing. In consequence it separates less readily than does the laminar boundary layer, i.e. it requires a much greater pressure rise to bring it to the separation condition.

However, two points must be stressed here. As already noted, the above discussion was confined to steady main stream flows in two dimensions. For unsteady flows it is by no means general that separation (in the sense of large scale convection of vorticity from the surface) occurs where the flow direction reverses close to the surface and the velocity gradient $(\partial u/\partial z)$ at the surface is zero. Indeed, the latter can occur with no other marked local changes of the flow behaviour. In three dimensional flow, separation is in general a more complex phenomenon than in two dimensions, and there is no simple local criterion to define it. It need not necessarily be adverse, aerodynamically speaking, nor need it be avoided as far as possible. On the contrary, we are learning how to design aircraft to exploit and control to advantage certain types of three dimensional separated flows which can be steady and stable. These points will be discussed in more detail later.

1.8 Drag

The resistance or *drag* of a body in motion in a fluid is the force component acting on it in the direction opposite to the direction of motion. It is a major aerodynamic characteristic of the body and for an aircraft the ratio of the drag to the lift is a performance parameter of prime importance. [The lift is the force component normal to the flight direction and in the plane of symmetry; in steady level flight the lift is equal to the aircraft's weight.]

The drag can be regarded as made up of (a) *skin friction drag* and (b) *pressure drag* or *form drag*. Skin friction drag is the resultant in the drag direction of the shear or frictional stresses acting on the body surface and tangential to it. The pressure drag is the resultant in the drag direction of the pressures acting on the surface and normal to it. The *drag due to lift*, or *induced drag*, is associated with the rate of mechanical energy expended in generating the trailing vortex system that develops when the body (or wing) has lift. It is manifest as part of the pressure drag, as is the *wave drag* due to mechanical energy dissipation across any shock waves that may appear in flight at high enough Mach numbers. [The Mach number is the ratio of a representative velocity (e.g. flight speed) to a representative speed of sound (e.g. that in the undisturbed flow). It is a measure of the effects of the compressibility of the fluid, and such effects become important for Mach numbers approaching unity and above. Of particular importance is the development of shock waves at such Mach numbers.] The drag is manifest in the defect of momentum flux in the wake far downstream and can be determined from measurement of the latter.

In the absence of lift dependent drag and wave drag the pressure drag is essentially due to the way the boundary layers and wake modify the effective shape of the body as far as the external flow is concerned, and so change the pressure distribution that acts on the body from that of inviscid flow where the drag is zero. Hence, both the skin friction drag and the pressure drag can then be ascribed to the viscosity of the fluid as manifest in the development of the boundary layers and wake. The pressure drag minus induced drag and wave drag is often referred to as *the boundary layer pressure drag* whilst the sum of the skin friction drag and the boundary layer pressure drag is called *boundary layer drag* or *profile drag*.

For a streamline shape with attached boundary layers and a thin downstream wake, the drag is very largely skin friction drag in the absence of lift and wave drag, the boundary layer pressure drag is then an order smaller than the skin friction drag. On the other hand, for a bluff body with a broad eddying wake, the drag is largely pressure drag, reflecting the considerable rate of energy expenditure needed to generate the wake and its large eddies.

The drag D is conveniently expressed in non-dimensional form as a drag coefficient:

$$C_D = D/\tfrac{1}{2}\rho_0 U_0^2 S \qquad (1.13)$$

where ρ_0 and U_0 are representative values of the density and velocity, respectively, in the main stream flow, and S is a representative area, e.g. wing plan form area in three dimensional flow or wing chord $\times$ unit span in two dimensional flow.

To illustrate the difference in drag between a streamline and a bluff shape we may note that at a Reynolds number of 5×10^6 for a flat plate at zero incidence:

$C_D = 0.0015$ with the boundary layers fully laminar
$\simeq 0.007$ with the boundary layers fully turbulent

whilst for the same plate held normal to the main stream flow:

$C_D \simeq 1.0$

Another example of a bluff shape is a circular cylinder held with its axis normal to the main stream direction for which the drag coefficient at a Reynolds number of 5×10^6 (in terms of its diameter and unit span) is about 0.7 (see Fig. 1.11).

Both skin friction drag and pressure drag can vary with Reynolds number, but in different measure. The dependence of the skin friction drag on Reynolds number follows mainly from its direct dependence on the viscosity as well as from any changes in the free stream velocity distribution near the surface with Reynolds number. For attached flow, the boundary layer pressure drag follows the associated changes in boundary layer thickness and therefore follows a similar trend to the skin friction drag. However, for separated flow the wake dimensions, which determine the effective inner boundary of the external flow, are less dependent on Reynolds number, unless there are dramatic movements of transition or separation locations with changes of Reynolds number, in which case corresponding significant changes in pressure drag can be expected.

An interesting illustration of this is provided by the variation with Reynolds number of the drag coefficient of a smooth circular cylinder shown in Fig. 1.11. We see that C_D is roughly constant, about 1.2, for Reynolds numbers between 10^4 and about $2-3 \times 10^5$, the so-called critical Reynolds number, above which there is a steep fall in C_D to a value of about 0.3 at a Reynolds number of about 8×10^5. However, with further increase in Reynolds number there is a steady increase in the

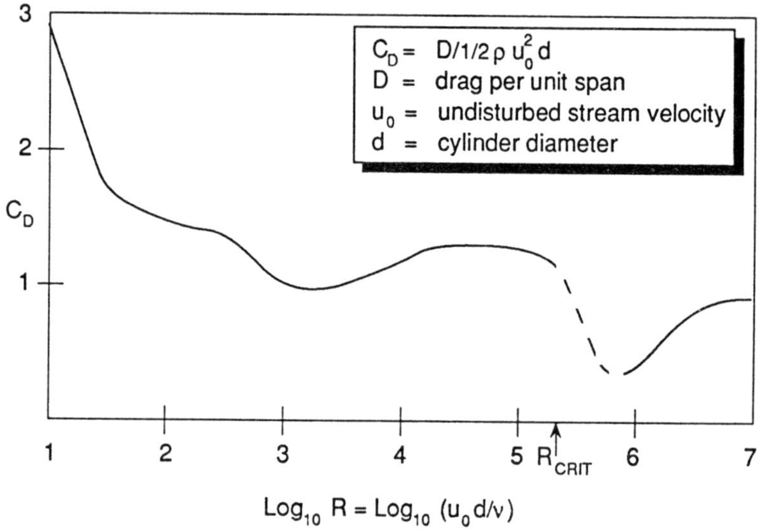

Fig. 1.11 Drag coefficient of a smooth circular cylinder with its axis normal to the stream.

drag coefficient which appears to reach another plateau at a Reynolds number of 10^7. Below the critical Reynolds number the boundary layers on the cylinder are laminar and separate relatively early from the surface where the angle to the forward stagnation point, ø, is about 80°, resulting in a wide wake and in consequence a high drag coefficient. At the critical Reynolds number, transition to turbulent flow occurs in the separated layer with subsequent reattachment to form a separation bubble and separation of the turbulent boundary layer occurs much further downstream at an angle ø of about 120°. The wake is in consequence much reduced and so is the drag coefficient. However, with further increase in the Reynolds number, transition moves forward and the turbulent boundary layer thickens and it separates somewhat earlier with a consequent broadening of the wake – hence the drag coefficient increases again. The evidence of a plateau at a Reynolds number of 10^7 suggests that the forward movement of transition has largely ceased by then or become very slow, probably because it has become coincident with the point of minimum pressure, and a wholly favourable pressure distribution, stabilising the laminar boundary layer, lies ahead of it.

Similar patterns of development and drag behaviour with changes of Reynolds number are found with many other bluff shapes, e.g. spheres and ellipsoids.

References

1.1 Prandtl, L. (1904) 'Über Flüssigkeitsbewegung bei sehr kleiner Reibung'. Proc. III Intern. Math. Congress, Heidelberg.

1.2 Schlichting, H. (1979) *Boundary Layer Theory*, 7th Ed., McGraw-Hill.

Chapter 2

Theoretical Foundations

2.1 Introduction

In this chapter we shall derive the main conservation equations for a viscous fluid in motion, viz. mass – the equation of continuity, momentum – the Navier–Stokes equations, and energy. We shall derive these in some generality but for cartesian axes. The boundary layer approximations of Prandtl will then be introduced leading to the boundary layer equations. However, in accordance with the limited aims and modest size of this book the emphasis at this stage will be mainly on two dimensional flow. A brief discussion of three dimensional boundary layers will be given in Chapter 10.

The present chapter will conclude with the development of the momentum and energy integral equations. These equations provide the basis for a wide class of approximate methods for predicting boundary layer characteristics, the so-called *integral methods*.

2.2 The conservation of mass – the equation of continuity

Consider a small box with sides parallel to the axes and of length δx, δy, δz with the point $O(x, y, z)$ at one corner, as illustrated in Fig. 2.1. Let the velocity components be (u, v, w). Then across the face OBFC in the yz plane there is a flow of mass into the box

$$= \rho u \,\delta y\, \delta z \qquad \text{per unit time}$$

whilst across the opposite face ADGE the flow out of the box

$$= \left[\rho u + \frac{\partial}{\partial x}(\rho u)\delta x\right]\delta y\, \delta z \qquad \text{per unit time}$$

to the first order in $\delta V = \delta x\, \delta y\, \delta z$. Hence across these two faces there is a net mass flow out of the box per unit volume and per unit time

$$= \frac{\partial}{\partial x}(\rho u)$$

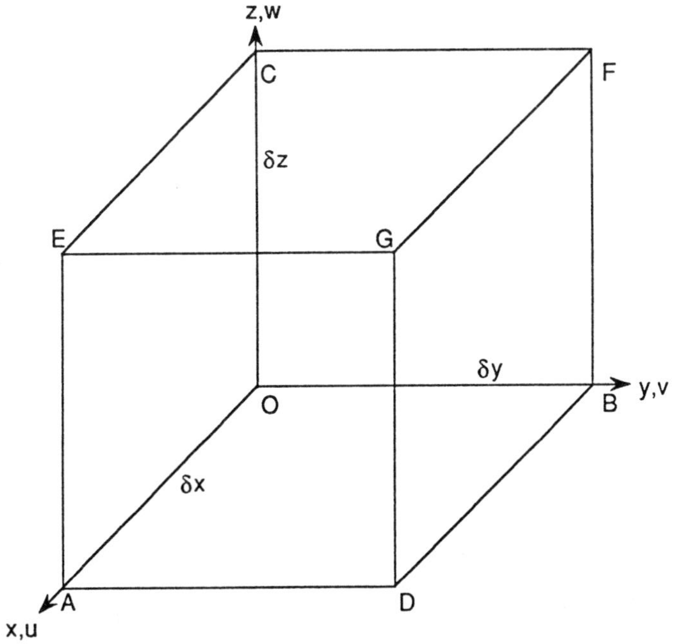

Fig. 2.1 Elementary fluid box for derivation of the conservation equations.

to the first order in δV. Similarly across the pairs of faces normal to the y and z axes we get net transports of mass out of the box, per unit volume and per unit time,

$$= \frac{\partial}{\partial y}(\rho v) \qquad \text{and} \qquad \frac{\partial}{\partial z}(\rho w)$$

respectively. If the flow is unsteady and the density is changing with time then we get an increase of mass per unit time and unit volume

$$= \frac{\partial \rho}{\partial t}$$

These expressions become increasingly exact as $\delta V \rightarrow 0$. Since mass must be conserved it follows that

$$\frac{\partial \rho}{\partial t} + \frac{\partial}{\partial x}(\rho u) + \frac{\partial}{\partial y}(\rho v) + \frac{\partial}{\partial z}(\rho w) = 0 \tag{2.1}$$

This equation can be written in the form

$$\left.\begin{aligned} &\frac{D\rho}{Dt} + \rho\Delta = 0 \\ \text{where}\qquad &\frac{D}{Dt} \equiv \frac{\partial}{\partial t} + u\frac{\partial}{\partial x} + v\frac{\partial}{\partial y} + w\frac{\partial}{\partial z} \\ \text{and}\qquad &\Delta = \frac{\partial u}{\partial x} + \frac{\partial v}{\partial y} + \frac{\partial w}{\partial z} \end{aligned}\right\} \tag{2.2}$$

where Δ is the *dilatation*. The operator D/Dt can be interpreted as the rate of change of a variable for a fluid particle moving with the fluid. We note that in incompressible flow ρ is constant and $\Delta = 0$. In vector notation equation (2.1) is

$$\frac{\partial\rho}{\partial t} + \nabla \cdot (\rho\,\boldsymbol{U}) = 0 \tag{2.3}$$

where $\boldsymbol{U} = (u, v, w)$ and $\nabla = \left(\frac{\partial}{\partial x}, \frac{\partial}{\partial y}, \frac{\partial}{\partial z}\right)$. In cartesian tensor notation this equation is

$$\frac{\partial\rho}{\partial t} + \frac{\partial}{\partial x_\alpha}(\rho u_\alpha) = 0 \tag{2.4}$$

2.3 Equations of motion

Consider again the elementary box of Fig. 2.1. We see that the flow of momentum in the x direction across OBFC is $(\rho u^2)\delta y\,\delta z$, per unit time into the box, whilst the corresponding flow in the x direction across ADGE is

$$\left[(\rho u^2) + \frac{\partial}{\partial x}(\rho u^2)\delta x\right]\delta y\,\delta z$$

out of the box, neglecting terms of order $\delta V\,\delta x$. Hence the net transport of momentum in the x direction out of the box per unit time and volume across the faces normal to the x direction

$$= \frac{\partial}{\partial x}(\rho u^2)$$

neglecting terms of the order of $\delta V\,\delta x$. Likewise, across the other pairs of faces normal to the y and z axes we find a net transport of momentum in the x direction out of the box, per unit time and volume

$$= \frac{\partial}{\partial y}(\rho uv) \qquad \text{and} \qquad \frac{\partial}{\partial z}(\rho uw)$$

respectively. If the motion is unsteady then the momentum in the x direction inside the box will have a consequent rate of increase per unit volume

$$= \frac{\partial}{\partial t}(\rho u)$$

By Newton's laws of motion the sum of these rates of change of momentum must equal the sum of the force components per unit volume in the x direction acting on the fluid in the box. On the faces normal to the x axis we readily find by the same arguments as were used above that the net force component is

$$\frac{\partial}{\partial x}\tau_{xx} \qquad \text{per unit volume}$$

and likewise on the faces normal to the y and z axes we derive net force components in the x direction per unit volume

$$= \frac{\partial}{\partial y}\tau_{xy} \qquad \text{and} \qquad \frac{\partial}{\partial z}\tau_{xz}$$

respectively. If there is a body force acting on the fluid $\boldsymbol{F} = (X, Y, Z)$, per unit mass, then this has a component in the x direction acting on the box per unit volume $= \rho X$. As before, these expressions become increasingly exact as $\delta V \to 0$. Hence, we obtain

$$\frac{\partial}{\partial t}(\rho u) + \frac{\partial}{\partial x}(\rho u^2) + \frac{\partial}{\partial y}(\rho uv) + \frac{\partial}{\partial z}(\rho uw) = \rho X + \frac{\partial}{\partial x}\tau_{xx} + \frac{\partial}{\partial y}\tau_{xy} + \frac{\partial}{\partial z}\tau_{xz}$$

Making use of the continuity equation (2.1) we can reduce this to

$$\rho\left[\frac{\partial u}{\partial t} + u\frac{\partial u}{\partial x} + v\frac{\partial u}{\partial y} + w\frac{\partial u}{\partial z}\right] \equiv \rho\frac{\mathrm{D}u}{\mathrm{D}t} = \rho X + \frac{\partial}{\partial x}\tau_{xx} + \frac{\partial}{\partial y}\tau_{xy} + \frac{\partial}{\partial z}\tau_{xz}$$

Similarly, we obtain for the y and z directions:

$$\rho\frac{\mathrm{D}v}{\mathrm{D}t} = \rho Y + \frac{\partial}{\partial x}\tau_{xy} + \frac{\partial}{\partial y}\tau_{yy} + \frac{\partial}{\partial z}\tau_{yz}$$

$$\rho\frac{\mathrm{D}w}{\mathrm{D}t} = \rho Z + \frac{\partial}{\partial x}\tau_{xz} + \frac{\partial}{\partial y}\tau_{yz} + \frac{\partial}{\partial z}\tau_{zz} \qquad (2.5)$$

or in tensor notation

$$\rho\frac{\mathrm{D}u_\alpha}{\mathrm{D}t} = \rho X_\alpha + \frac{\partial}{\partial x_\beta}\tau_{\alpha\beta}$$

With the aid of equation (2.4) this can also be written

$$\frac{\partial}{\partial t}(\rho u_\alpha) = \rho X_\alpha + \frac{\partial}{\partial x_\beta}(\tau_{\alpha\beta} - \rho u_\alpha u_\beta)$$

2.4 The Navier–Stokes equations

If we now substitute in equation (2.5) the relations between the stress and rate of strain tensors of equation (1.12) we get:

$$\rho \frac{\mathrm{D}u_\alpha}{\mathrm{D}t} = \rho X_\alpha - \frac{\partial}{\partial x_\beta}(p + 2\mu\Delta/3)\delta_{\alpha\beta} + \frac{\partial}{\partial x_\beta}(\mu\, e_{\alpha\beta})$$

or
$$\rho \frac{\mathrm{D}u}{\mathrm{D}t} = \rho X - \frac{\partial}{\partial x}(p + 2\mu\Delta/3) + \frac{\partial}{\partial x}(\mu\, e_{xx}) + \frac{\partial}{\partial y}(\mu\, e_{xy}) + \frac{\partial}{\partial z}(\mu\, e_{xz}) \tag{2.6}$$

plus two similar equations obtained by cyclic interchange of x, y, z. If we now make use of equations (1.8) and (1.9) we obtain:

$$\left.\begin{aligned}
\rho \frac{\mathrm{D}u}{\mathrm{D}t} &= \rho X - \frac{\partial p}{\partial x} + \mu\,\nabla^2 u + \mu\frac{\partial \Delta/3}{\partial x} - 2\Delta\frac{\partial \mu/3}{\partial x} + 2\frac{\partial \mu}{\partial x}\frac{\partial u}{\partial x}\\
&\quad + \frac{\partial \mu}{\partial y}\left(\frac{\partial u}{\partial y} + \frac{\partial v}{\partial x}\right) + \frac{\partial \mu}{\partial z}\left(\frac{\partial u}{\partial z} + \frac{\partial w}{\partial x}\right)\\
\rho \frac{\mathrm{D}v}{\mathrm{D}t} &= \rho Y - \frac{\partial p}{\partial y} + \mu\,\nabla^2 v + \mu\frac{\partial \Delta/3}{\partial y} - 2\Delta\frac{\partial \mu/3}{\partial y} + 2\frac{\partial \mu}{\partial y}\frac{\partial v}{\partial y}\\
&\quad + \frac{\partial \mu}{\partial z}\left(\frac{\partial v}{\partial z} + \frac{\partial w}{\partial y}\right) + \frac{\partial \mu}{\partial x}\left(\frac{\partial v}{\partial x} + \frac{\partial u}{\partial y}\right)\\
\rho \frac{\mathrm{D}w}{\mathrm{D}t} &= \rho Z - \frac{\partial p}{\partial z} + \mu\,\nabla^2 w + \mu\frac{\partial \Delta/3}{\partial z} - 2\Delta\frac{\partial \mu/3}{\partial z} + 2\frac{\partial \mu}{\partial z}\frac{\partial w}{\partial z}\\
&\quad + \frac{\partial \mu}{\partial x}\left(\frac{\partial w}{\partial x} + \frac{\partial u}{\partial z}\right) + \frac{\partial \mu}{\partial y}\left(\frac{\partial w}{\partial y} + \frac{\partial v}{\partial z}\right)
\end{aligned}\right\} \tag{2.7}$$

In tensor notation these equations are:

$$\rho \frac{\mathrm{D}u_\alpha}{\mathrm{D}t} = \rho X_\alpha - \frac{\partial p}{\partial x_\alpha} + \mu\,\nabla^2 u_\alpha + \frac{\mu}{3}\frac{\partial}{\partial x_\alpha}\left(\frac{\partial u_\beta}{\partial x_\beta}\right) - \frac{2}{3}\frac{\partial \mu}{\partial x_\alpha}\cdot\frac{\partial u_\beta}{\partial x_\beta} + \frac{\partial \mu}{\partial x_\beta}\left(\frac{\partial u_\alpha}{\partial x_\beta} + \frac{\partial u_\beta}{\partial x_\alpha}\right)$$

These are the *Navier–Stokes equations* for three dimensional compressible flow. For incompressible flow ρ and μ are constants and $\Delta = \operatorname{div} \boldsymbol{U} = 0$, and then:

$$\left.\begin{aligned}
\rho \frac{\mathrm{D}u_\alpha}{\mathrm{D}t} &= \rho X_\alpha - \frac{\partial p}{\partial x_\alpha} + \mu\,\nabla^2 u_\alpha\\
\text{or}\quad \rho \frac{\mathrm{D}\boldsymbol{u}}{\mathrm{D}t} &= \rho \boldsymbol{F} - \nabla p + \mu\,\nabla^2 \boldsymbol{u}
\end{aligned}\right\} \tag{2.8}$$

2.5 The energy equation

We shall find it especially convenient in this section to make greater use of tensor notation than in the previous sections. However, the reader should have little difficulty in translating the material into ordinary cartesian notation if required. Any reader unversed in the elements of compressible flow is advised as a preliminary to consult an introductory text book dealing with the subject e.g. Reference 2.1.

We shall denote the sum of the kinetic and internal energies per unit mass by Σ, i.e. $\Sigma = \frac{1}{2}(u^2 + v^2 + w^2) + e$, with

$$e = \int^{T} c_{\text{v}}\, \mathrm{d}T$$

for a perfect gas, where c_{v} is the specific heat at constant volume.

If we consider again the flow through the elementary box of Fig. 2.1 then it follows from the first law of thermodynamics that:

the rate of work done on the fluid in the box by the external forces (W_{F})
+ the rate of work done on the fluid by the stresses on the faces of the box (W_{S})
+ the rate of conduction of heat energy into the box (Q_1)
+ the rate of convection of kinetic and internal energy into the box (Σ_1)
= the rate of change of the kinetic and internal energy inside the box
$= \dfrac{\partial}{\partial t}(\Sigma p) \, . \, \delta V$

Now $$W_{\text{F}} = (\rho\, X_\alpha\, u_\alpha)\delta V \tag{2.9}$$

and $$W_{\text{S}} = \frac{\partial}{\partial x_\alpha}(\tau_{\alpha\beta} \, . \, u_\beta) \, . \, \delta V = \left(\frac{\partial}{\partial x_\alpha}\tau_{\alpha\beta} \, . \, u_\beta + \tau_{\alpha\beta} \, . \, \frac{\partial u_\beta}{\partial x_\alpha}\right)\delta V$$

But since

$$\rho\frac{\mathrm{D}u_\alpha}{\mathrm{D}t} = \rho X_\alpha + \frac{\partial}{\partial x_\beta}\tau_{\alpha\beta}$$

it follows that

$$\frac{\rho}{2}\frac{\mathrm{D}u_\alpha^2}{\mathrm{D}t} = \rho\, u_\alpha X_\alpha + \frac{\partial}{\partial x_\beta}\tau_{\alpha\beta} \, . \, u_\alpha$$

Hence,
$$\begin{aligned} W_{\text{S}} &= \left(\tau_{\alpha\beta}\frac{\partial u_\beta}{\partial x_\alpha} + \frac{\rho}{2}\frac{\mathrm{D}}{\mathrm{D}t}u_\alpha^2 - \rho\, u_\alpha X_\alpha\right)\delta V \\ &= \left(\frac{\tau_{\alpha\beta} \, . \, e_{\alpha\beta}}{2} + \frac{\rho}{2}\frac{\mathrm{D}u_\alpha^2}{\mathrm{D}t} - \frac{W_{\text{F}}}{\delta V}\right)\delta V \\ &= \left(\phi_1 + \frac{\rho}{2}\frac{\mathrm{D}u_\alpha^2}{\mathrm{D}t} - \frac{W_{\text{F}}}{\delta V}\right)\delta V \end{aligned} \tag{2.10}$$

Here $\quad \phi_1 = \frac{1}{2}\tau_{\alpha\beta} \cdot e_{\alpha\beta}$

Also $\quad Q_1 = \frac{\partial}{\partial x_\alpha}\left(k\,\frac{\partial T}{\partial x_\alpha}\right) \cdot \delta V$

where k = coefficient of heat conductivity, and

$$\Sigma_1 = -\frac{\partial}{\partial x_\alpha}\,(\rho u_\alpha \Sigma)\,.\,\delta V$$

Hence, in the limit as $\delta V \to 0$ we have

$$\frac{\partial}{\partial t}\,(\rho\Sigma) = \phi_1 + \frac{\rho}{2}\,\frac{\mathrm{D}u_\alpha^2}{\mathrm{D}t} + \frac{\partial}{\partial x_\alpha}\left(k\,\frac{\partial T}{\partial x_\alpha}\right) - \frac{\partial}{\partial x_\alpha}\,(\rho u_\alpha \Sigma)$$

or, using equation (2.4)

$$\rho\,\frac{\mathrm{D}e}{\mathrm{D}t} = \phi_1 + \frac{\partial}{\partial x_\alpha}\left(k\,\frac{\partial T}{\partial x_\alpha}\right) \qquad (2.11)$$

Now from the first law of thermodynamics the total rate of heat absorption per unit volume of the fluid in the box is

$$\rho\left[\frac{\mathrm{D}e}{\mathrm{D}t} + p\,\frac{\mathrm{D}}{\mathrm{D}t}\left(\frac{1}{\rho}\right)\right]$$

$$= \phi_1 + \frac{\partial}{\partial x_\alpha}\left(k\,\frac{\partial T}{\partial x_\alpha}\right) + p\rho\,.\,\frac{\mathrm{D}}{\mathrm{D}t}\left(\frac{1}{\rho}\right)$$

and with the aid of equation (2.2) this becomes

$$\phi_1 + \frac{\partial}{\partial x_\alpha}\left(k\,\frac{\partial T}{\partial x_\alpha}\right) + p\Delta = \phi_1 + \frac{Q_1}{\delta V} + p\Delta \qquad (2.12)$$

But this must equal the rate of heat conducted from outside the box per unit volume ($Q_1/\delta V$) plus the heat generated internally by irreversible dissipation of mechanical energy. If we denote the latter by ϕ, then

$$\phi = \phi_1 + p\Delta = \tau_{\alpha\beta}\,.\,e_{\alpha\beta}/2 + p\,.\,\Delta$$

With the aid of equation (1.12) this reduces to

$$\phi = \mu\, e_{\alpha\beta}\,.\,e_{\alpha\beta}/2 - 2\mu\,\Delta^2/3 \qquad (2.13)$$

If we write this out in full we obtain

$$\phi = \mu(e_{yz}^2 + e_{zx}^2 + e_{xy}^2) + \tfrac{1}{6}\mu[(e_{yy} - e_{zz})^2 + (e_{zz} - e_{xx})^2 + (e_{xx} - e_{yy})^2]$$

It is clear that $\phi \geqslant 0$, which is to be expected of a dissipation function. Hence equation (2.11) can be written

$$\rho\,\frac{\mathrm{D}e}{\mathrm{D}t} = \phi - p\Delta + \frac{\partial}{\partial x_\alpha}\left(k\,\frac{\partial T}{\partial x_\alpha}\right) \qquad (2.14)$$

The enthalpy per unit mass is $i = e + p/\rho$, and so we deduce that

$$\rho \frac{Di}{Dt} = \phi + \frac{Dp}{Dt} + \frac{\partial}{\partial x_\alpha}\left(k \frac{\partial T}{\partial x_\alpha}\right) \tag{2.15}$$

making use of the equation of continuity, equation (2.2). Also, if s is the entropy per unit mass, $T\, ds = de + p\, d(1/\rho)$, and so

$$\rho T \frac{Ds}{Dt} = \rho \frac{De}{Dt} + p\Delta = \phi + \frac{\partial}{\partial x_\alpha}\left(k \frac{\partial T}{\partial x_\alpha}\right) \tag{2.16}$$

We recall that if the gas is perfect then its equation of state is

$$p = \rho\, \mathcal{R}\, T$$

where $\mathcal{R}$ is a constant, also

$$e = \int_0^T c_v \mathrm{d}T \quad \text{and} \quad i = \int_0^T c_p \mathrm{d}T$$

where c_v and c_p are the specific heats at constant volume and constant pressure, respectively. Further,

$$c_p - c_v = \mathcal{R}$$

If we can assume that c_v and c_p are constant, then the two forms of the energy equation (2.14) and (2.15) can be written

$$\rho \frac{De}{Dt} = \rho \frac{D}{Dt}(c_v T) = \phi - p\Delta + \frac{\partial}{\partial x_\alpha}\left(k \frac{\partial T}{\partial x_\alpha}\right) \tag{2.17}$$

$$\rho \frac{Di}{Dt} = \rho \frac{D}{Dt}(c_p T) = \phi + \frac{Dp}{Dt} + \frac{\partial}{\partial x_\alpha}\left(k \frac{\partial T}{\partial x_\alpha}\right) \tag{2.18}$$

2.6 An exact solution of the Navier–Stokes equations

Exact analytic solutions of the Navier–Stokes equations are rare and are only obtained for relatively simple cases (see references 2.2 and 2.3 for reviews of such cases). One group is characterised by the welcome fact that the non-linear convection terms on the LHS of the equations vanish. We shall here refer to only one such case, because of its relevance to boundary layers. It is that of an infinite plate which at time $t = 0$ is started impulsively from rest into motion in its own plane with velocity U_1, in the x direction. The fluid, assumed incompressible, in which the plate is immersed, is therefore set in motion although far from the plate the fluid remains at rest. This problem is generally called the Rayleigh problem as he dealt with it in a paper in 1911,[2.4] but Stokes dealt with it many years before in 1851[2.5]. It is clear that for this problem $\partial/\partial x = 0$, so u is a function of z and t only, where z is measured normal to the plate.

The equation of continuity is $\partial u/\partial x + \partial w/\partial z = 0$ and so w = function of t only $= 0$, since w must vanish at the plate. The boundary conditions for the fluid are therefore:

$$\begin{aligned} t \leqslant 0, \quad & u = 0, && \text{all } z \\ t > 0, \quad & u = U_1, && z = 0 \\ & u = 0, && z = \infty \end{aligned}$$

The equations of motion reduce to

$$\rho\, \partial u/\partial t = \mu\, \partial^2 u/\partial z^2$$

or

$$\partial u/\partial t = \nu\, \partial^2 u/\partial z^2 \tag{2.19}$$

This form of equation is well known in the theory of heat conduction. Since there is no characteristic length in this problem we can expect that the distributions of u as a function of z at different times are similar, i.e. they can be derived from each other by a simple scaling of z by a function of t. Hence we seek solutions in which $u/U_1 = f(\eta)$, say, where $\eta = z/[\sqrt{\nu}\ \psi\ (t)]$, and $f(\eta)$ and $\psi(t)$ are functions to be determined. Then, since

$$\frac{\partial}{\partial z} = \frac{\partial \eta}{\partial z} \cdot \frac{\partial}{\partial \eta} = \frac{1}{\sqrt{\nu}\,.\,\psi} \cdot \frac{\partial}{\partial \eta}$$

it follows from equation (2.19) that

$$\frac{-f'z\,.\,\psi'}{\sqrt{\nu}\,.\,\psi^2} = \frac{\nu f''}{\nu\psi^2}$$

or $\quad \psi\,.\,\psi' = f''/(f'\eta)$

Now the LHS of this equation is a function of t only, whilst the RHS is a function of η only. It follows that both must be a constant K, say. Therefore,

$$\psi^2/2 = Kt + \text{const.}$$

and there is no loss of generality if we take the constant as zero and put $K = 2$. Then

$$\eta = z/2(\nu t)^{1/2} \tag{2.20}$$

and $\quad f'' + 2f'\eta = 0$

Therefore, $\quad \ln f' = -\eta^2 + \text{const.}$

or $\quad f' = C\,\mathrm{e}^{-\eta^2}, \quad$ say,

and so $\quad f = A + B \int_0^{\eta} \mathrm{e}^{-\eta^2}\, d\eta, \quad$ say

Now $f = 1$ when $\eta = 0$, and hence $A = 1$. Also $f = 0$ when $\eta = \infty$, and since

$$\int_0^{\infty} \mathrm{e}^{-\eta^2}\, d\eta = \pi^{1/2}/2$$

it follows that $B = -2/\pi^{1/2}$. Therefore

$$f = \frac{u}{U_1} = 1 - \frac{2}{\sqrt{\pi}} \int_0^{\eta} e^{-\eta^2} \, d\eta$$
$$= 1 - \operatorname{erf} \eta \qquad (2.21)$$

where the so-called error function

$$\operatorname{erf} \eta = \frac{2}{\sqrt{\pi}} \int_0^{\eta} e^{-\eta^2} \, d\eta$$

This result is illustrated in Fig. 2.2. We see that the plate's impulsive motion induces a motion in the fluid in a layer adjacent to the plate. If we measure the thickness of the layer δ at any time t by the value of z for which u/U_1 is some specified small quantity then the thickness grows with time at a rate proportional to $(\nu t)^{1/2}$, e.g. if we specify that $u/U_1 = 0.01$ at the outer edge of the layer, then $\delta \simeq 4(\nu t)^{1/2}$.

One is tempted at this stage to assume that this solution can be used to infer, at least qualitatively, what happens in a steady flow along a semi-infinite flat plate, by arguing that in the latter case the perturbation induced by the plate diffuses outwards at the same rate as for the impulsive

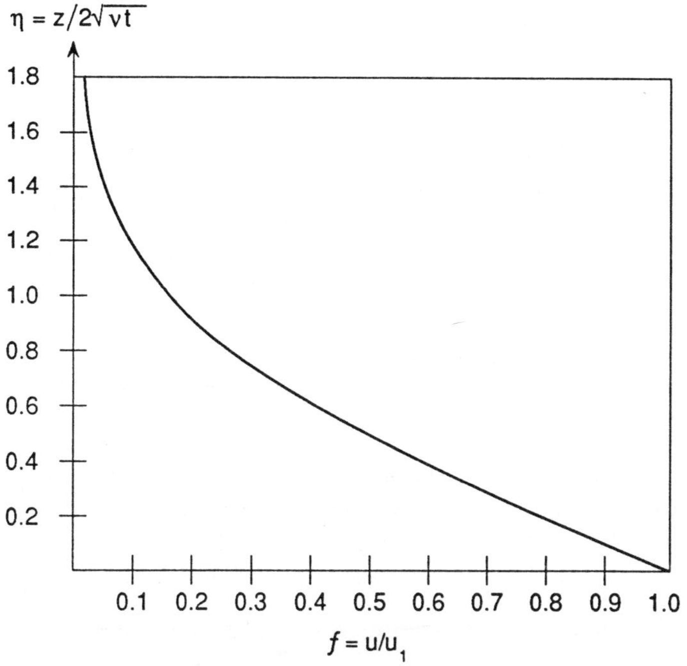

Fig. 2.2 Velocity distribution above an infinite flat plate impulsively started in motion from rest.

motion but it is also swept downstream by the flow at a rate U_1. It then follows that we can replace t in the above solution by x/U_1, so that in the steady flow the layer thickness becomes proportional to $(\nu x/U_1)^{1/2}$, i.e.

$$\delta/x \propto 1/R_x^{1/2}$$

where $R_x = U_1 x/\nu$. As we shall see this result can be proved more rigorously. If we venture further, we can deduce that the velocity distribution for the steady flow problem is given by

$$\frac{u}{U_1} = \text{erf}\left(\frac{z}{2}\sqrt{\frac{U_1}{\nu x}}\right)$$

and the frictional stress at the surface is

$$\tau = \mu\left(\frac{\partial u}{\partial z}\right)_0 = \rho\, U_1^2\left(\frac{\nu}{\pi\, U_1 x}\right)^{1/2}$$

or the skin friction coefficient

$$c_f = \tau/\tfrac{1}{2}\rho\, U_1^2 = 2/(\pi\, R_x)^{1/2} = 1.128\, R_x^{-1/2}$$

Neither of these two results is accurate enough to be of more than qualitative guidance – for example, we shall find that the correct value of the skin friction coefficient is given by

$$c_f = 0.664\, R_x^{1/2}$$

The approximations involved are in effect the replacement of $(u\partial u/\partial x + w\partial u/\partial z)$ in the equation of motion by $U_1\partial u/\partial x$ and the neglect of the term $\mu\, \partial^2 u/\partial x^2$. The latter assumpton is, as we shall see, a key feature of boundary layer theory and can be readily justified, but the former cannot be expected to yield reliable results.

2.7 The boundary layer equations; two dimensional flow over a plane surface

At this stage we shall confine our attention to the flow along a flat plate, the x axis will be taken parallel to the flow direction with the plate leading edge as origin, and the z axis taken normal to the plate. The equations of continuity, momentum and energy for a perfect gas with zero body forces then become:

$$\frac{\partial \rho}{\partial t} + u\frac{\partial \rho}{\partial x} + w\frac{\partial \rho}{\partial z} + \rho\left(\frac{\partial u}{\partial x} + \frac{\partial w}{\partial z}\right) = 0 \qquad (2.22)$$

$$\rho\left(\frac{\partial u}{\partial t} + u\frac{\partial u}{\partial x} + w\frac{\partial u}{\partial z}\right) = -\frac{\partial p}{\partial x} + \mu\left(\frac{\partial^2 u}{\partial x^2} + \frac{\partial^2 u}{\partial z^2}\right) + \frac{\mu}{3}\frac{\partial}{\partial x}\left(\frac{\partial u}{\partial x} + \frac{\partial w}{\partial z}\right)$$
$$-\frac{2}{3}\frac{\partial \mu}{\partial x}\left(\frac{\partial u}{\partial x} + \frac{\partial w}{\partial z}\right) + 2\frac{\partial \mu}{\partial x}\frac{\partial u}{\partial x} + \frac{\partial \mu}{\partial z}\left(\frac{\partial u}{\partial z} + \frac{\partial w}{\partial x}\right) \tag{2.23}$$

$$\rho\left(\frac{\partial w}{\partial t} + u\frac{\partial w}{\partial x} + w\frac{\partial w}{\partial z}\right) = -\frac{\partial p}{\partial z} + \mu\left(\frac{\partial^2 w}{\partial x^2} + \frac{\partial^2 w}{\partial z^2}\right) + \frac{\mu}{3}\frac{\partial}{\partial z}\left(\frac{\partial u}{\partial x} + \frac{\partial w}{\partial z}\right)$$
$$-\frac{2}{3}\frac{\partial \mu}{\partial z}\left(\frac{\partial u}{\partial x} + \frac{\partial w}{\partial z}\right) + 2\frac{\partial \mu}{\partial z}\frac{\partial w}{\partial z} + \frac{\partial \mu}{\partial x}\left(\frac{\partial u}{\partial z} + \frac{\partial w}{\partial x}\right) \tag{2.24}$$

$$\rho\left[\frac{\partial}{\partial t}(c_p T) + u\frac{\partial}{\partial x}(c_p T) + w\frac{\partial}{\partial z}(c_p T)\right] - \left(\frac{\partial p}{\partial t} + u\frac{\partial p}{\partial x} + w\frac{\partial p}{\partial z}\right)$$
$$= \left[k\left(\frac{\partial^2 T}{\partial x^2} + \frac{\partial^2 T}{\partial z^2}\right) + \frac{\partial k}{\partial x}\frac{\partial T}{\partial x} + \frac{\partial k}{\partial z}\frac{\partial T}{\partial z}\right] + \phi \tag{2.25}$$

where

$$\phi = \mu\left[-\frac{2}{3}\left(\frac{\partial u}{\partial x} + \frac{\partial w}{\partial z}\right)^2 + 2\left(\frac{\partial u}{\partial x}\right)^2 + 2\left(\frac{\partial w}{\partial z}\right)^2 + \left(\frac{\partial w}{\partial x} + \frac{\partial u}{\partial z}\right)^2\right]$$

If the fluid is incompressible so that ρ, u and k are constants the equations of continuity and momentum simplify to:

$$\frac{\partial u}{\partial x} + \frac{\partial w}{\partial z} = 0 \tag{2.26}$$

$$\frac{\partial u}{\partial t} + u\frac{\partial u}{\partial x} + w\frac{\partial u}{\partial z} = -\frac{1}{\rho}\frac{\partial p}{\partial x} + \nu\left(\frac{\partial^2 u}{\partial x^2} + \frac{\partial^2 u}{\partial z^2}\right) \tag{2.27}$$

$$\frac{\partial w}{\partial t} + u\frac{\partial w}{\partial x} + w\frac{\partial w}{\partial z} = -\frac{1}{\rho}\frac{\partial p}{\partial z} + \nu\left(\frac{\partial^2 w}{\partial x^2} + \frac{\partial^2 w}{\partial z^2}\right) \tag{2.28}$$

Since ρ is then constant the energy equation does not need to be solved to solve these equations but it plays a direct part in problems of heat transfer and can be written

$$\rho\frac{\mathrm{D}i}{\mathrm{D}t} - \frac{\mathrm{D}p}{\mathrm{D}t} = k\left(\frac{\partial^2 T}{\partial x^2} + \frac{\partial^2 T}{\partial z^2}\right) + \phi \tag{2.29}$$

Reverting to compressible flow and equations (2.22) to (2.25) we now assume, in accordance with Prandtl's concept of the boundary layer that, in a layer of thickness δ_u adjacent the plate the gradients of velocity components with respect to z are large enough for the viscous terms in those equations to be significant whilst outside the layer they can be neglected. We further assume that at the Reynolds numbers of interest δ_u/x is small compared with unity, so that in the boundary layer

$$\frac{\partial}{\partial z} \gg \frac{\partial}{\partial x}$$

If we assume that the plate is impermeable, we have that on the plate surface ($z = 0$) $u = w = 0$. At the edge of the layer $u \simeq U_1$, say.

We suppose that U_1, c, ρ_1 and μ_1 can be taken as representative values of the velocity, length (e.g. plate chord), density and viscosity coefficient, respectively. Then if we examine the orders of magnitude of the terms in the equation of continuity (2.22) we see that $u\,\partial\rho/\partial x$ and $\rho\,\partial u/\partial x$ are of order $U_1\rho_1/c$ whilst the terms $w\partial\rho/\partial z$ and $\rho\partial w/\partial z$ are of order $w\rho_1/\delta_u$. Hence if the terms involving w are to be of the same order as those involving u it follows that:

$$w/U_1 = O(\delta_u/c) \ll 1 \tag{2.30}$$

Here, O( ) denotes 'of the order of'. We assume that in each of the equations the terms involving $\partial/\partial t$ are at most of the same order as the remaining terms. Turning now to the first equation of motion (2.23) we see that the terms on the LHS are of order $\rho_1 U_1^2/c$, whilst the terms of largest order of magnitude on the RHS are:

$$\mu \frac{\partial^2 u}{\partial z^2} + \frac{\partial \mu}{\partial z}\frac{\partial u}{\partial z} = \frac{\partial}{\partial z}\left(\mu \frac{\partial u}{\partial z}\right)$$

and these are of order $\mu_1 U_1/\delta_u^2$. For these orders of magnitude to be the same

$$\delta_u^2/c^2 = O(U_1 c/\nu_1)^{-1} \qquad \text{where } \nu_1 = \mu_1/\rho_1$$

i.e. $\quad \delta_u/c = O(R)^{-1/2}$ (2.31)

where the Reynolds number is $R = U_1 c/\nu_1$.

Proceeding to the second equation of motion (2.24) and making use of equation (2.30) we find that the order of magnitude of the terms on the LHS are at most $\rho_1 U_1^2\delta_u/c^2$ whilst that of the terms on the RHS is $\mu_1 U_1/\delta_u c$. These are consistent in view of equation (2.31). Further, if $\partial p/\partial z$ is to be of the same order then the change in p across the thickness of the boundary layer is of order $\rho_1 U_1^2\delta_u^2/c^2$, so that

$$\Delta p/\rho_1 U_1^2 \ll 1$$

Hence the change in p across any section of the boundary layer can be neglected and p can be regarded as a function of x and t only. Therefore, equation (2.23) becomes:

$$\rho\left(\frac{\partial u}{\partial t} + u\frac{\partial u}{\partial x} + w\frac{\partial u}{\partial z}\right) = -\frac{\partial p}{\partial x} + \frac{\partial}{\partial z}\left(\mu\frac{\partial u}{\partial z}\right) \tag{2.32}$$

Following the same argument it is likewise postulated that the gradients of temperature with z are large enough for the heat conduction terms in equation (2.25) to be significant only in a thin layer adjacent the surface, the temperature boundary layer, the thickness of which we will denote as δ_T, with $\delta_T/c \ll 1$. We shall further assume that δ_u and δ_T are of the same order. In the temperature boundary layer the temperature changes from a value, T_w, say, at the surface to a value in the external flow at the edge of the boundary layer of the order of T_1, say, and for flow quantities in the boundary layer we can again infer that $\partial/\partial z \gg \partial/\partial x$. Hence, retaining only the terms of largest order of magnitude in equation (2.25), it reduces to:

$$\rho\left[\frac{\partial}{\partial t}(c_p T) + u\frac{\partial}{\partial x}(c_p T) + w\frac{\partial}{\partial z}(c_p T)\right] - \left(\frac{\partial p}{\partial t} + u\frac{\partial p}{\partial x}\right) = \frac{\partial}{\partial z}\left(k\frac{\partial T}{\partial z}\right) + \mu\left(\frac{\partial u}{\partial z}\right)^2 \qquad (2.33)$$

The terms on the LHS are of order $\rho_1 U_1 c_p \Delta T/c$, where ΔT is the change in T across the boundary layer. The heat conduction term on the RHS is of order $k_1 \Delta T/\delta_T^2$, where k_1 is a representative value of the coefficient of thermal conductivity. Hence

$$\delta_T^2/c^2 = O(k_1/\rho_1 c_p U_1 c)$$

We had $\quad \delta_u^2/c^2 = O(\mu_1/\rho_1 U_1 c)$

therefore $\quad \delta_u/\delta_T = O(\mu_1 c_p/k_1)^{1/2} = O(\sigma_1^{1/2}) \qquad (2.34)$

where $\sigma = \mu c_p/k$ is the Prandtl number.

Hence our assumption that δ_u and δ_T are of the same order of magnitude is equivalent to assuming that the Prandtl number is of order unity. For air, as for gases in general, the Prandtl number is about 0.7 and is remarkably constant and insensitive to changes of temperature in ranges away from dissociation and liquefaction. It will be evident that the Prandtl number is a measure of the ratio of the diffusion of momentum to the diffusion of heat by molecular movements.

To recapitulate, the assumptions of Prandtl's boundary layer theory lead to the boundary layer equations (2.22), (2.32) and (2.33) with the pressure a function only of x and t. These equations are still non-linear but they are very much simpler than the Navier–Stokes equations from which they were derived.

An alternative and illuminating way of regarding the approximations introduced by boundary layer theory is as follows. We define a non-dimensional set of variables as:

$$x' = x/c, \quad z' = zR^{1/2}/c, \quad t' = tU_1/c, \quad u' = u/U_1$$
$$w' = wR^{1/2}/U_1, \quad p' = p/\rho_1 U_1^2, \quad \rho' = \rho/\rho_1, \quad T' = T/T_1$$

Here, R is the Reynolds number $= U_1 c/\nu_1$; and we must note that the introduction of the factor $R^{1/2}$ in the non-dimensional forms of z and w has been made in the light of the results of the above order of magnitude analysis.

We then express the equations of continuity, motion and energy [equations (2.22)–(2.25)] in terms of these non-dimensional variables and assume that in any one of these equations terms containing the same power of $1/R$ are of comparable magnitude. We then find that as R tends to infinity the equations tend to the non-dimensional forms of the above boundary layer equations. Thus, the boundary layer equations can be regarded as the limit of the full equations of viscous flow for $R \to \infty$.

It can also be inferred from the non-dimensional equations that for complete dynamic and thermodynamic similarity of the flows past two geometrically similar bodies, the corresponding values of R, M_1 (a representative Mach number), $\gamma = c_p/c_v$ and σ (the Prandtl number) must be the same. In addition, μ must be the same function of T and the temperature boundary conditions must be similar.

This approach has led to work being done on the development of second order theories of the boundary layer in which terms of higher order in $1/R^{1/2}$ are evaluated (see reference 2.6). These terms can be significant for R less than about 10^4, but we shall be generally concerned with Reynolds numbers higher than this.

2.8 Two dimensional flow over a curved surface

For the flow over a two dimensional curved surface, such as that of an aerofoil, we can conveniently adopt orthogonal curvilinear coordinates where x is now taken along the surface in the direction of the flow and z is normal to the surface.

If the curvature of the surface is denoted by κ there must be a pressure gradient normal to the wall to balance the centrifugal forces, i.e.

$$\partial p/\partial z = \kappa \rho u^2$$

As a result there will be a pressure change Δp across the boundary layer for a given x such that

$$\Delta p \simeq O(\kappa \rho_1 U_1^2 \delta_u)$$

This can be neglected if $\kappa \delta_u \ll 1$.

Also it will be evident from Fig. 2.3 that if δx_0 is an element of x along the surface then at a height z above the surface the length of the corresponding element is

$$\delta x_0(1 + z\kappa)$$

Hence, if $\kappa \delta_u \ll 1$ the change of element length with z can be ignored, i.e. the metric $h_1 = 1 + \kappa z \simeq 1$. Also, in the Navier–Stokes equation of motion in the direction normal to the surface the viscous term involves $\partial h_1/\partial x$, which can be neglected for large Reynolds numbers provided $\delta_u(c/R)\mathrm{d}\kappa/\mathrm{d}x \ll 1$.

If these limitations on the curvature and its gradient with respect to x

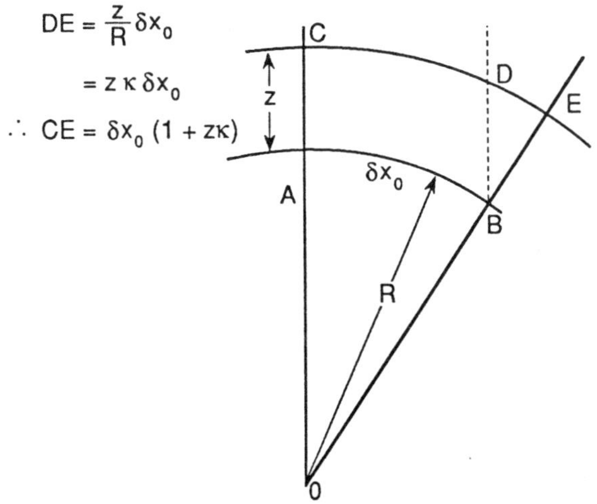

Fig. 2.3 Change of δx with z on a curved surface.

are met then the boundary layer equations (2.22), (2.32) and (2.33) that we have derived for flow over a flat plate, will apply unchanged in form to the flow over a curved surface. However, we must note that the conditions in the external flow at the edge of the boundary layer will be a function of x. If we denote flow quantities there by suffix e then since the viscous terms are negligible there:

$$\frac{\partial \rho_e}{\partial t} + \frac{\partial}{\partial x}(\rho_e u_e) = 0 \tag{2.35}$$

$$\rho_e\left(\frac{\partial u_e}{\partial t} + u_e\frac{\partial u_e}{\partial x}\right) = -\frac{\partial p}{\partial x} \tag{2.36}$$

$$\rho_e\left(\frac{\partial i_e}{\partial t} + u_e\frac{\partial i_e}{\partial x}\right) = -\left(\frac{\partial p}{\partial t} + u_e\frac{\partial p}{\partial x}\right) \tag{2.37}$$

2.9 Axi-symmetric flow over a body of revolution

For this case we can again use orthogonal curvilinear coordinates with x parallel to the surface in a meridian plane and z normal to the surface. Again, it is found, following the same reasoning as in Section 2.8, that the flat plate boundary layer equations of motion and energy apply provided $\delta_u\kappa$ and $\delta_u(c/R)\mathrm{d}\kappa/\mathrm{d}x$ are small compared to unity, where κ is the curvature of the meridian profile. However, the equation of continuity is now:

$$\frac{\partial}{\partial t}(\rho r) + \frac{\partial}{\partial x}(\rho r u) + \frac{\partial}{\partial z}(\rho r w) = 0$$

where r is the distance from the axis. This can be written:

$$\frac{\partial \rho}{\partial t} + \frac{\partial}{\partial x}(\rho u) + \frac{\partial}{\partial z}(\rho w) + \frac{\rho u}{r}\frac{\partial r}{\partial x} + \frac{\rho w}{r}\frac{\partial r}{\partial z} = 0$$

But, see Fig. 2.4, $\partial r/\partial z = \cos\chi$, and $\partial r/\partial x = \sin\chi$, where χ is the angle between the axis and the tangent to the surface in the meridian plane at the point $(x, 0)$. In general, since $w \ll u$, $(\rho w/r)\partial r/\partial z$ can be neglected compared with $(\rho u/r)\partial r/\partial x$. Hence the equation of continuity becomes:

$$\frac{\partial \rho}{\partial t} + \frac{\partial}{\partial x}(\rho u) + \frac{\partial}{\partial z}(\rho w) + \frac{\rho u}{r}\frac{\partial r}{\partial x} = 0 \qquad (2.38)$$

If δ_u/r is small, as is usual over most of a streamline shape with an attached boundary layer, except over the rearmost portion, then the last term in equation (2.38) can be replaced by

$$\frac{\rho u}{r_0}\frac{\mathrm{d}r_0}{\mathrm{d}x}$$

where r_0 is the local radius of cross section of the body. In such cases Mangler[2.7] has shown that it is possible to transform the flow over the axi-symmetric body to that over a related flat plate.

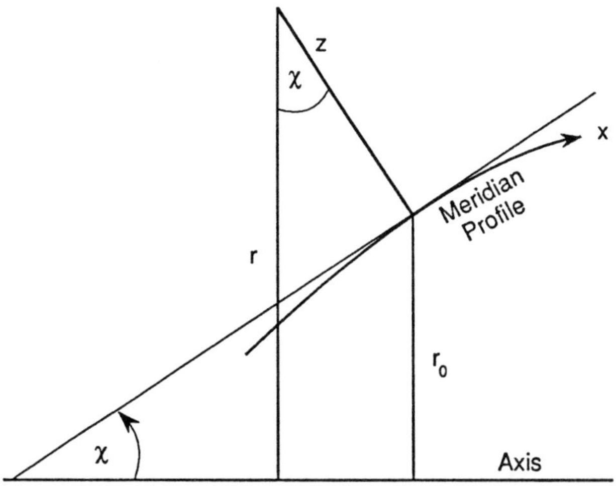

Fig. 2.4 Geometrical relations between (x, z) and (r, χ) for body of revolution.

2.10 Boundary conditions, boundary layer thicknesses

If the flow is over a surface which is impermeable, then the relative motion of the fluid and surface is zero at the surface (no-slip condition), and so with axes fixed in the surface we have:

when $\quad z = 0, \qquad u = w = 0$

Also the temperature of the surface (or wall) may be given, or the condition of zero heat transfer there may be prescribed, i.e.:

when $\quad z = 0, \qquad T = T_w$, say, $\qquad$ or $(\partial T/\partial z)_w = 0$

We note from equations (2.32) and (2.36) that at the wall for steady flow

$$\frac{\partial}{\partial z}\left(\mu\,\frac{\partial u}{\partial z}\right)_w = \frac{dp}{dx} = -\rho_e u_e\,\frac{du_e}{dx}$$

Hence when the pressure is constant and the flow is incompressible

$$(\partial^2 u/\partial z^2)_w = 0$$

We see that if $dp/dx > 0$ then $(\partial^2 u/\partial z^2)_w > 0$, and since near the boundary layer outer edge $(\partial^2 u/\partial z^2) < 0$ it follows that somewhere in the boundary layer $(\partial^2 u/\partial z^2) = 0$, i.e. the velocity profile has a point of inflection.

At the outer edge of the boundary layer we require the boundary layer and the external flow to merge smoothly. Let us denote conditions in the outer flow at the edge of the boundary layer by suffix e, so that there we have:

$$u = u_e, \quad \rho = \rho_e, \quad T = T_e, \quad \partial^n u/\partial z^n = \partial^n T/\partial z^n = 0, \quad n = 1, 2, \ldots$$

As far as the boundary layer equations are concerned these conditions apply strictly for $z = \infty$. However, as we have seen, and as is implied by the approximations of boundary layer theory, these conditions are approached very rapidly if asymptotically within a relatively small distance from the surface, δ. We have not attempted as yet to give an unambiguous definition of δ, but with any practically based definition, e.g. the value of z for which $u = 0.995u_e$, or the value of z for which the measured total (i.e. pitot) pressure is indistinguishable within the accuracy of the instruments used from that of the external flow, we find δ/c to be a very small quantity at the Reynolds numbers of aeronautical interest. For example, on a flat plate at zero incidence, with the outer edge of the laminar boundary layer defined as that for which $u/u_e = 0.995$, then $\delta/c \simeq 5/R^{1/2}$, whilst for $u/u_e = 0.999$, $\delta/c \simeq 6/R^{1/2}$, so that at the trailing edge of such a plate for which $R = 10^6$, say, $(\delta/c)_{0.999} = 0.006$.

If, therefore, a finite thickness δ_u is postulated for the velocity boundary layer, then the corresponding outer boundary conditions are:

$$u = u_e, \quad \partial u/\partial z = \partial^2 u/\partial z^2 = \ldots = 0, \quad \text{at } z = \delta_u$$

with as many of the derivatives put equal to zero as practical requirements and computational difficulties permit. Likewise, with a finite temperature boundary layer thickness δ_T we have:

$$T = T_e, \quad \rho = \rho_e, \quad \partial T/\partial z = \partial^2 T/\partial z^2 = \ldots = 0, \quad \text{at } z = \delta_T$$

where again we seek to satisfy as many of the latter conditions involving higher order derivatives as practical considerations show are desirable.

However, we can with greater precision and less ambiguity define certain quantities which can be regarded as measures of the thickness of the boundary layer and which have the merit of having considerable physical significance.

Displacement thickness – We shall denote this as δ^* where

$$\delta^* = \int_0^h \left(1 - \frac{\rho u}{\rho_e u_e}\right) dz \tag{2.39}$$

where h can be any value of z a little greater than δ_u or δ_T. We see that $\rho_e u_e \delta^*$ is the total defect in rate of mass flow in the boundary layer as compared with that which would occur in the absence of the boundary layer. This defect must result in a movement of fluid out from the boundary into the external flow. Its effect on the external flow is therefore equivalent to displacing the surface outwards and normal to itself by the distance δ^*. This is an important concept in helping us to understand the interaction between the boundary and the external flow.

For axi-symmetric flow past a body of revolution we likewise define the displacement thickness as (see Fig. 2.4):

$$\delta^* = \int_0^h \left(1 + \frac{z}{r_0} \cos \chi\right)\left(1 - \frac{\rho u}{\rho_e u_e}\right) dz \tag{2.40A}$$

provided $r_0/\delta^* \gg 1$. This is generally true over much of a typical streamline body with attached flow. However, near a pointed rear of the body where r_0 tends to zero and $\delta^*/r_0 \gg 1$ we have instead:

$$\pi \delta^{*2} \rightarrow \int_0^h 2\pi(r_0 + z \cos \chi) \left(1 - \frac{\rho u}{\rho_e u_e}\right) dz \tag{2.40B}$$

Momentum thickness – This we denote as θ where

$$\theta = \int_0^h \frac{\rho u}{\rho_e u_e} \left(1 - \frac{u}{u_e}\right) dz \tag{2.41}$$

Here we see that $\rho_e u_e^2 \theta$ is the defect in the rate of streamwise momentum transport in the boundary layer as compared with the rate in the absence of the boundary layer. It therefore is related to the contributions to the drag of the surface upstream, both frictional and pressure. Again, for axi-symmetric flow and $r_0/\theta \gg 1$,

$$\theta = \int_0^h \left(1 + \frac{z}{r_0} \cos \chi\right) \frac{\rho u}{\rho_e u_e} \left(1 - \frac{u}{u_e}\right) dz \tag{2.42A}$$

but where $r_0/\theta \ll 1$

$$\pi\theta^2 = \int_0^h 2\pi(r_0 + z \cos \chi) \frac{\rho u}{\rho_e u_e} \left(1 - \frac{u}{u_e}\right) dz \tag{2.42B}$$

Kinetic energy thickness – We denote this as δ_E, where

$$\delta_E = \int_0^h \frac{\rho u}{\rho_e u_e} \left[1 - \left(\frac{u}{u_e}\right)^2\right] dz \tag{2.43}$$

In this case we see that $\rho_e u_e^3 \delta_E/2$ is the defect in the rate of transport of kinetic energy in the boundary layer as compared with the rate in the absence of the boundary layer. It is related to the dissipation of mechanical energy in the boundary layer due to viscous effects.

For axi-symmetric flow and $r_0/\delta_E \gg 1$:

$$\delta_E = \int_0^h \left(1 + \frac{z}{r_0} \cos \chi\right) \frac{\rho u}{\rho_e u_e} \left[1 - \left(\frac{u}{u_e}\right)^2\right] dz \tag{2.44A}$$

but where $r_0/\delta_E \ll 1$

$$\pi\delta_E^2 = \int_0^h 2\pi(r_0 + z \cos \chi) \frac{\rho u}{\rho_e u_e} \left[1 - \left(\frac{u}{u_e}\right)^2\right] dz \tag{2.44B}$$

2.11 Integral equations for steady flow

If we integrate the boundary layer equation of motion (2.32) with respect to z from $z = 0$ to the outer edge of the boundary layer for any station x, we obtain an equation which relates the overall rate of flux of momentum across the cross-section of the boundary layer to the local pressure gradient and surface frictional stress. This is the so-called *momentum integral equation* (MIE), which was first derived by Von Karman.[2.8] Its value lies primarily in the fact that it provides a mean description of the equation of motion averaged over the boundary layer section. It can therefore be used as the basis of a class of approximate methods of solution; these are characterised by the use of assumed forms of velocity (and temperature) profiles across the boundary layer determined by a small number of postulated parameters. The values of these parameters are then obtained by satisfying the momentum integral equation together with some auxiliary relations, the number of which depends on the number of parameters postulated. In the derivation of the equation that follows the method adopted is a variant of that described above.

Other mean or integral equations can be obtained by multiplying the boundary layer equation of motion by $u^m z^n$, where m and n are integers, before integration. One such equation, for $m = 1.0$ and $n = 0$, is the so-called *kinetic energy* (or *dissipation*) *integral equation*. It can likewise be used as a basis for approximate methods of solution either as an alternative to the momentum integral equation or together with it.

Similarly, we can obtain a series of energy integral equations by integrating the boundary layer energy equation (2.33) either alone or in combination with the equation of motion over a boundary layer section with or without multiplying factors of the kind described above. A simple but illuminating example, the *total energy integral equation*, is derived below.

2.11.1 Momentum integral equation

In Fig. 2.5 AE and DF are two adjacent sections of the boundary layer, AD being a small element of the surface Δx. AB = DC = h, where h is a little greater than the boundary layer thickness. We assume that the sections of the boundary layer are of unit depth normal to the paper. Then we have:

Rate of mass flow across DC − rate of mass flow across AB

$$= \frac{\mathrm{d}}{\mathrm{d}x}\left[\int_0^h \rho u \,\mathrm{d}z\right]\Delta x + \text{terms of order } \Delta x^2$$

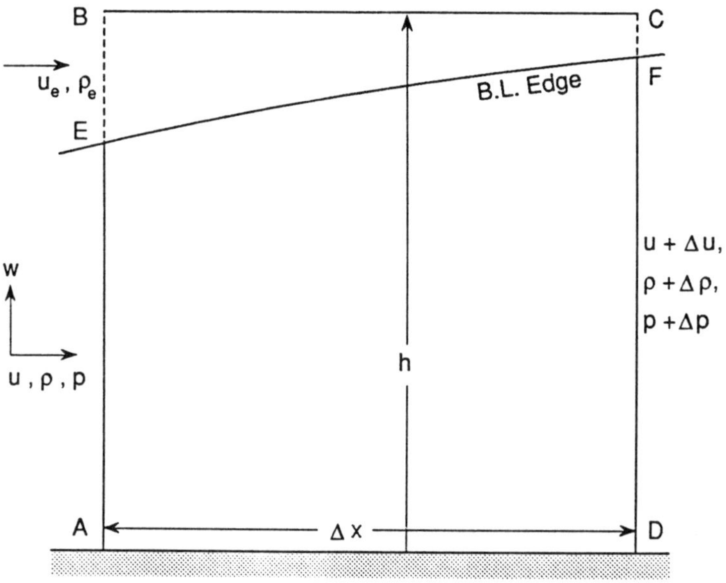

Fig. 2.5 Elementary boundary layer section for deriving the momentum integral equation in two dimensional flow.

Rate of mass flow across BC $= \rho_e w_h . \Delta x$. Hence continuity requires that

$$\rho_e w_h = -\frac{d}{dx}\left[\int_0^h \rho u \, dz\right] + \text{terms of order } \Delta x \qquad (2.45)$$

Similarly, the rate of transport of momentum in the x direction across DC − the rate of transport of momentum across AB

$$= \frac{d}{dx}\left[\int_0^h \rho u^2 dz\right] + \text{terms of order } \Delta x^2$$

The rate of transport of momentum in the x direction across BC

$$= \rho w_h \, . \, u_e \Delta x = -u_e \frac{d}{dx}\left[\int_0^h \rho u \, dz\right]\Delta x + O(\Delta x)^2$$

The force on the sectional element of the boundary layer due to the pressures

$$= -h\Delta p = -h(dp/dx)\Delta x + O(\Delta x)^2$$

whilst the force on the element due to the frictional stress on the surface

$$= -\tau_w \, . \, \Delta x$$

Hence, in the limit as $\Delta x \to 0$

$$\frac{d}{dx}\left(\int_0^h \rho u^2 dz\right) - u_e \frac{d}{dx}\left(\int_0^h \rho u \, dz\right) = -h\frac{dp}{dx} - \tau_w \qquad (2.46)$$

This is the momentum integral equation but it can be usefully expressed in terms of the displacement and momentum thicknesses as follows.

From equation (2.36) we have for steady flow

$$-dp/dx = \rho_e u_e du_e/dx \qquad (2.47)$$

and so (2.46) can be written

$$\frac{d}{dx}\left[\int_0^h \rho u(u - u_e) dz\right] + \frac{du_e}{dx}\left[\int_0^h \rho u \, dz\right] = h\rho_e u_e \frac{du_e}{dx} - \tau_w$$

or $$\frac{d}{dx}\left[\int_0^h \rho u(u_e - u) dz\right] + \frac{du_e}{dx}\left[\int_0^h \rho_e u_e - \rho u) dz\right] = \tau_w$$

or $$\frac{d}{dx}(\rho_e u_e^2 \theta) + \frac{du_e}{dx}\rho_e u_e \delta^* = \tau_w \qquad (2.48)$$

The ratio δ^*/θ is usually denoted by H, and is referred to as a form parameter of the boundary layer since its value is related to the shape of the velocity profile. Hence, equation (2.48) can be written

$$\frac{d\theta}{dx} + \frac{1}{u_e}\frac{du_e}{dx}\theta(H + 2) + \frac{\theta}{\rho_e}\frac{d\rho_e}{dx} = \frac{\tau_w}{\rho_e u_e^2} \qquad (2.49)$$

This is the most commonly used form of the MIE. In incompressible flow ρ is constant and the equation reduces to

$$\frac{d\theta}{dx} + \frac{1}{u_e}\frac{du_e}{dx}\theta(H+2) = \frac{\tau_w}{\rho_e u_e^2} \tag{2.50}$$

The same argument applied to the boundary layer section in axi-symmetric flow illustrated in Fig. 2.6 finally results in

$$\frac{d\theta}{dx} + \frac{\theta}{u_e}\left[(H+2)\frac{du_e}{dx} + \frac{1}{\rho_e}\frac{d\rho_e}{dx} + \frac{1}{r_0}\frac{dr_0}{dx}\right] = \frac{\tau_w}{\rho_e u_e^2} \tag{2.51}$$

where $H = \delta^*/\theta$ as defined in equation (2.40) and (2.42).

In the form presented these equations apply to both laminar and turbulent boundary layers, but certain provisos apply to their use for turbulent boundary layers which are discussed later.

2.11.2 Kinetic energy integral equation

If we multiply the boundary layer equation of motion (2.32) by u and then integrate it with respect to z from $z = 0$ to $z = h$ we obtain for steady flow

$$\int_0^h \rho u^2 \frac{\partial u}{\partial x}dz + \int_0^h \rho u w\frac{\partial u}{\partial z}dz = -\int_0^h u\frac{dp}{dx}dz + \int_0^h u\frac{\partial}{\partial z}\left(\mu\frac{\partial u}{\partial z}\right)dz$$

We also have the equation of continuity

$$\frac{\partial}{\partial x}(\rho u) + \frac{\partial}{\partial z}(\rho w) = 0$$

and so $$\rho_e w_e = -\int_0^h \frac{\partial}{\partial z}(\rho u)dz$$

and from equation (2.47)

$$dp/dx = -\rho_e u_e du_e/dx$$

Now
$$\int_0^h \rho u w\frac{\partial u}{\partial z}dz = [\rho u^2 w]_0^h - \int_0^h u\frac{\partial}{\partial z}(\rho u w)dz$$
$$= [\rho u^2 w]_0^h - \int_0^h u^2\frac{\partial}{\partial z}(\rho w)dz - \int_0^h u\rho w\frac{\partial u}{\partial z}dz$$

and so

$$\int_0^h \rho u w\frac{\partial u}{\partial z}dz = \tfrac{1}{2}\rho_e u_e^2 w_h + \int_0^h \tfrac{1}{2}u^2\frac{\partial}{\partial x}(\rho u)dz$$
$$= \int_0^h \tfrac{1}{2}(u^2 - u_e^2)\frac{\partial}{\partial x}(\rho u)dz$$

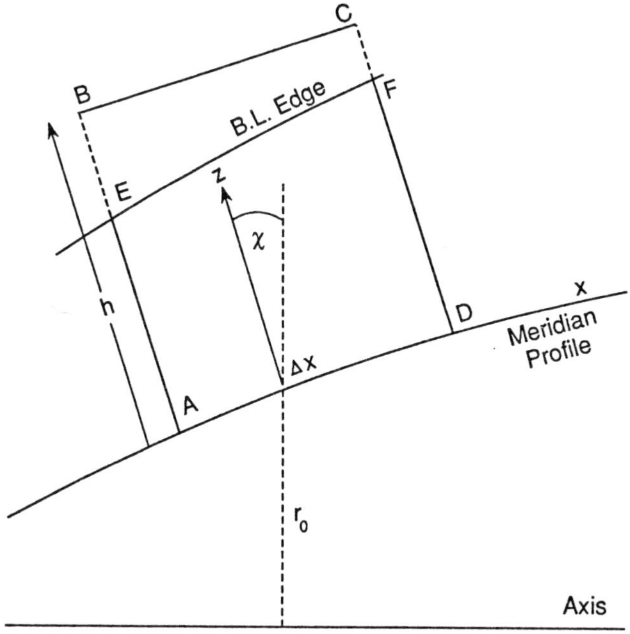

Fig. 2.6 Elementary boundary layer section for deriving the momentum integral equation in axi-symmetric flow.

If we substitute this relation as well as that above for dp/dx in the integral equation we get

$$\int_0^h \left[\rho u \frac{\partial}{\partial x}\left(\frac{u^2}{2}\right) + \tfrac{1}{2}(u^2 - u_e^2)\frac{\partial}{\partial x}(\rho u)\right]dz - \rho_e u_e \frac{du_e}{dx}\int_0^h u\,dz$$
$$= \left[\mu u \frac{\partial u}{\partial z}\right]_0^h - \int_0^h \mu\left(\frac{\partial u}{\partial z}\right)^2 dz$$

or $$\frac{d}{dx}\int_0^h \rho u\,\tfrac{1}{2}(u_e^2 - u^2)dz + u_e\frac{du_e}{dx}\int_0^h u(\rho_e - \rho)dz = \int_0^h \mu\left(\frac{du}{dz}\right)^2 dz$$

i.e. $$\frac{d}{dx}[\rho_e \tfrac{1}{2}u_e^3\delta_E] + \rho_e u_e^2\frac{du_e}{dx}\int_0^h \frac{u}{u_e}\left(1 - \frac{\rho}{\rho_e}\right)dz = \int_0^h \mu\left(\frac{\partial u}{\partial z}\right)^2 dz$$

If we write

$$\delta^{*i} = \int_0^h \left(1 - \frac{u}{u_e}\right)dz$$

i.e. the expression for δ^* as if the flow were incompressible, then

$$\int_0^h \frac{u}{u_e}\left(1 - \frac{\rho}{\rho_e}\right).\,dz = \delta^* - \delta^{*i}$$

Hence, the energy integral equation becomes

$$\frac{d}{dx}\left(\tfrac{1}{2}\rho_e u_e^3 \delta_E\right) + \rho_e u_e^2 \frac{du_e}{dx}\left[\delta^* - \delta^{*i}\right] = \int_0^h \mu \left(\frac{\partial u}{\partial z}\right)^2 dz \qquad (2.52)$$

$$= \int_0^h \tau_{xz} \frac{\partial u}{\partial z} dz$$

This is the rate of energy dissipation by the shear stress τ_{xz}.

Note that with the RHS expressed in this latter form the equation is applicable to both laminar and turbulent flow. In incompressible flow the second term vanishes and we have

$$\frac{d}{dx}\left(\tfrac{1}{2}u_e^3 \delta_E\right) = \nu \int_0^h \left(\frac{\partial u}{\partial z}\right)^2 dz = \frac{1}{\rho}\int_0^h \tau_{xz}\frac{\partial u}{\partial z} dz \qquad (2.53)$$

2.11.3 The total energy equation

The boundary layer equations of motion and energy (2.32) and (2.33) are

$$\rho \frac{Du}{Dt} = -\frac{dp}{dx} + \frac{\partial}{\partial z}\left(\mu \frac{\partial u}{\partial z}\right)$$

and $$\rho \frac{Di}{Dt} = \frac{Dp}{Dt} + \frac{\partial}{\partial z}\left(k \frac{\partial T}{\partial z}\right) + \mu\left(\frac{\partial u}{\partial z}\right)^2$$

where $$i = \int^T c_p dT = c_p T$$

for a perfect gas with constant specific heats.

If we now multiply the first of these equations by u and add to the second, we get for steady flow

$$\frac{\rho D(c_p T_H)}{Dt} = \frac{\partial}{\partial z}\left\{\frac{k}{c_p}\left[\frac{\partial}{\partial z}\left(c_p T + \tfrac{1}{2}\sigma u^2\right)\right]\right\} \qquad (2.54)$$

where $T_H = T + u^2/2c_p$ and is sometimes called the *total temperature*, whilst $c_p T_H$ is called the *total energy per unit mass*.

If we now integrate this equation with respect to z between 0 and h and take note of the boundary conditions $\partial T/\partial z = \partial u/\partial z = 0$, when $z = h$, we get

$$\int_0^h [\rho D(c_p T_H)/Dt]dz = -k_w(\partial T/\partial z)_w \qquad (2.55)$$

where suffix w denotes wall (or surface values). Using the equation of continuity (2.22) the LHS of equation (2.55) can be written

$$\frac{d}{dx}\int_0^h (\rho u c_p T_H)dz + \int_0^h \frac{\partial}{\partial z}(\rho w c_p T_H)dz$$

$$= \frac{d}{dx}\left\{c_p\left[\int_0^h \rho u (T_H - T_{He})dz\right]\right\} \qquad (2.56)$$

If we now integrate equation (2.55) with respect to x from x_1 to x_2, say, then we get

$$\left[\int_0^h \rho u c_p(T_H - T_{He})\mathrm{d}z\right]_{x_1}^{x_2} = -\int_{x_1}^{x_2} k_w(\partial T/\partial z)_w \,\mathrm{d}x$$
$$= -\int_{x_1}^{x_2} k_w(\partial T_H/\partial z)_w \,\mathrm{d}x \qquad (2.57)$$

The RHS is the rate at which heat energy is being transferred from the surface between x_1 and x_2 to the fluid, and the LHS is the difference between the flux of total energy increment across normals to the surface through the boundary layer at x_1 and x_2. If the surface is insulated and so no heat is transferred from it to the fluid then it follows that

$$\int_0^h \rho u c_p(T_H - T_{He})\mathrm{d}z \text{ is a constant } = 0$$

since at the start of the boundary layer $T_H = T_{He}$. It then follows that

$$\int_0^h \rho u c_p T_H \mathrm{d}z \Big/ \int_0^h \rho u \mathrm{d}z = c_p T_{He} \qquad (2.58)$$

Thus, the mean rate of flow of total energy across a surface normal through the boundary layer divided by the mean rate of mass flow is equal to the total energy per unit mass outside the boundary layer.

If we can also assume $\sigma = 1.0$ then equation (2.54) simplifies to

$$\rho \frac{\mathrm{D}T_H}{\mathrm{D}t} = \frac{\partial}{\partial z}\left(\mu \frac{\partial T_H}{\partial z}\right) \qquad (2.59)$$

of which a solution is clearly

$$T_H = \text{const.} = T_{He} \qquad (2.60)$$

It is readily apparent that this solution must be that for zero heat transfer at the wall since it requires that $(\partial T_H/\partial z)_w = 0$.

Another interesting deduction from equation (2.59) is obtained when we compare it with the equation of motion for the case with zero pressure gradient, namely,

$$\rho \frac{\mathrm{D}u}{Dt} = \frac{\partial}{\partial z}\left(\mu \frac{\partial u}{\partial z}\right)$$

We see that a relation of the form

$$T_H = K_1 u + K_2 \qquad (2.61)$$

is admissible when $\sigma = 1.0$, where K_1 and K_2 are constants determined by the temperature boundary conditions.

2.12 A simple example of the application of the momentum integral equation

To illustrate at a very simple level the use of the MIE let us consider a steady laminar boundary layer over a flat plate at zero incidence in incompressible flow. Here p and ρ are constant and the MIE [equation (2.50)] becomes

$$d\theta/dx = \tau_w/\rho u_e^2 \tag{2.62}$$

We now assume that the velocity profile across the boundary layer for any value of x is of the form

$$u/u_e = \text{a function of } \eta = z/\delta \tag{2.63}$$

This assumption implies similarity of the velocity profiles for all x, i.e. the profiles for different values of x differ only in a scaling factor on the z ordinate, where the scaling factor is a function of x only. In this instance the scaling factor is $1/\delta$, where δ as a function of x is to be determined by satisfying equation (2.62). The assumption of similarity is consistent with the fact that the flow has no characteristic geometric length.

The function of η in equation (2.63) is a matter of choice and depends on the number of boundary conditions one wishes to satisfy. In the earliest use of the MIE by Pohlhausen[2.9] a quartic was assumed but other polynomials or suitable transcendental functions can be chosen. Here, we shall postulate

$$u/u_e = \sin(\pi\eta/2) \tag{2.64}$$

We see that this satisfies the boundary conditions (see section 2.10):

$$u = 0, \qquad \partial^2 u/\partial z^2 = 0 \qquad \text{at } z = 0$$

and $\quad u = u_e, \qquad \partial u/\partial z = 0 \qquad \text{at } z = \delta$

Hence $\quad \tau_w = \mu(\partial u/\partial z)_w = \mu u_e \pi/2\delta$

and

$$\delta^* = \int_0^\delta \left(1 - \frac{u}{u_e}\right) dz = \delta\left(1 - \frac{2}{\pi}\right)$$

$$\theta = \int_0^\delta \frac{u}{u_e}\left(1 - \frac{u}{u_e}\right) dz = \frac{\delta(4 - \pi)}{2\pi}$$

Substituting these expressions for τ_w and θ into equation (2.62) we get

$$(4 - \pi)\frac{d\delta}{dx} = \frac{2\pi^2 \nu}{2u_e\delta} \qquad \text{or} \qquad \frac{d\delta^2}{dx} = \frac{2\pi^2\nu}{(4 - \pi)u_e}$$

Since $\delta = 0$ when $x = 0$ it follows that

$$\delta/x = [2\pi^2/(4 - \pi)]^{1/2}/R_x^{1/2} = 4.795/R_x^{1/2} \tag{2.65}$$

Also, the local skin friction coefficient is

$$c_{\mathrm{f}} = 2\tau_{\mathrm{w}}/\rho u_{\mathrm{e}}^2 = [(4 - \pi)/2R_x]^{1/2} = 0.6551/R_x^{1/2} \tag{2.66}$$

and the overall skin friction coefficient for one surface of chord c is

$$C_{\mathrm{F}} = \int_0^c c_{\mathrm{f}}\,\mathrm{d}x/c = 1.310/R_c^{1/2} \tag{2.67}$$

where $R_c = u_{\mathrm{e}}c/\nu$. One can further deduce that

$$\delta^*/x = 1.702/R_x^{1/2}, \qquad \theta/x = 0.655/R_x^{1/2} \qquad H = \delta^*/\theta = 2.660 \tag{2.68}$$

As we shall see (Section 3.2) these results compare quite well with the corresponding results of the exact solution of the boundary layer equations (the Blasius solution) which are

$$c_{\mathrm{f}} = 0.664/R_x^{1/2}, \qquad C_{\mathrm{F}} = 1.328/R_c^{1/2}$$

$$\delta^*/x = 1.721/R_x^{1/2}, \qquad \theta/x = 0.664/R_x^{1/2}, \qquad H = 2.59 \tag{2.69}$$

The exact solution does not strictly yield a finite value of δ but we may note that for

$z/x = 5/R_x^{1/2}, \qquad u/u_{\mathrm{e}} = 0.9916$

whilst for $z/x = 6/R_x^{1/2}, \qquad u/u_{\mathrm{e}} = 0.9990$

It is of interest to record that even as crude an assumpton as a linear velocity profile (i.e. $u/u_{\mathrm{e}} = z/\delta$) yields the following surprisingly good results

$$c_{\mathrm{f}} = \theta/x = 0.557/R_x^{1/2}, \qquad C_{\mathrm{F}} = 1.155/R_c^{1/2}$$

$$\delta^*/x = 1.732/R_x^{1/2}, \qquad H = 3.00 \tag{2.70}$$

whilst the classical quartic profile of Pohlhausen ($u/u_{\mathrm{e}} = 2\eta - 2\eta^2 + \eta^4$) gives

$$c_{\mathrm{f}} = \theta/x = 0.686/R_x^{1/2}, \qquad C_{\mathrm{F}} = 1.372/R_c^{1/2}$$

$$\delta^*/x = 1.752/R_x^{1/2}, \qquad H = 2.55 \tag{2.71}$$

We see that for all the above sets of results

$$\theta/x = c_{\mathrm{f}} = C_{\mathrm{F}}/2 \tag{2.72}$$

(with $c = x$). These relations are quite general for the flat plate boundary layer with zero pressure gradient and can be derived directly from the result that $\theta = Kx^{1/2}$, where K is independent of x. For then, from equation (2.62),

$$c_{\mathrm{f}}/2 = \tau_{\mathrm{w}}/\rho u_{\mathrm{e}}^2 = \mathrm{d}\theta/\mathrm{d}x = K/2x^{1/2} = \theta/2x$$

so $\quad c_{\mathrm{f}} = \theta/x$

and $\quad C_F(x) = \frac{1}{x}\int_0^x c_f \mathrm{d}x = \frac{1}{x}\int_0^x K/x^{1/2}\mathrm{d}x = 2K/x^{1/2} = 2c_f = 2\theta/x$

Another illuminating way of arriving at this last result is as follows. Consider a control volume ABCD in the form of a box of unit depth as sketched in Fig. 2.7 of which AB is one side of the plate of length c. AD and BC are of length l where $l \gg \delta$. There will be a mass flow rate across DC = defect in the mass flow rate in the boundary layer across BC

$$= \int_0^l (\rho_e u_e - \rho u)\mathrm{d}z$$

This will transport momentum in the x direction

$$= \int_0^l u_e(\rho_e u_e - \rho u)\mathrm{d}z$$

out of the box. The momentum flux across AD minus that across BC

$$= \int_0^l (\rho_e u_e^2 - \rho u^2)\mathrm{d}z$$

and so the net rate of loss of momentum in the x direction

$$= \int_0^l \rho u(u_e - u)\mathrm{d}z$$

Since the pressure is constant there are no forces acting on the control volume other than the frictional force F acting on the plate due to the frictional stresses there. We can therefore write

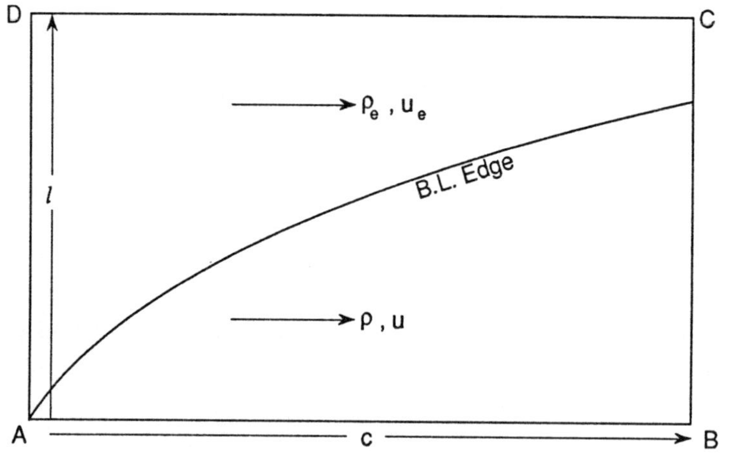

Fig. 2.7 Control volume for deriving the relation $C_F = 2\theta_c/c$ for the boundary layer on a flat plate at zero incidence.

$$F = \int_0^l \rho u(u_e - u)dz$$

and hence

$$C_F = 2F/\rho_e u_e^2 c = 2\theta_c/c$$

where θ_c is the value of θ at $x = c$.

References

2.1 Duncan, W.J., Thom, A.S., and Young A.D. (1970) *Mechanics of Fluids*, 2nd Ed., Ed. Arnold., Ch. 9.
2.2 Schlichting, H. (1979) *Boundary Layer Theory*, 7th Ed., McGraw-Hill.
2.3 Rosenhead, L. (Ed.) (1963) *Laminar Boundary Layers*, OUP.
2.4 Rayleigh, Lord (1911) 'On the motion of solid bodies through viscous fluids'. *Phil. Mag.* (6), **21**, p. 697.
2.5 Stokes, G.G. (1851) 'On the effect of internal friction of fluids on the motion of pendulums'. *Camb. Phil. Trans.* IX, 8.
2.6 Gersten, K. and Gross, J.F. (1975) 'Higher order boundary layer theory'. *Fluid Dynamics Transactions.*
2.7 Mangler, K.W. (1948) 'Zusammenhang zwischen ebenen und rotationssymmetrischen Grenzschichten in kompressiblen Flüssigkeiten' *ZAMM*, **28**, p. 97.
2.8 Von Karman, Th. (1921) 'Über laminare und turbulente Reibung' *ZAMM* **1**, p. 233.
2.9 Pohlhausen, K. (1921) 'Zur näherungsweisen Integration der Differentialgleichung der laminaren Reibungsschicht'. *ZAMM* **1**, p. 252.

Chapter 3

Some Basic Solutions of the Steady Laminar Boundary Layer Equations in Two Dimensions

3.1 Introduction

In general, the laminar boundary layer equations form a system of partial differential equations which in principle can be solved numerically for any particular case given the external pressure distribution. Modern computers have made this task much easier than in the early days of the development of the subject. Nevertheless, a number of basic solutions that were then developed are still of fundamental importance in revealing essential physical features and in providing guidance and benchmarks for the development of relatively simple approximate methods of solution. The latter are needed because of their speed and low computing costs particularly in the early stages of engineering design.

A characteristic of these basic solutions is that they refer to cases for which the boundary layer equations reduce to ordinary differential equations which are much more readily solved either numerically or analytically and their solutions have been obtained to a high degree of accuracy. In this chapter we will discuss a number of these basic solutions for both incompressible and compressible flow.

3.2 Incompressible flow over a flat plate at zero incidence

3.2.1 The classical Blasius solution over a solid surface

For this case the pressure is constant and the boundary layer equations of continuity and momentum become:

$$\left.\begin{aligned} &\frac{\partial u}{\partial x} + \frac{\partial w}{\partial z} = 0 \\ \text{and} \quad &u\frac{\partial u}{\partial x} + w\frac{\partial u}{\partial z} = \nu\partial^2 u/\partial z^2 \end{aligned}\right\} \tag{3.1}$$

The boundary conditions are $u = w = 0$ at $z = 0$; $u = u_e$ at $z = \infty$.

We have already noted (Section 2.12) that in this case there is no characteristic geometric length to determine the flow and hence the velocity profiles for different values of x are similar and can be made identical by imposing scaling factors on z which are functions of x only. We therefore seek for a solution in the form

$$u/u_e = f(\zeta), \qquad \text{where} \quad \zeta = z/\phi(x), \quad \text{say} \tag{3.2}$$

so that the lines $z = \text{const.}\ \phi(x)$ are loci of constant u. We have seen that measures of the boundary layer thickness vary as $x/R_x^{1/2}$, so we are led to consider

$$\zeta = \text{const.}\ \frac{z}{x}\left(\frac{u_e x}{\nu}\right)^{1/2} = \frac{z}{2}\left(\frac{u_e}{\nu x}\right)^{1/2} \tag{3.3}$$

where for later numerical convenience we have taken the constant as $\frac{1}{2}$.

From the equation of continuity it follows that there is a stream function ψ such that

$$u = \partial\psi/\partial z \qquad w = -\partial\psi/\partial x \tag{3.4}$$

We can therefore replace equations (3.1) by one equation for ψ which we seek to express non-dimensionally as a function of ζ only. We write therefore

$$\psi = (u_e x \nu)^{1/2}\ .\ F(\zeta) \tag{3.5}$$

and our problem then reduces to determining the function $F(\zeta)$ that satisfies the momentum equation in (3.1). We see that

$$u = \frac{\partial\psi}{\partial z} = \frac{\partial\psi}{\partial\zeta}\frac{\partial\zeta}{\partial z} = (u_e x\nu)^{1/2}\frac{F'}{2}\left(\frac{u_e}{\nu x}\right)^{1/2} = \frac{u_e F'}{2} \tag{3.6}$$

where the prime denotes differentiation with respect to ζ. We note that this relation is in accordance with the similarity condition, equation (3.2). Also

$$w = -\partial\psi/\partial x = \tfrac{1}{2}(u_e\nu/x)^{1/2}\,(F'\zeta - F) \tag{3.7}$$

Hence $$\frac{\partial u}{\partial x} = \frac{u_e}{2}F''\frac{z}{4x}\left(\frac{u_e}{\nu x}\right)^{1/2},\quad \frac{\partial u}{\partial z} = u_e\frac{F''}{4}\left(\frac{u_e}{\nu x}\right)^{1/2}$$

$$\frac{\partial^2 u}{\partial z^2} = \frac{u_e^2 F'''}{8\nu x}$$

The momentum equation then reduces to

$$F\,F'' + F''' = 0 \tag{3.8}$$

with the boundary conditions $F = 0$, $F' = 0$ at $\zeta = 0$; and $F' = 2$ at $\zeta = \infty$.

Thus, the self-consistency of the assumption of similarity is demonstrated since we have reduced the two partial differential equations in x and z with which we started to a single ordinary differential in ζ. Our problem therefore reduces to solving equation (3.8). This is a non-linear equation of the third order and it does not lend itself to straightforward analytic solution but must be solved numerically. This was first done by Blasius[3.1] in 1908 and the resulting solution is generally associated with his name. Subsequent authors have improved on the accuracy or speed of his solution. It is a boundary value or two point problem, i.e. the boundary conditions that must be satisfied are at the extreme values of ζ, namely 0 and ∞. The details of Blasius's mode of solution will not be given here, suffice it to say that it involved an expression for F in a power series in ζ for small ζ which satisfied the boundary conditions at $\zeta = 0$ in terms of the unknown $F''(0)$, the value of which was then determined by making the solution compatible with an asymptotic solution for $\zeta \to \infty$ which satisfied the boundary condition there.

However, it was subsequently noted that near $\zeta = 0$, F can be expressed as the series

$$F = \frac{\alpha\zeta^2}{2!} - \frac{\alpha^2\zeta^5}{5!} + 11\,\frac{\alpha^3\zeta^8}{8!} + 375\,\frac{\alpha^4\zeta^{11}}{11!} + \ldots$$

where $\alpha = F''(0)$, and this series as well as equation (3.8) are consistent with

$$F/\alpha^{1/3} = \text{function of } (\alpha^{1/3}\zeta) = G(\alpha^{1/3}\zeta), \quad \text{say}$$

Hence if we take $\alpha = 1$ we can obtain a numerical integration of equation (3.8), for which this series is used to provide starting values near $\zeta = 0$, and this yields at $\zeta = \infty$

$$G'(\infty) = F'(\infty)/\alpha^{2/3} = 2/\alpha^{2/3}$$

so $$\alpha = [G'(\infty)/2]^{-2/3}$$

However, with the widespread use of digital computers the solution of the Blasius equation is today relatively straightforward. Details and programs will be found in various books, notably Cebeci and Bradshaw,[3.2] and need not be reproduced here. Essentially, the equation is replaced by a system of first order ordinary differential equations treating F, F', F'' as the unknown variables and these equations are then rewritten in terms of finite differences with appropriately small intervals in ζ. An iterative shooting process based on Newton's method is then used to ensure that the outer boundary condition is satisfied to a specified accuracy. Some resulting values to four decimals of F and $\frac{1}{2}F'$ $(= u/u_e)$ are given in Table 3.1.

The comments in Section 2.10 about the effective thickness of the boundary layer are readily borne out by Table 3.1. A plot of u/u_e and $(w/u_e)R_x^{1/2}$ as functions of ζ are given in Fig. 3.1. It will be seen that as ζ increases w rapidly approaches the asymptotic value:

$$w/u_e = 0.865/R_x^{1/2} \qquad (3.9)$$

Table 3.1

$\zeta = \frac{1}{2}z(u_e/\nu x)^{1/2}$	F	$\frac{1}{2}F'$ (u/u_e)
0	0	0
0.1	0.0066	0.0664
0.2	0.0266	0.1328
0.3	0.0597	0.1989
0.4	0.1061	0.2647
0.5	0.1656	0.3298
0.6	0.2380	0.3938
0.8	0.4203	0.5168
1.0	0.6500	0.6298
1.2	0.9223	0.7290
1.4	1.2310	0.8115
1.6	1.5691	0.8761
1.8	1.9295	0.9233
2.0	2.3058	0.9555
2.5	3.2833	0.9916
3.0	4.2976	0.9990
3.5	5.2793	0.9999
4.0	6.2792	1.0000

This reflects the displacement effect of the boundary layer on the external flow which is associated with a non-zero normal velocity component at the edge of the boundary layer into the external flow.

The value of $\alpha = F''(0)$ is found to be 1.328 and since $\tau_w = \mu(\partial u/\partial z)_w = \frac{1}{4}\mu u_e(u_e/\nu x)^{1/2}F''(0)$ it follows that the local skin friction coefficient is

$$c_f = 2\,\frac{\tau_w}{\rho u_e^2} = \frac{1}{2}\left(\frac{\nu}{u_e x}\right)^{1/2} F''(0) = \frac{0.664}{R_x^{1/2}} \tag{3.10}$$

Hence, the total skin friction coefficient for one side of the plate of chord c is

$$C_F = \frac{1}{c}\int_0^c c_f \, dx = \frac{1.328}{R^{1/2}} \tag{3.11}$$

where $R = u_e c/\nu$, and the drag coefficient taking both sides of the plate into account is

$$C_D = 2.656/R^{1/2} \tag{3.12}$$

The displacement thickness δ^* is found to be given by

$$\delta^*/x = 1.721/R_x^{1/2} \tag{3.13}$$

and since $d\theta/dx = \frac{1}{2}c_f$, the momentum thickness θ is given by

$$\theta/x = 0.664/R_x^{1/2} \tag{3.14}$$

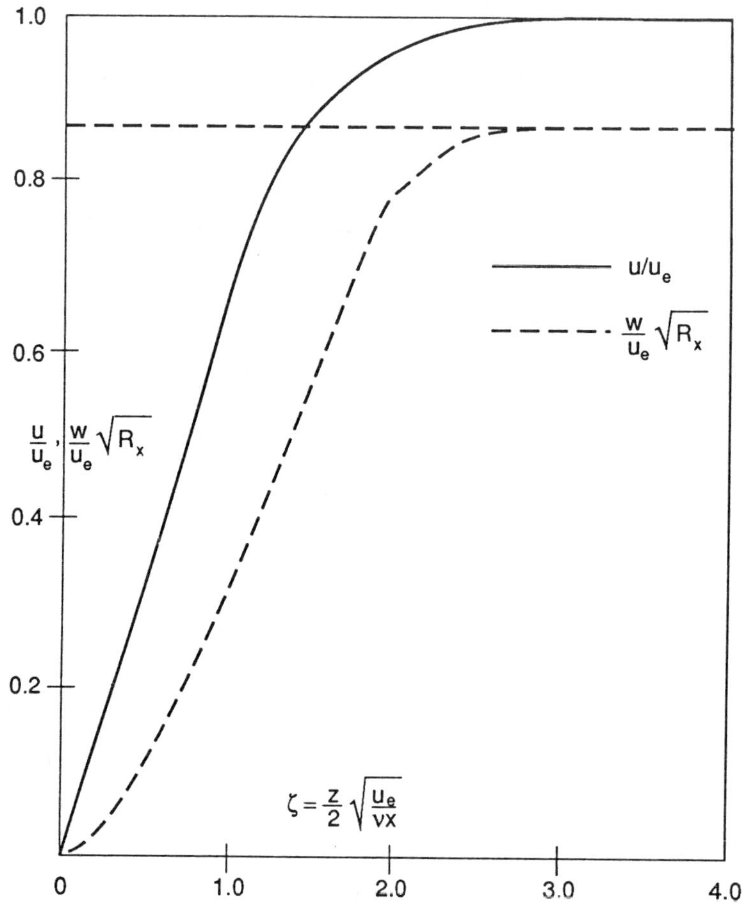

Fig. 3.1 Velocity profile in the laminar boundary layer on a flat plate at zero incidence (Blasius solution).

3.2.2 Asymptotic solution for uniform wall suction

A case of classical simplicity for which the solution was first derived by Griffith and Meredith in 1936[3.3] is the asymptotic boundary layer far downstream from the leading edge of a flat plate with uniform suction at the plate surface which is assumed permeable. The rate of change of quantities with x can then be taken as zero for large x and the velocity component w at the surface is negative. The equations of continuity and momentum reduce to

$$\left.\begin{aligned} \partial w/\partial z &= 0 \\ w\,\partial u/\partial z &= \nu \partial^2 u/\partial z^2 \end{aligned}\right\} \qquad (3.15)$$

and the boundary conditions are $u = 0$, $w = w_w = \text{const.}$ at $z = 0$; $u = u_e$ at $z = \infty$.

It is to be noted that these equations are the asymptotic form as $x \to \infty$ of the Navier–Stokes equations for incompressible flow with uniform pressure and uniform suction.

From the first equation of (3.15) we see that $w = \text{const.} = w_w$, for all z; and the second equation can be integrated to give

$$u/u_e = 1 - \exp(w_w z/\nu) \tag{3.16}$$

It follows that

$$\delta^* = \int_0^\infty (1 - u/u_e)\mathrm{d}z = -\nu/w_w \tag{3.17}$$

$$\theta = \int_0^\infty \frac{u}{u_e}\left(1 - \frac{u}{u_e}\right)\mathrm{d}z = -\frac{\nu}{2w_w} = \tfrac{1}{2}\delta^* \tag{3.18}$$

and

$$c_f = 2\mu(\mathrm{d}u/\mathrm{d}z)_w/(\rho u_e^2) = -2w_w/u_e$$
$$= 1/R_\theta = 2/R_\delta^* \tag{3.19}$$

where $R_\theta = u_e\theta/\nu$, and $R_\delta^* = u_e\delta^*/\nu$.

It is of interest to compare these relations with those of the Blasius solution where

$$c_f = 0.441/R_\theta = 1.143/R_\delta^*$$

We see that we can write the velocity profile as

$$u/u_e = 1 - \exp(-z/2\theta)$$

Not surprisingly, this profile as a function of z/θ is considerably fuller than the Blasius profile since the suction continuously removes the slower innermost regions of the layer.

3.2.3 The Pohlhausen solution[3.4] for the temperature boundary layer

In incompressible flow the momentum equation is uncoupled from the energy equation and for the case of constant pressure the velocity boundary layer given by the Blasius solution of the previous section applies whatever the temperature distribution. To determine the temperature boundary layer we must then solve the energy equation which for this case can be written

$$\rho c_p(u\,\partial T/\partial x + w\,\partial T/\partial z) = k\,\partial^2 T/\partial z^2 + \mu(\partial u/\partial z)^2 \tag{3.20}$$

The conduction term on the RHS is of order kT_e/δ_T^2 and the dissipation term is of order $\mu(u_e/\delta_u)^2$ and so the ratio of the latter to the former is of order $u_e^2/c_p T_e$, since $\sigma = \mu\, c_p/k$ and $\delta_u/\delta_T = O(\sigma^{1/2})$. We have

$$c_p T_e = \frac{\gamma}{\gamma - 1}\frac{p_e}{\rho_e} = \frac{a_e^2}{\gamma - 1}$$

for a perfect gas with constant specific heats, here a_e is the speed of sound in the external flow. Hence the above ratio $= O[M_e^2(\gamma - 1)]$, where M_e is the flow Mach number at the outer edge of the boundary layer. Therefore, at low speeds we may neglect the dissipation term as compared with the conduction term in equation (3.20) and we have

$$\rho c_p(u\,\partial T/\partial x + w\,\partial T/\partial z) = k\,\partial^2 T/\partial z^2 \tag{3.21}$$

with the boundary conditions $T = T_w$, say, at $z = 0$; and $T = T_e$ at $z = \infty$. Just as for the velocity profiles the condition of similarity should hold for the temperature profiles and so we write

$$(T - T_w)/(T_e - T_w) = \phi(\zeta), \quad \text{say} \tag{3.22}$$

where ζ is given in equation (3.3). Then, after a little algebra equation (3.21) reduces to

$$\phi'' + \phi' F\sigma = 0 \tag{3.23}$$

where F is the solution of the Blasius equation (3.8). The boundary conditions for ϕ are $\phi = 0$ at $\zeta = 0$; $\phi = 1$ at $\zeta = \infty$. This equation can be integrated to give

$$\phi' = C \exp\left(\int_0^\zeta -F\sigma\, d\zeta\right)$$

where C is a constant. Integrating again we get

$$\phi = C \int_0^\zeta \left\{\exp\left[\int_0^\zeta -F(\zeta_1)\sigma\, d\zeta_1\right] d\zeta\right\}$$

Since $\phi = 1$ at $\zeta = \infty$

$$C = 1\Big/\int_0^\infty \exp\left(\int_0^\zeta -F\sigma\, d\zeta\right) d\zeta = C(\sigma), \quad \text{say} \tag{3.24}$$

From equation (3.8) we have $F = -F'''/F''$, and we deduce that

$$\phi = C(\sigma) \int_0^\zeta \left[\frac{F''(\zeta)}{F''(0)}\right]^\sigma d\zeta \tag{3.25}$$

Pohlhausen's calculations of $C(\sigma)$ showed that it was closely given by

$$C(\sigma) = 0.664\,\sigma^{1/3} \tag{3.26}$$

The rate of heat transfer from one side of a plate of length c and unit span is

$$Q = -k \int_0^c (\partial T/\partial z)_w\, dx$$

$$= \tfrac{1}{2}k \int_0^c (T_w - T_e)\, \phi'(0) \,.\, (u_e/\nu x)^{1/2}\, dx$$
$$= k(T_w - T_e)\, C(\sigma)\, (u_e c/\nu)^{1/2}$$

We define a Nusselt number Nu, an overall heat transfer coefficient, as

$$\mathrm{Nu} = Ql/(Sk\Delta T)$$

where l is a representative length, S is the surface area and ΔT is the driving temperature difference. In this case we take $l = c$, $S = c\,.\,1$, and $\Delta T = T_w - T_e$. Then

$$\begin{aligned} \mathrm{Nu} &= C(\sigma)R^{1/2}, \qquad \text{where } R = u_e c/\nu \\ &= 0.664\,\sigma^{1/3}\,R^{1/2} \end{aligned} \tag{3.27}$$

Equations (3.11) and (3.27) show that there is a simple relation between Nu and the overall skin friction coefficient C_F, namely

$$\mathrm{Nu} = \tfrac{1}{2}C_F\, R\, \sigma^{1/3} \simeq 0.448\, C_F\, R \text{ for air } (\sigma = 0.72) \tag{3.28}$$

Alternatively, we can define a Stanton number St by

$$\mathrm{St} = Q/(S\rho u_e c_p \Delta T),$$

which leads to

$$\begin{aligned} \mathrm{St} &= k\, C(\sigma)\, R^{1/2}/(\rho u_e c_p c) = C(\sigma)/(\sigma\, R^{1/2}) \\ &= 0.664\,\sigma^{-2/3}/R^{1/2} = \tfrac{1}{2}C_F\,\sigma^{-2/3} \end{aligned} \tag{3.29}$$

Like the Nusselt number this Stanton number is an overall heat transfer coefficient. However, we can define a Stanton number which is a local heat transfer coefficient, namely

$$\mathrm{St}_l = (\mathrm{d}Q/\mathrm{d}S)/(\rho u_e c_p \Delta T)$$

Now, per unit span,

$$\begin{aligned} \mathrm{d}Q/\mathrm{d}S &= \mathrm{d}Q/\mathrm{d}x \\ &= -k(\partial T/\partial z)_w \\ &= \tfrac{1}{2}k(T_w - T_e)\phi'(0)(u_e/\nu x)^{1/2} \end{aligned}$$

Hence

$$\left.\begin{aligned} \mathrm{St}_l(x) &= \phi'(0)\, R_x^{-1/2}/2\sigma = C(\sigma)\, R_x^{-1/2}/2\sigma \\ &= 0.332\,\sigma^{-2/3}\, R_x^{-1/2} = \tfrac{1}{2}c_f\,\sigma^{-2/3} \end{aligned}\right\} \tag{3.30}$$
$$\simeq 0.622\, c_f \text{ for air}$$

Thus, we have established a simple relationship between the local Stanton number and the local skin friction coefficient. Such relations reflect the fact that there is a close analogy between heat and momentum transfer in the restrictive conditions postulated. In the presence of a pressure gradient the analogy weakens, essentially because energy is a scalar quantity whereas momentum is a vector directly influenced by the pressure gradient.

3.2.4 The wall temperature for zero heat transfer – the thermometer problem

To determine the wall temperature when there is zero heat transfer we can no longer ignore the dissipation term in equation (3.20), as it is the conversion of the kinetic energy of the fluid into heat by frictional effects that causes the wall temperature to differ from that of the external flow. For this reason we can expect that $T - T_e$ is proportional to u_e^2/c_p times a function of ζ, and so we write

$$T - T_e = \frac{u_e^2}{2c_p}\,\theta(\zeta), \quad \text{say} \tag{3.31}$$

Equation (3.20) then reduces to

$$\theta'' + \sigma F \theta' = -\tfrac{1}{2}\sigma F''^2 \tag{3.32}$$

and the boundary conditions are $\theta' = 0$ for $\zeta = 0$; $\theta = 0$ for $\zeta \to \infty$.

If we now multiply equation (3.32) by the integrating factor

$$I = \exp\left[\int_0^\zeta \sigma F \, d\zeta\right]$$

we obtain

$$d(\theta' I)/d\zeta = -\sigma F''^2 I/2$$

hence
$$\theta(\zeta) = \tfrac{1}{2}\sigma \int_\zeta^\infty \frac{1}{I}\left[\int_0^\zeta F''^2 I \, d\zeta\right] d\zeta$$

when account is taken of the boundary conditions. But since $F = -F'''/F''$ from equation (3.8), we have

$$I = [F''(\zeta)/F''(0)]^{-\sigma}$$

hence
$$\theta(\zeta) = \tfrac{1}{2}\sigma \int_\zeta^\infty (F'')^\sigma \left[\int_0^\zeta (F'')^{2-\sigma} \, d\zeta\right] d\zeta \tag{3.33}$$

We note that when $\sigma = 1$, $\theta = 1 - F'^2/4 = 1 - (u/u_e)^2$, and so

$$T + \frac{u^2}{2c_p} = T_e + \frac{u_e^2}{2c_p}$$

and hence the total temperature T_H is constant across the boundary layer as anticipated in Section 2.11, equation (2.60).

Pohlhausen[3.4] has evaluated $\theta(\zeta)$ from equation (3.33) for a range of values of σ and some of his results are illustrated in Fig. 3.2. Of particular interest is his result that a good approximation to $\theta(0)$ as a function of σ is $\theta(0) \simeq \sigma^{1/2}$, so that the wall temperature is

$$T_w \simeq T_e + u_e^2 \, \sigma^{1/2}/2c_p \tag{3.34}$$

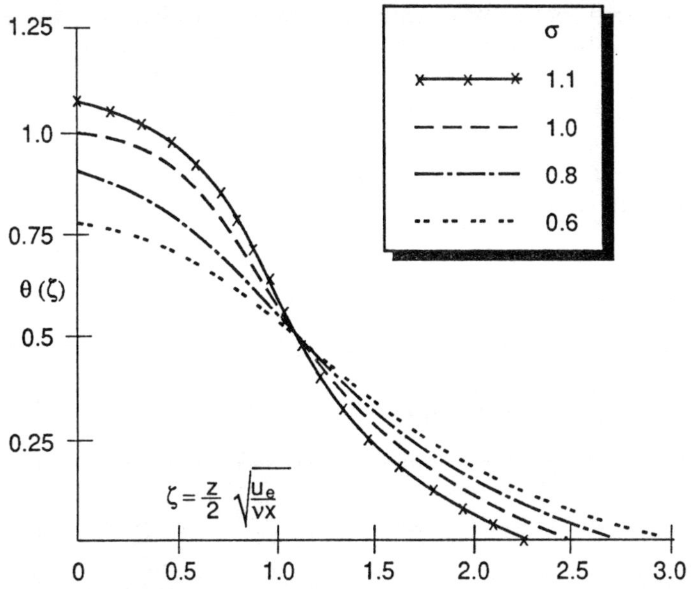

Fig. 3.2 Temperature distribution in the boundary layer of an insulated flat plate – Pohlhausen solution for the 'thermometer problem'.

The factor $\sigma^{1/2}$ is referred to as a *recovery factor*. It is a measure of the external kinetic energy that is recovered as heat energy at the wall where the flow has been brought to rest by the frictional stresses in the boundary layer.

3.3 Steady compressible flow over a flat plate at zero incidence

3.3.1 Introduction

Notable early attempts at dealing with the basic problem of compressible flow over an impermeable surface with zero pressure gradient are associated with the names of Busemann[3.5], von Karman and Tsien[3.6], Emmons and Brainerd[3.7], Hantsche and Wendt[3.8], and Crocco[3.9]. They differed somewhat in approach and in the assumptions made, and the details of all their methods and results cannot be gone into here. We shall confine ourselves to presenting the salient features of Crocco's work because of its general scope and its elegance.

However, some key results anticipated by the earlier workers are worth briefly noting at this stage. Thus, Emmons and Brainerd found that equation (3.34) for the wall temperature with zero heat transfer applied with good accuracy for compressible flow as well as in the case considered by Pohlhausen. With zero heat transfer they all noted a thickening of the boundary layer and a tendency for the velocity profile to become more linear with increase of Mach number. Wall cooling tended to reduce the boundary layer thickness. These points are illustrated by some velocity and temperature profiles obtained by von Karman and Tsien shown in Fig. 3.3.

3.3.2 *The Crocco transformation*[3.9]

The boundary layer equations (Section 2.7) reduce in this case to:

$$\left.\begin{aligned} \frac{\partial}{\partial x}(\rho u) + \frac{\partial}{\partial z}(\rho w) &= 0 \\ \rho u \frac{\partial u}{\partial x} + \rho w \frac{\partial u}{\partial z} &= \frac{\partial}{\partial z}\left(\mu \frac{\partial u}{\partial z}\right) \\ \rho u \frac{\partial i}{\partial x} + \rho w \frac{\partial i}{\partial z} &= \frac{\partial}{\partial z}\left(k \frac{\partial T}{\partial z}\right) + \mu \left(\frac{\partial u}{\partial z}\right)^2 \\ &= \frac{1}{\sigma}\frac{\partial}{\partial z}\left(\mu \frac{\partial i}{\partial z}\right) + \mu\left(\frac{\partial u}{\partial z}\right)^2 \end{aligned}\right\} \tag{3.35}$$

Here $\quad i = \int^T c_p \, \mathrm{d}T = c_p T$

for a perfect gas, and we take σ as constant. The boundary conditions are $u = w = 0$ at $z = 0$; $u = u_e$ at $z = \infty$. Also, at $z = 0$, either i_w or zero heat transfer $[(\partial T/\partial z)_w = (\partial i/\partial z)_w = 0]$ is specified; and at $z = \infty$, $i = i_e$.

Following Crocco we change the independent variables from (x, z) to (x, u) and we make the shear stress τ and the enthalpy i the dependent variables. Now if q is a dependent variable and we express δq in terms of small increments in the two sets of independent variables we readily find that

$$\left(\frac{\partial q}{\partial x}\right)_u = \left(\frac{\partial q}{\partial x}\right)_z + \left(\frac{\partial q}{\partial z}\right)_x \left(\frac{\partial z}{\partial x}\right)_u, \qquad \left(\frac{\partial q}{\partial u}\right)_x = \left(\frac{\partial q}{\partial z}\right)_x \left(\frac{\partial z}{\partial u}\right)_x$$

where the suffix denotes the variable kept constant in the partial differentiation. With the aid of these relations equations (3.35) transform in terms of the variables (x, u) to:

$$\left.\begin{aligned} \frac{\partial z}{\partial u}\frac{\partial}{\partial x}(\rho u) - \frac{\partial z}{\partial x}\frac{\partial}{\partial u}(\rho u) + \frac{\partial}{\partial u}(\rho w) &= 0 \\ -\rho u \frac{\partial z}{\partial x} + \rho w &= \partial \tau/\partial u \\ (1-\sigma)\frac{\partial \tau}{\partial u}\frac{\partial i}{\partial u} + \left(\frac{\partial^2 i}{\partial u^2} + \sigma\right)\tau - \rho u \sigma \frac{\partial i}{\partial x}\frac{\partial z}{\partial u} &= 0 \end{aligned}\right\} \tag{3.36}$$

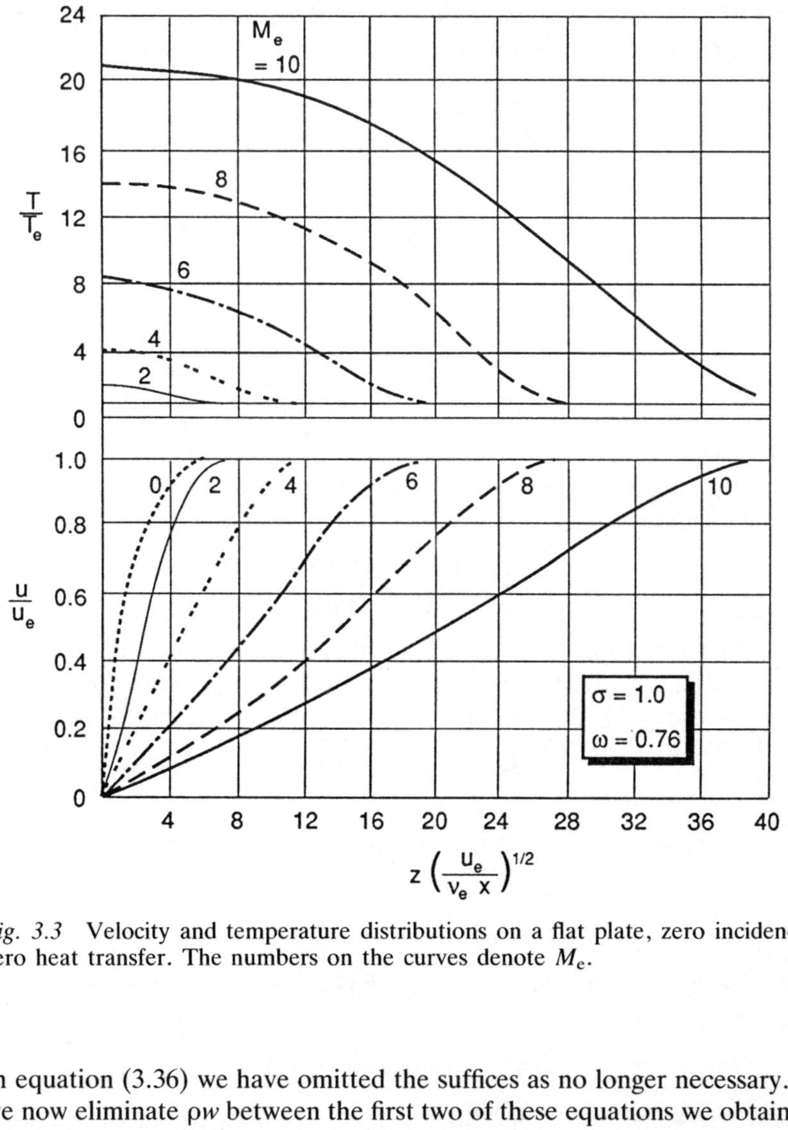

Fig. 3.3 Velocity and temperature distributions on a flat plate, zero incidence, zero heat transfer. The numbers on the curves denote M_e.

In equation (3.36) we have omitted the suffices as no longer necessary. If we now eliminate ρw between the first two of these equations we obtain:

$$u\frac{\partial}{\partial x}(\rho\mu/\tau) + \frac{\partial^2\tau}{\partial u^2} = 0$$

and the energy equation can be written

$$(1-\sigma)\tau\frac{\partial\tau}{\partial u}\frac{\partial i}{\partial u} + \left(\frac{\partial^2 i}{\partial u^2} + \sigma\right)\tau^2 - \rho u\sigma\mu\frac{\partial i}{\partial x} = 0 \qquad (3.37)$$

As in the case of incompressible flow we now seek for a solution in which both the temperature and velocity profiles form similar families for different values of x, and the same argument for such similarity applies in view of the absence of a geometric length scale controlling the flow. In that case both i and u must be functions of a single variable ζ of the form $z\phi(x)$, say. Hence i can be expressed as a function of u only and is independent of x, so $\partial i/\partial x = 0$. Also, since ζ is expressible as a function of u only as is μ, we have

$$\tau = \mu \Big/ \frac{\partial z}{\partial u} = \phi(x) \text{ times a function of } u = \phi(x) \,.\, f(u), \text{ say} \tag{3.38}$$

Therefore equations (3.37) become

$$\phi(x) \,.\, f''(u) - \frac{\rho u \mu}{f(u)} \frac{\phi'(x)}{\phi(x)^2} = 0$$

$$\phi(x)^2 f(u) \left[(1 - \sigma) \frac{di}{du} f'(u) + \left(\frac{d^2 i}{du^2} + \sigma \right) . f(u) \right] = 0$$

where the primes denote differentiation with respect to the appropriate variable. Hence, we see from the first of these equations that

$$\frac{\phi'(x)}{\phi(x)^3} = \frac{f''(u) \,.\, f(u)}{u \rho \mu}$$

Since the LHS of this equation is a function of x only, whilst the RHS is a function of u only, both sides must be constant so that $\phi(x) = \text{const.}/x^{1/2}$, as in incompressible flow, where x is measured from the leading edge. We can therefore write without loss of generality

$$\tau = f(u)/(2x)^{1/2} \tag{3.39}$$

The above equations for f and i as functions of u then become

$$\left. \begin{aligned} ff'' + u\rho\mu &= 0 \\ (i'' + \sigma)f + (1 - \sigma)i'f' &= 0 \end{aligned} \right\} \tag{3.40}$$

We now make these equations non-dimensional by writing $u/u_e = \eta$, $i/i_e = \theta$, $\rho/\rho_e = r$, $\mu/\mu_e = m$ and

$$G(\eta) = f(u) \,.\, (2/\rho_e \mu_e u_e^3)^{1/2}$$

Then equations (3.40) become

$$\left. \begin{aligned} GG'' + 2\eta r m &= 0 \\ (\theta'' + \sigma u_e^2/i_e)G + (1 - \sigma)\theta' G' &= 0 \end{aligned} \right\} \tag{3.41}$$

The boundary conditions are $G' = 0$, $\theta = \theta_w = i_w/i_e$ at $\eta = 0$ and $G = 0$, $\theta = 1$ at $\eta = 1$.

Now $\quad u_e^2/i_e = u_e^2/c_p T_e = (\gamma - 1)u_e^2/a_e^2 = (\gamma - 1)M_e^2 \qquad (3.42)$

for a perfect gas with constant specific heats, where $\gamma = c_p/c_v$, a_e and M_e are the speed of sound and the Mach number, respectively, in the external flow. The equation of state for a perfect gas with constant pressure leads to ρ/T = const. and hence $r\theta = 1$. If we assume that the relation between viscosity and temperature is of the form $\mu \propto T^\omega$, then $m = \theta^\omega$, and so the first of equations (3.41) can be written

$$GG'' + 2\eta\theta^{\omega-1} = 0 \tag{3.43}$$

The local skin friction coefficient is

$$c_f = 2\tau_w/\rho_e u_e^2 = 2f(0)/[(2x)^{1/2}\rho_e u_e^2] = G(0)/R_x^{1/2} \tag{3.44}$$

where $R_x = u_e x/\nu$.

Also, since $\quad \tau = \mu/(\partial z/\partial u)$

$$z = \int_0^u (\mu/\tau)du = (4x/\rho_e u_e^3 \mu_e)^{1/2} u_e \mu_e \int_0^\eta (m/G)\mathrm{d}\eta$$

and so $\quad (z/x)R_x^{1/2} = 2\int_0^\eta (m/G)\mathrm{d}\eta \quad (3.45)$

This equation enables the distance from the wall to be determined for a given η once the profiles $\theta(\eta)$ and $G(\eta)$ are determined.

We see from equations (3.41) that considerable simplifications can arise when either $\sigma = 1$ or $rm = 1$. The latter condition implies $\rho\mu$ = const. and is satisfied if μ = const. T (i.e. $\omega = 1$) or ρ and μ are constant as in incompressible flow. We will now consider these special cases further.

3.3.3 *Case when* $\sigma = 1$

Here the second of equations (3.41) becomes

$$\theta'' + u_e^2/i_e = 0 \tag{3.46}$$

and we infer that θ is a quadratic function of η, independent of the form of the non-dimensional shear stress profile $G(\eta)$. This result was in effect anticipated in Section 2.11. The coefficients of the quadratic are readily determined from the boundary conditions. Having θ as a function of η we can then determine $2\eta rm$ as a function of $\eta = h(\eta)$, say. The first equation of (3.41) then takes the form

$$GG'' + h = 0 \tag{3.47}$$

with the boundary conditions $G' = 0$, when $\eta = 0$; $G = 0$ when $\eta = 1$. If we integrate (3.47) twice we then get

$$G(\eta) = \int_\eta^1 \left[\int_0^\eta (h/G)\mathrm{d}\eta\right]\mathrm{d}\eta \tag{3.48}$$

This equation lends itself to an iterative process of solution, starting with some approximate solution on the RHS, but it can be readily solved numerically with a modern computer using a trial and error process to satisfy the boundary conditions at the end values of η.

3.3.4 *Case when* $\rho\mu$ = *const.*

In this case $rm = 1$ and the first equation of (3.41) becomes

$$GG'' + 2\eta = 0 \tag{3.49}$$

and hence G is independent of θ and of σ. Indeed, this equation can be readily transformed into that for incompressible flow, i.e. the Blasius equation (3.8) by writing

$$G = \tfrac{1}{2}d^2F/\mathrm{d}\zeta^2, \qquad \eta = \tfrac{1}{2}\mathrm{d}F/\mathrm{d}\zeta$$

as can be readily verified.

Since $\eta = 0$ when $\zeta = 0$, $G(0)$ is given by its value for incompressible flow. Therefore, from equation (3.39)

$$c_f R_x^{1/2} = \text{const.} = 0.664 \tag{3.50}$$

as in incompressible flow and this holds whatever the values of σ *and* i_w.

Some values of $G(\eta) = \tau R_x^{1/2}/\tfrac{1}{2}\rho_e u_e^2$ are given in Table 3.2.

Having determined the function $G(\eta)$ we can proceed to solve the second equation of (3.41) for θ. Thus, it can be written

$$\frac{\partial}{\partial\eta}(\theta' G^{1-\sigma}) = -\sigma\frac{u_e^2}{i_e}G^{1-\sigma}$$

Table 3.2

$\eta = u/u_e$	$G(\eta)$
0	0.6641
0.1	0.6636
0.2	0.6601
0.3	0.6505
0.4	0.6317
0.5	0.6001
0.6	0.5518
0.7	0.4816
0.8	0.3819
0.9	0.2388
0.95	0.1410
1.0	0

and so $\theta' = C\,G^{\sigma-1} - \sigma\,G^{\sigma-1}\dfrac{u_e^2}{i_e}\displaystyle\int_0^{\eta} G^{1-\sigma}\mathrm{d}\eta$

where C is a constant. Integrating again, and taking note of the wall boundary condition for θ, we obtain

$$\theta = \theta(0) + \theta'(0)\,A(\eta,\sigma) - \sigma\frac{u_e^2}{i_e}B(\eta,\sigma) \qquad (3.51)$$

where $A(\eta,\sigma) = \displaystyle\int_0^{\eta} [G(\eta_1)/G(0)]^{\sigma-1}\mathrm{d}\eta_1$

and $B(\eta,\sigma) = \displaystyle\int_0^{\eta} [G(\eta_1)/G(0)]^{\sigma-1}\left\{\int_0^{\eta_1} [G(\eta_2)/G(0)]^{1-\sigma}\mathrm{d}\eta_2\right\}\mathrm{d}\eta_1$

We see that the enthalpy distribution can be regarded as made up of two fields, one arising from the temperature and heat transfer at the wall [the first and second terms on the RHS of equation (3.51)] and the other from the heat generated by the dissipation of mechanical energy by the frictional stresses [the third term on the RHS of equation (3.51)]. The latter is proportional to M_e^2, where M_e is the Mach number of the external flow. The functions A and B were calculated by Crocco for a range of values of σ, and some of the results are illustrated in Fig. 3.4.

Since $\theta = 1$ when $\eta = 1$, it follows from equation (3.51) that

$$1 = \theta(0) + \theta'(0)\,[A(1,\sigma) - \sigma\frac{u_e^2}{i_e}B(1,\sigma) \qquad (3.52)$$

and so

$$\theta = 1 - \theta'(0)\,[A(1,\sigma) - A(\eta,\sigma)] + \sigma\frac{u_e^2}{i_e}[B(1,\sigma) - B(\eta,\sigma)]$$

If the heat transfer at the wall is zero then $\theta'(0) = 0$ and we have

$$\theta(0) = 1 + \sigma\frac{u_e^2}{i_e}B(1,\sigma)$$

since $B(0,\sigma) = 0$. The wall temperature for zero heat transfer is sometimes called the *recovery temperature* and is denoted T_r, and so we can write

$$T_r/T_e = 1 + \tfrac{1}{2}(\gamma - 1)M_e^2 \,.\, g(\sigma), \quad \text{say,}$$

where $g(\sigma) = \sigma \,.\, B(1,\sigma)$

To a close approximation we find that $g(\sigma) = \frac{1}{2}\sigma^{1/2}$, in good agreement with Pohlhausen's result for low speed flow [equation (3.34)]. Hence, we can write

$$T_r/T_e = 1 + \tfrac{1}{2}(\gamma - 1)M_e^2\sigma^{1/2} \qquad (3.53)$$

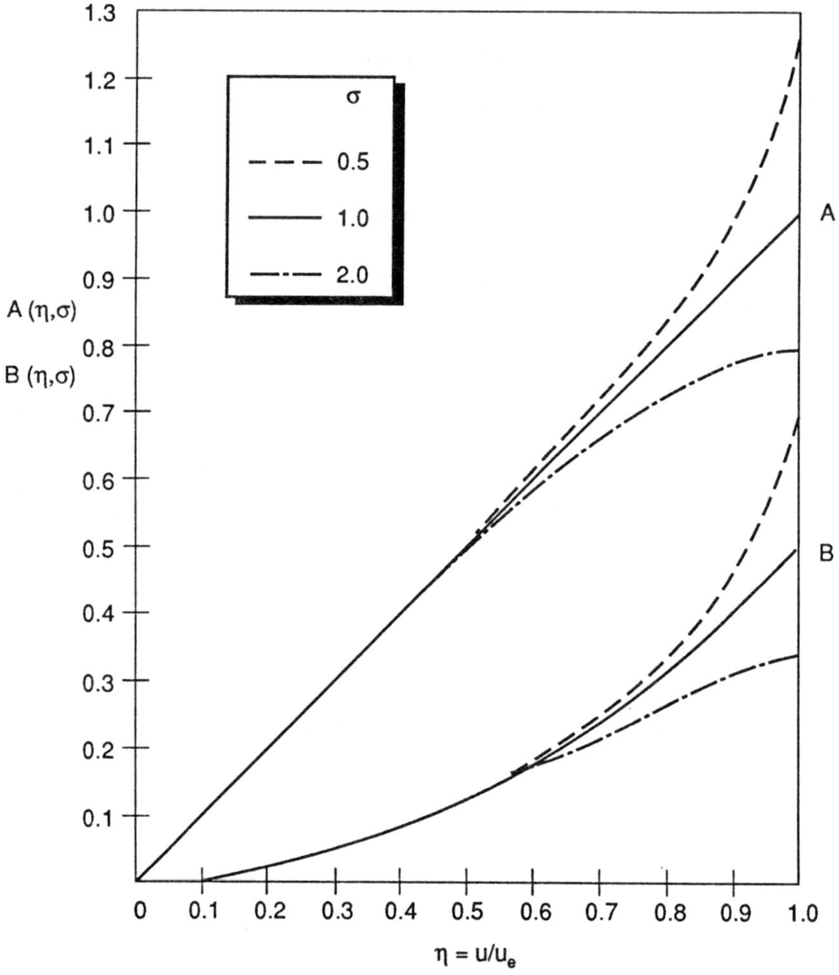

Fig. 3.4 The functions $A(\eta, \sigma)$ and $B(\eta, \sigma)$.

It therefore follows that the recovery factor is given by $\sigma^{1/2}$ as it was at low speeds. In a non-dissipative (or isentropic) flow, the temperature at points where the flow is brought to rest is given by

$$T_H = T_e + u_e^2/2c_p = T_e[1 + \tfrac{1}{2}(\gamma - 1)M_e^2]$$

and so again we see that the difference between the recovery factor and unity arises from the effect of viscous dissipation in the boundary layer in bringing the flow to rest at the wall.

We have seen from equation (3.45) that we can determine z for given x and η once the functions $G(\eta)$ and $\theta(\eta)$ are known. From equation (3.45) we can write

$$\mathrm{d}z/\mathrm{d}\eta = 2xm/[R_x^{1/2}G(\eta)]$$

and for incompressible flow

$$\mathrm{d}z_i/\mathrm{d}\eta = 2x/[R_x^{1/2}G(\eta)]$$

where suffix i refers to incompressible flow. We have seen that $G(\eta)$ is independent of Mach number when $\rho\mu$ = const. and so

$$z = \int_0^{z_i} m\,\mathrm{d}z_i = \int_0^{z_i} (\mu/\mu_e)\mathrm{d}z_i = \int_0^{z_i} (i/i_e)\mathrm{d}z_i = \int_0^{z_i} (\rho_e/\rho)\mathrm{d}z_i \qquad (3.54)$$

Likewise $\quad z_i = \int_0^{z} (\mu_e/\mu)\mathrm{d}z$

Thus, as a function of

$$z_i = \int_0^{z} (\mu_e/\mu)\mathrm{d}z$$

the velocity distribution in the boundary layer is independent of Mach number. For zero heat transfer $\mu > \mu_e$ inside the boundary layer so it follows that $z/z_i \geqslant 1$ and increases with Mach number for a given $\eta = u/u_e$.

3.3.5 General case, but with σ near unity

Crocco numerically solved equations (3.41) for values of ω ranging from 0.5 to 1.25, a range of Mach numbers up to 5 and with $\sigma = 0.725$. Later, he used the Sutherland relation for the viscosity–temperature relation with a corresponding range of values of the parameter C (see Section 1.3). His results led to the interesting conclusion that the enthalpy as a function of velocity, i.e. $\theta(\eta)$ was practically independent of the viscosity–temperature relation, and was closely given by the solution for the case $\rho\mu$ = const. or $\omega = 1$.

By way of explanation, it can be argued that in equation (3.51), which is generally applicable, the possible variation of $G(\eta)/G(0)$ cannot be large, and since it appears to the power $(1 - \sigma)$ its variation with ω or C is still further reduced provided σ is near unity. With the same proviso equation (3.53) is also found to be of wide validity.

Hence the shear stress function G is determined by the solution of the equation

$$GG'' + h(\eta) = 0$$

where $h(\eta)$ is known. We have already discussed the solution of this equation in Section 3.3.3 when considering the case of $\sigma = 1$.

Starting with an iterative solution of this equation and making use of Crocco's results as well as of earlier calculations by Emmons and Brainerd[3.7], Karman and Tsien[3.6] and Cope and Hartree,[3.10] Young[3.11] deduced that a very close fit to the skin friction distribution is given by

$$c_f R_x^{1/2} = 0.664[0.45 + 0.55(T_w/T_e) + 0.09(\gamma - 1)M_e^2\sigma^{1/2}]^{-(1-\omega)/2} \quad (3.55)$$

With the aid of equation (3.48) we find that for zero heat transfer ($T_w = T_r$)

$$c_f R_x^{1/2} = 0.664[1 + 0.365(\gamma - 1)M_e^2\sigma^{1/2}]^{-(1-\omega)/2} \quad (3.56)$$

An illustration of the close fit of these formulae to the results of numerical calculations is given in Fig. 3.5.

We note that equation (3.55) is in conformity with the result already demonstrated that when $\omega = 1$ the skin friction is independent of Mach number. If $\omega < 1$, as is the case with air, then we see that c_f decreases with increase of M_e for a constant value of T_w/T_e.

We can define an *intermediate temperature* T_m of the boundary layer given by

$$T_m = T_e[0.45 + 0.55(T_w/T_e) + 0.09(\gamma - 1)M_e^2\sigma^{1/2}] \quad (3.57)$$

and we shall define

$$c_{fm} = \tau_w/\tfrac{1}{2}\rho_m u_e^2, \qquad R_{xm} = u_e x \rho_m/\mu_m$$

where ρ_m and μ_m are determined at the temperature T_m. Then we find that

$$\begin{aligned} c_{fm} R_{xm}^{1/2} &= c_f(\rho_e/\rho_m)R_x^{1/2}(\rho_m\mu_e/\rho_e\mu_m)^{1/2} \\ &= c_f R_x^{1/2}(T_m/T_e)^{(1-\omega)/2} = 0.664 \end{aligned} \quad (3.58)$$

Hence in terms of the intermediate temperature we recover the incompressible relation between c_{fm} and R_{xm}.

The local rate of heat transfer from the plate to the fluid is

$$\begin{aligned} q = \mathrm{d}Q/\mathrm{d}S &= -k_w(\partial T/\partial z)_w = -k_w(\partial i/\partial u)_w(\partial u/\partial z)_w \\ &= -\theta'(0)\, i_e \tau_w/(\sigma u_e) \end{aligned}$$

Since $\theta'(0)$ is independent of x it follows that

$$q/\tau_w = Q/F_s = -i_e\theta'(0)/(\sigma u_e) \quad (3.59)$$

where F_s is the frictional force on unit span of the plate of chord c

$$= \int_0^c c_f \tfrac{1}{2}\rho_e u_e^2 \mathrm{d}x = 2G(0)\tfrac{1}{2}\rho_e u_e^2/R^{1/2}$$

Then the Nusselt number is

$$\begin{aligned} \mathrm{Nu} = Q/[k_e(T_w - T_e)] &= -i_e\theta'(0)F_s/[\sigma u_e k_e(T_w - T_e)] \\ &= R^{1/2}G(0)\theta'(0)/[1 - \theta(0)] \end{aligned} \quad (3.60)$$

With the aid of equation (3.52) we deduce from equation (3.59) that

$$q/\tau_w = Q/F_s = -\frac{-i_e}{\sigma\, u_e A(1, \sigma)}\left[1 - \theta(0) + \sigma\frac{u_e^2}{i_e}B(1, \sigma)\right]$$

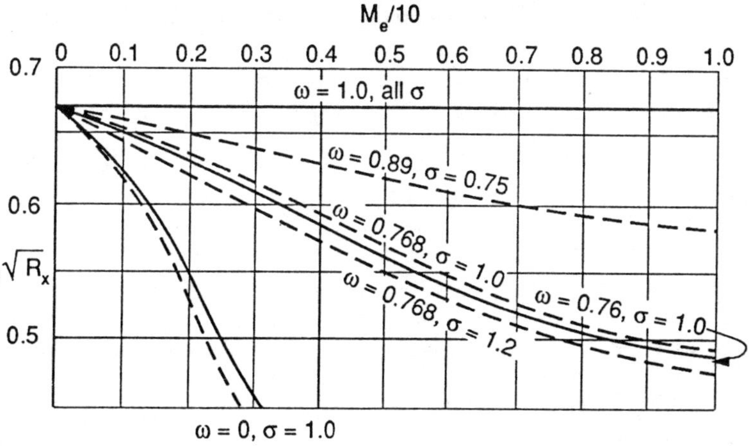

Fig. 3.5 Illustration of the fit of equation (3.55) to predictions of various workers for skin friction on a flat plate with no heat transfer.
——— calculated by various workers
– – – – – equation (3.55).

$$= -\frac{i_e}{u_e}\left\{\gamma(\sigma)[1 - \theta(0)] + \frac{u_e^2}{i_e}\,\delta(\sigma)\right\}$$

where $\gamma(\sigma) = 1/[\sigma A(1, \sigma)]$ and $\delta(\sigma) = B(1, \sigma)/A(1, \sigma)$.

Then
$$\mathrm{Nu} = R^{1/2}G(0)\left\{\sigma\gamma(\sigma) + \frac{u_e^2\sigma\delta(\sigma)}{i_e[1 - \theta(0)]}\right\} \qquad (3.61)$$

With zero heat transfer $Q = \theta'(0) = \mathrm{Nu} = 0$, and then

$$1 - \theta(0) = -u_e^2 \,.\, \delta(\sigma)/[i_e \,.\, \gamma(\sigma)]$$

or
$$T_r/T_e = 1 + u_e^2 \,.\, \delta(\sigma)/[i_e \,.\, \gamma(\sigma)] \qquad (3.62)$$

where T_r is the recovery temperature (wall temperature for zero heat transfer).

Hence

$$Q/k_e = \mathrm{Nu}(T_w - T_e) = R^{1/2}G(0)\sigma\gamma(\sigma)(T_w - T_r)$$
$$= \tfrac{1}{2}C_F R\sigma\gamma(\sigma)(T_w - T_r) \qquad (3.63)$$

At low speeds $T_r = T_e$ and then

$$Q/k_e = \tfrac{1}{2}C_F R\sigma\gamma(\sigma)(T_w - T_e)$$

Since $\gamma(\sigma)$ is independent of Mach number it follows that *the form of the relation between heat transfer and skin friction remains independent of*

Mach number if in the low speed relation we replace the driving temperature difference $(T_w - T_e)$ *by* $(T_w - T_r)$.

We have seen that the low speed calculations of Pohlhausen showed that to a good approximation $\gamma(\sigma) \simeq \sigma^{-2/3}$. Hence we can write

$$Q/k_e = \tfrac{1}{2}C_F R\sigma^{1/3}(T_w - T_r) = F_s\sigma^{1/3}(T_w - T_r)/(\mu_e u_e)$$

and so $$Q/F_s = q/\tau_w = c_p(T_w - T_r)/(\sigma^{2/3} \, . \, u_e) \qquad (3.64)$$

If, in the light of the above we define $\mathrm{Nu} = Q/[k_e(T_w - T_r)]$ then

$$\mathrm{Nu} = \tfrac{1}{2}C_F R\sigma^{1/3} \qquad (3.65)$$

as for low speed flow. Likewise if we define the local Stanton number as

$$\mathrm{St}_l = q/[\rho_e u_e c_p(T_w - T_r)] \qquad (3.66)$$

then $$\mathrm{St}_l = \tfrac{1}{2}c_f \, . \, \sigma^{-2/3} \qquad (3.67)$$

again as in low speed flow. We have already noted that $\delta(\sigma)/\gamma(\sigma) = \sigma B(1, \sigma) \simeq \sigma^{1/2}/2$, and so again

$$T_r/T_e = 1 + \tfrac{1}{2}(\gamma - 1)M_e^2\sigma^{1/2} \qquad (3.53)$$

3.3.6 *The asymptotic solution for large x and uniform suction*

The extension to compressible flow of the analysis of Griffith and Meredith described in Section 3.2.2 for the asymptotic solution with uniform suction at the wall is relatively simple with the aid of the Crocco transformation and was first presented by Young[3.12].

We confine ourselves to the case of zero heat transfer by conduction at the plate surface and since we are considering a station far downstream we can again assume all quantities in the boundary layer are independent of x. Then equations (3.37) reduce to

$$\tau_{uu} = 0 \qquad (1 - \sigma)\tau_u i_u + \tau(i_{uu} + \sigma) = 0 \qquad (3.68)$$

where the suffix now denotes differentiation with respect to u, i.e. $\tau_u = d\tau/du$, etc. From the equation of continuity $\partial(\rho w)/\partial z = 0$ and hence

$$\rho w = \text{const.} = \rho_w w_w = \rho_e w_e \qquad (3.69)$$

Hence, from equation (3.36)

$$\tau_u = \rho w = \rho_e w_e$$

so that at $z = 0$, $u = 0$, $\tau_u = \rho_e w_e$ and $i_u = 0$; and at $z = \infty$, $u = u_e$ and $i = i_e$. Therefore, from equation (3.68)

$$\left.\begin{aligned} \tau &= -\rho_e w_e(u_e - u) \\ \text{and} \quad (1 - \sigma)i_u &- (u_e - u)(i_{uu} + \sigma) = 0 \end{aligned}\right\} \qquad (3.70)$$

This last equation can be integrated to give

$$i_e - i = \frac{\sigma u_e^2}{2(2 - \sigma)}\left[(1 - u/u_e)^2 - \frac{2}{\sigma}(i - u/u_e)^\sigma\right] \tag{3.71}$$

The first equation of (3.70) and the relation $\tau = \mu\, du/dz$ yield

$$u/u_e = 1 - \exp \xi \qquad \text{where } \xi = \rho_e w_e \int_0^z dz/\mu \tag{3.72}$$

Note the similarity of this relation with the corresponding one for incompressible flow in Section 3.2.2. Equation (3.71) then yields

$$(i/i_e) - 1 = \tfrac{1}{2}\sigma(\gamma - 1)M_e^2\, H(\xi)/(2 - \sigma) \tag{3.73}$$

where $$H(\xi) = \frac{2}{\sigma}\exp(\sigma\xi) - \exp(2\xi)$$

Hence, we can express u/u_e and $[(i/i_e) - 1]/(\gamma - 1)M_e^2$ as functions of ξ, independent of Mach number. To obtain the distributions as functions of z we infer from equation (3.72) that

$$w_e z/\nu_e = \int_0^\xi (i/i_e)^\omega d\xi = \int_0^\xi \left[1 + \frac{(\gamma - 1)M_e^2}{2(2 - \sigma)}\sigma H(\xi)\right]^\omega d\xi \tag{3.74}$$

In general this integral must be evaluated numerically, but when $\omega = 1$ it can be integrated analytically to yield

$$w_e z/\nu_e = \xi + \frac{(\gamma - 1)M_e^2\sigma}{2(2 - \sigma)}\left[\frac{2}{\sigma^2}\exp(\sigma\xi) - \tfrac{1}{2}\exp(2\xi) - \frac{2}{\sigma^2} + \tfrac{1}{2}\right] \tag{3.75}$$

Some results are illustrated in Fig. 3.6.

3.4 Similar solutions: incompressible flow

We have noted that for the flow with the pressure uniform the laminar boundary layer equations yield a similar set of velocity profiles for different values of x, so that we can write u/u_e = function of a single variable ζ, say, where ζ is of the form z/(function of x). The question then naturally arises as to whether families of similar solutions also exist when the pressure varies with x, and if so what forms this variation may take.

The details need not be gone into here but it can be readily shown that for incompressible flow such solutions exist when either:

(a) u_e = const. x^m, where m is a constant, or
(b) u_e = const. $\exp(\alpha x)$, where α is a constant $\geqslant 0$.

We will confine ourselves to case (a).

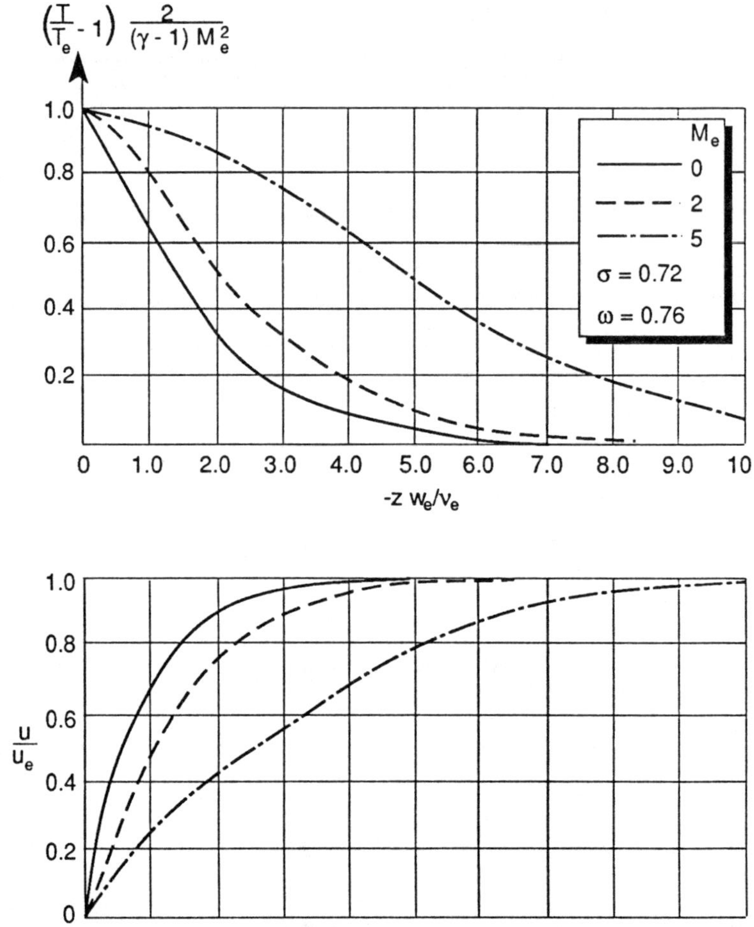

Fig. 3.6 Asymptotic velocity and temperature distributions on an infinite flat plate with suction and zero heat transfer.

Let us then consider

$$u_e = u_0(x/L)^m \tag{3.76}$$

where u_0 and L are a reference velocity and length, respectively. The boundary layer equations of continuity and momentum are

$$\partial u/\partial x + \partial w/\partial z = 0$$

and $$u\,\partial u/\partial x + w\,\partial u/\partial z = u_e \mathrm{d}u_e/\mathrm{d}x + \nu\,\mathrm{d}^2u/\partial z^2$$

With the uniform pressure case as guide we take

$$\zeta = \tfrac{1}{2}z(u_e/\nu x)^{1/2} = \tfrac{1}{2}z(u_0/\nu)^{1/2}x^{(m-1)/2}/L^{m/2} \tag{3.77}$$

and we write the stream function, implied by the equation of continuity, as

$$\psi = (u_e\nu x)^{1/2}F(\zeta) = (u_0\nu)^{1/2}x^{(m+1)/2}F(\zeta)/L^{m/2} \tag{3.78}$$

Then
$$u = \partial\psi/\partial z = u_e F'(\zeta)/2 \tag{3.79}$$

where the prime denotes differentiation with respect to ζ. This equation is clearly consistent with the postulate of similarity. If we now express w, $\partial u/\partial z$, $\partial u/\partial x$, and $\partial^2 u/\partial z^2$ in terms of F and its derivatives with respect to ζ, the boundary layer equation of motion after a little algebra then reduces to the ordinary differential equation:

$$F''' + (m + 1)FF'' - 2mF'^2 + 8m = 0 \tag{3.80}$$

The boundary conditions are $F = F' = 0$, when $\zeta = 0$; $F' = 2$, when $\zeta = \infty$.

An alternative form of this equation is obtained by taking as the independent variable ξ, say, where

$$\xi = \zeta[2(m + 1)]^{1/2} = z[\tfrac{1}{2}(m + 1)u_e/\nu x]^{1/2} \tag{3.81}$$

and writing

$$f(\xi) = [\tfrac{1}{2}(m + 1)]^{1/2}F(\zeta)$$

Then $u/u_e = f'(\xi)$ and equation (3.80) becomes

$$f''' + ff'' + \beta(1 - f'^2) = 0 \tag{3.82}$$

where $\beta = 2m/(m + 1)$. The boundary conditions are $f = f' = 0$, when $\xi = 0$; $f' = 1$, when $\xi = \infty$. Primes here denote differentiation with respect to ξ. The equation is often referred to as the Falkner–Skan equation.

The first set of numerical solutions for this equation was provided by Falkner and Skan[3.13], and Hartree[3.14] later gave solutions of greater accuracy. A plot of $u/u_e = f'(\xi)$ for some values of m is shown in Fig. 3.7, and we readily find that the local skin friction coefficient is

$$\begin{aligned} c_f = 2\tau_w/\rho u_e^2 &= 2\nu(\partial u/\partial z)_w/u_e^2 = \tfrac{1}{2}F''(0)/R_x^{1/2} \\ &= [2(m + 1)]^{1/2}f''(0)/R_x^{1/2} \end{aligned} \tag{3.83}$$

Also the displacement and momentum thicknesses are given by

$$\left.\begin{aligned} (\delta^*/x)R_x^{1/2} &= [2/(m + 1)]^{1/2}\int_0^\infty (1 - f')\mathrm{d}\xi \quad \text{and} \\ (\theta/x)R_x^{1/2} &= [2/(m + 1)]^{1/2}\int_0^\infty f'(1 - f')\mathrm{d}\xi \end{aligned}\right\} \tag{3.84}$$

The values of these boundary layer characteristics for a range of values of m are set out in Table 3.3.

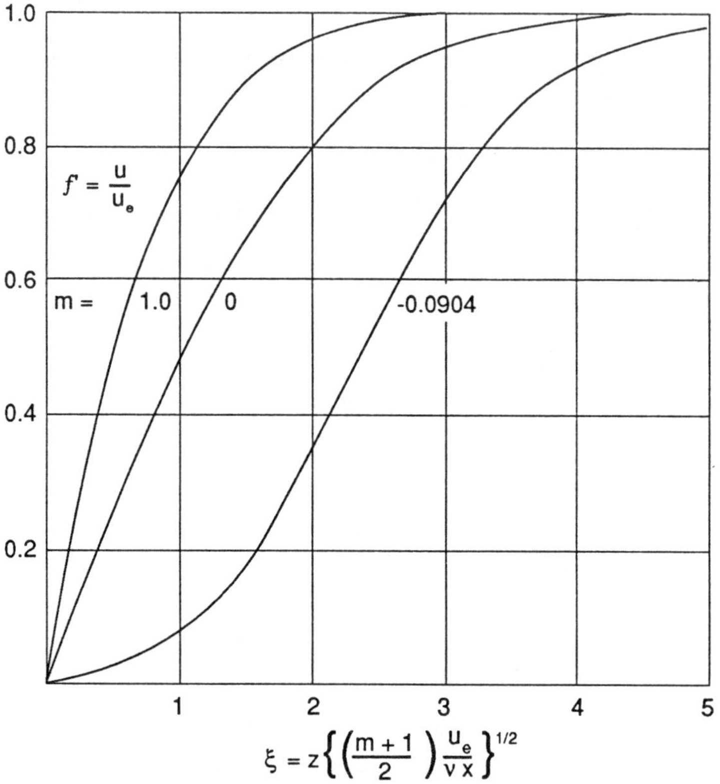

Fig. 3.7 Representative velocity profiles obtained from the Falkner–Skan similar solutions equation (3.82). $m = -0.0904$ is for a boundary layer on the verge of separation.

Table 3.3

m	$c_f R_x^{1/2}$	$(\delta^*/x)R_x^{1/2}$	$(\theta/x)R_x^{1/2}$	$H = \delta^*/\theta$
1	1.2326	0.6479	0.2923	2.216
1/3	0.7575	0.9854	0.4290	2.297
0.1	0.4966	1.3478	0.5566	2.422
0	0.3321	1.7207	0.6641	2.591
−0.01	0.3115	1.7800	0.6789	2.622
−0.05	0.2135	2.1174	0.7515	2.818
−0.0904	0	3.4277	0.8680	3.949

For any $m > 0$, where the pressure is falling with x (i.e. favourable pressure gradient), the solution is unique; but for $m < 0$, with the pressure rising (i.e. adverse pressure gradient), there are an infinite number of solutions with the given boundary solutions. However, Hartree argued that the solution corresponding to the physical situation was that which approached most rapidly from below the condition $f' = 1$ at the outer edge of the boundary layer. This solution is the only one that has an exponential approach to that condition and thus has continuity with the unique solutions for $m > 0$.

We note that in the case $m = 0$ when u_e, and therefore p, are constant, we recover the Blasius solution. For $m < 0$, the velocity profiles are inflected as was anticipated in Section 2.10. For the case $m = -0.0904$ the shear stress at the surface is zero $[f''(0) = 0]$ and the boundary layer is on the point of separating for all x.

Of special interest is the fact that the inviscid flow past a wedge of semi-angle $\pi\beta/2$ at zero incidence is given by

$$u_e = \text{const. } x^{\beta/(2-\beta)}$$

where x is the distance aft of the nose,

$$= \text{const. } x^m$$

Hence any solution for $m > 0$ corresponds to the boundary layer near the leading edge of such a wedge.

The case $\beta = m = 1$ corresponds to the flow in the neighbourhood of a forward stagnation point, such as that of a round nosed aerofoil, and we note that in this case ζ and ξ are independent of x. Hence, the velocity profile u/u_e is a function of z only in that region and the boundary layer thickness there is independent of the distance from the stagnation point.

Another case of some interest is the case $m = -1.0$ but with $u_0 < 0$. It corresponds to the flow in a straight sided channel converging towards a sink at $x = 0$ from positive x. Here the solution can be obtained in a closed analytic form, for further details see Schlichting (Reference 2.2, Chapter IX).

Stewartson[3.15] has identified another set of solutions for the range of m between 0 and -0.0904 which have profiles with reversed flow close to the surface, i.e. they can be used to describe flows downstream of separation. The above discussion refers to a non-permeable surface. If we consider a porous surface permitting suction or blowing through it then additional similar solutions are possible. Such solutions arise when

$$u_e \propto x^m = u_0(x/L)^m$$

and we have a value of w at the wall given by

$$w_w = -Cu_0(x/L)^{(m-1)/2} \tag{3.85}$$

where C is a constant. With $w_w < 0$ there is suction at the wall, and with $w_w > 0$ there is blowing. Equation (3.82) applies as before but now the boundary conditions include

$$f_w = f(0) = -w_w\{2x/[(m + 1)u_e v]\}^{1/2} = C\left[\frac{2}{(m + 1)}\frac{u_0 L}{v}\right]^{1/2}$$

Therefore, we see that with the prescribed suction distribution f_w is independent of x and the similarity condition holds.

A set of solutions was determined by Koh and Hartnett[3.16] for a range of values of $m > 0$, and later another set of solutions for negative values of m ranging from -0.1 to -0.9 and a wide range of values of f_w was determined by Zamir and Young.[3.17] Fig. 3.8 shows the suction required to just maintain a separation profile for different values of $m < 0$; not surprisingly it increases as m decreases. Fig. 3.9 shows the local skin friction coefficient as a function of m. We note that for a given m, c_f increases as the suction is increased; this is because the boundary layer thickness is decreased and the velocity profile becomes fuller with suction. Also, c_f increases as m increases, because a falling pressure gradient speeds up the flow in the boundary layer and also makes it thinner and so increases the rate of shear and hence the shear stress close to the surface.

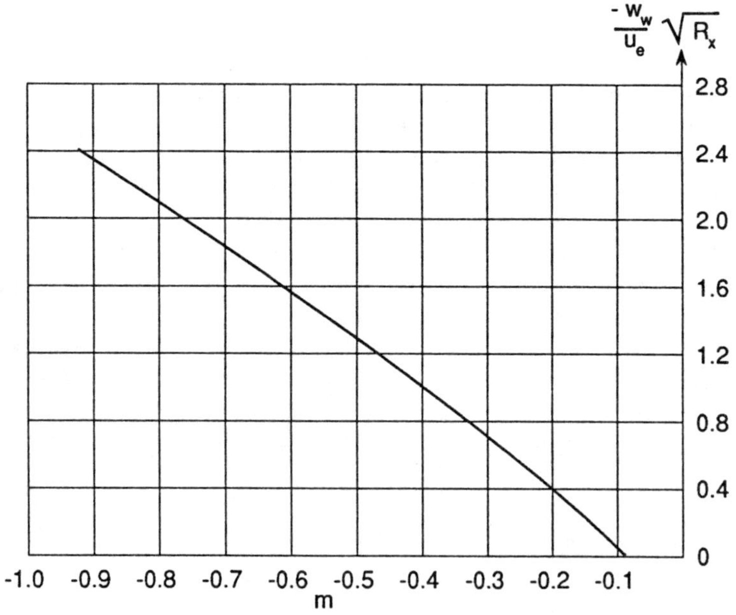

Fig. 3.8 Suction ($-w_w$) required to just avoid separation in the presence of an adverse pressure gradient ($m < 0$) for the family of similar profiles of reference 3.17.

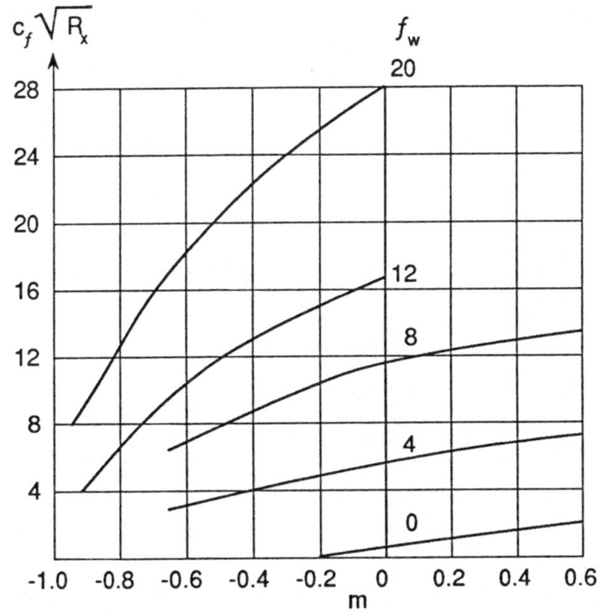

Fig. 3.9 Skin friction with various suction quantities and pressure gradients for the family of similar profiles of reference 3.17.
$f_w = -(w_w/u_e)\,[2R_x/(m+1)]^{1/2}$ $\quad u_e \propto x^m \quad w_w \propto x^{(m-1)/2}$
$R_x = u_e x/\nu \quad c_f = 2\tau_w/\rho u_e^2$

3.5 Similar solutions: steady compressible flow; Stewartson–Illingworth transformation

The question naturally follows as to what conditions must be satisfied to obtain similarity of velocity and temperature distributions for different values of x when the flow is compressible. This question was first considered by Illingworth[3.18] and he concluded that without additional limitations such solutions only existed if the external velocity (or pressure) was uniform – the case we have already discussed in Section 3.3. This is of course a much more restrictive condition than when the flow is incompressible.

However, if we can assume that the Prandtl number $\sigma = 1.0$ and that μ/T is constant (or $\omega = 1.0$) then Stewartson[3.19] and Illingworth[3.20] have independently shown that there exists a transformation from a compressible flow boundary layer to a related incompressible flow boundary layer. In that case the similar solutions of the latter that we have discussed in Section 3.4 can be transformed to yield families of similar solutions for compressible flow, provided there is zero heat transfer between the surface and the fluid. However, in the case when the wall temperature T_w = const. and there is therefore heat transfer at the wall, the boundary layer equations can be transformed to equations which are closely similar to the incompressible flow equations and which permit of relatively easy numerical solution. The basic analytic steps leading to the transformation are as follows:

We write
$$\mu/\mu_0 = CT/T_0 \tag{3.86}$$

where C is a constant and suffix 0 now refers to stagnation conditions in the external flow, i.e.

$$T_0 = T_e + u_e^2/2c_p = T_e[1 + \tfrac{1}{2}(\gamma - 1)M_e^2]$$

The equations of motion and energy (see Section 2.7) for steady flow are:

$$\left.\begin{aligned} \rho\left(u\frac{\partial u}{\partial x} + w\frac{\partial u}{\partial z}\right) &= -\frac{dp}{dx} + \frac{\partial}{\partial z}\left(\mu\frac{\partial u}{\partial z}\right) \\ \rho c_p\left(u\frac{\partial T}{\partial x} + w\frac{\partial T}{\partial z}\right) &= u\frac{dp}{dx} + \frac{\partial}{\partial z}\left(k\frac{\partial T}{\partial z}\right) + \mu\left(\frac{\partial u}{\partial z}\right)^2 \end{aligned}\right\} \tag{3.87}$$

From the equation of continuity we infer a stream function ψ exists such that

$$\rho u/\rho_0 = \partial\psi/\partial z \quad \text{and} \quad \rho w/\rho_0 = -\partial\psi/\partial x \tag{3.88}$$

We now tranform to new coordinates X and Z such that

$$\left.\begin{aligned} X &= \int_0^x Cp_e a_e/(p_0 a_0)\mathrm{d}x = \int_0^x C(a_e/a_0)^{(3\gamma-1)/(\gamma-1)}\mathrm{d}x \\ Z &= (a_e/a_0)\int_0^z (\rho/\rho_0)dz \end{aligned}\right\} \tag{3.89}$$

[*Note* that in insentropic flow $p_e/p_0 = (T_e/T_0)^{\gamma/(\gamma-1)}$, and $a_e/a_0 = (T_e/T_0)^{1/2}$ = function of x.]

Then
$$\begin{aligned} \partial/\partial x &= (\partial X/\partial x)\partial/\partial X + (\partial Z/\partial x)\partial/\partial Z \\ &= (Cp_e a_e/p_0 a_0)\partial/\partial X + (\partial Z/\partial x)\partial/\partial Z \end{aligned}$$
and
$$\partial/\partial z = (\partial Z/\partial z)\partial/\partial Z = (a_e\rho/a_0\rho_0)\partial/\partial Z$$

After a little algebra we then find

$$\begin{aligned} u\,\partial u/\partial x + w\,\partial u/\partial z = C(p_e a_e^3/p_0 a_0^3)&\left[\frac{\partial\psi}{\partial Z}\frac{\partial^2\psi}{\partial Z\partial X} - \frac{\partial\psi}{\partial X}\frac{\partial^2\psi}{\partial Z^2}\right. \\ &\left. + \frac{1}{a_e}\frac{\mathrm{d}a_e}{\mathrm{d}X}\left(\frac{\partial\psi}{\partial Z}\right)^2\right] \end{aligned}$$

In the external flow $a_e^2 + \frac{1}{2}(\gamma - 1)u_e^2 = a_0^2$, and so

$$\frac{1}{a_e}\frac{\mathrm{d}a_e}{\mathrm{d}X} = \frac{1}{a_e}\frac{\mathrm{d}a_e}{\mathrm{d}x}C\frac{p_0 a_0}{p_e a_e} = -\tfrac{1}{2}(\gamma - 1)\frac{u_e}{a_e^2}\frac{\mathrm{d}u_e}{\mathrm{d}x}C\frac{p_0 a_0}{p_e a_e}$$

Also $\quad (\partial\psi/\partial Z)^2 = (\partial\psi/\partial z)^2\,(\partial z/\partial Z)^2 = (a_0/a_e)^2\,u^2$

$$\frac{1}{\rho}\frac{\partial}{\partial z}\left(\mu\frac{\partial u}{\partial z}\right) = \left(\frac{a_e}{a_o}\right)^3 C\,\frac{p_e}{p_0}\frac{\partial^3\psi}{\partial Z^3}, \qquad \frac{1}{\rho}\frac{dp}{dx} = -\frac{T}{T_e}\,u_e\frac{du_e}{dx}$$

If we now introduce

$$S = \frac{T_H}{T_{H0}} - 1 = \frac{c_p T + \frac{1}{2}u^2}{c_p T_e + \frac{1}{2}u_e^2} - 1 \tag{3.90}$$

then $\quad T/T_e = (1 + S)a_0^2/a_e^2 - \frac{1}{2}(\gamma - 1)u^2/a_e^2$
and so

$$\frac{1}{\rho}\frac{dp}{dx} = -u_e\frac{du_e}{dx}\,\{(1 + S)(a_0/a_e)^2 - \tfrac{1}{2}(\gamma - 1)u^2/a_e^2\}$$

Hence the equation of motion becomes

$$\frac{\partial\psi}{\partial Z}\frac{\partial^2\psi}{\partial Z\,\partial X} - \frac{\partial\psi}{\partial X}\frac{\partial^2\psi}{\partial Z^2} = (1 + S)(a_0/a_e)^5\,\frac{p_0}{p_e}\,u_e\frac{du_e}{dx} + \nu_0\frac{\partial^3\psi}{\partial Z^3} \tag{3.91}$$

If we now define

$$U = \partial\psi/\partial Z, \qquad W = -\partial\psi/\partial X \tag{3.92}$$

then $U = u\,a_0/a_e$; and we denote $U_e = u_e a_0/a_e$. Taking note of the variation of a_e with x we find after a little algebra

$$u_e\,du_e/dx = (U_e\,dU_e/dX)(a_e/a_0)^5\,C(p_e/p_0)$$

The equation of motion then takes the form

$$U\,\partial U/\partial X + W\,\partial U/\partial Z = (1 + S)U_e\frac{dU_e}{dX} + \nu_0\frac{\partial^3\psi}{\partial Z^3} \tag{3.93}$$

This is the same as the boundary layer equation for incompressible flow apart from the factor $(1 + S)$ in the first term on the RHS (the pressure gradient term). The velocity boundary conditions are also formally the same. In the case of zero heat transfer and with $\sigma = 1$ we recall that $T_H =$ const. $= T_{H0}$ and so $S = 0$. The correspondence with an incompressible flow for the external velocity distribution $U_e(x)$ is then complete.

For the transformed energy equation we turn to the total energy form [Equation (2.54)]

$$\rho u\frac{\partial T_H}{\partial x} + \rho w\frac{\partial T_H}{\partial z} = \frac{\partial}{\partial z}\left[\frac{k}{c_p}\frac{\partial}{\partial z}(c_p T + \tfrac{1}{2}\sigma u^2)\right]$$

or

$$\rho u\frac{\partial S}{\partial x} + \rho w\frac{\partial S}{\partial z} = \frac{\partial}{\partial z}\left[\frac{\mu}{\sigma}\frac{\partial S}{\partial z} + \frac{(\sigma - 1)}{\sigma}\frac{\partial}{\partial z}(\tfrac{1}{2}u^2/c_p T_0)\right]$$

If we now transform to the coordinates (X, Z) this becomes, after some algebra,

$$U\frac{\partial S}{\partial X}+W\frac{\partial S}{\partial Z}=\nu_0\left[\frac{1}{\sigma}\frac{\partial^2 S}{\partial Z^2}+\frac{(\sigma-1)}{\sigma}\frac{\frac{1}{2}(\gamma-1)M_e^2}{1+\frac{1}{2}(\gamma-1)M_e^2}\frac{\partial^2(U/U_e)^2}{\partial Z^2}\right] \tag{3.94}$$

With $\sigma = 1$, this reduces to

$$U\frac{\partial S}{\partial X}+W\frac{\partial S}{\partial Z}=\nu_0\frac{\partial^2 S}{\partial Z^2} \tag{3.95}$$

In incompressible flow we can write $T = T_e(1 + S)$, with T_e = const. so that equation (3.95) is the same as equation (3.21) for the temperature boundary layer in incompressible flow. The boundary conditions are: at $Z = 0$, $U = W = 0$ with either $S = S_w$, say, or $\partial S/\partial Z = 0$ (zero heat transfer); at $Z = \infty$, $U = U_e(x)$, $S = 0$.

If $S_w > 0$ then heat is transferred from the wall to the fluid, and $S_w < 0$ implies the converse. In the case of $S = 0$ (zero heat transfer) T_H = const. and then

$$T = T_e + \tfrac{1}{2}(u_e^2 - u^2)/c_p \tag{3.96}$$

If the pressure is uniform then $du_e/dx = dU_e/dX = 0$ and equations (3.93) and (3.95) become identical in form with similar boundary conditions for U and $(S - S_w)$ and therefore permit the relation

$$(S - S_w) = \text{const.}\ U, \qquad \text{or} \quad S/S_w = 1 - U/U_e = 1 - u/u_e \tag{3.97}$$

This yields a quadratic relation between T and u/u_e:

$$\frac{T}{T_e}=\frac{T_w}{T_e}-\frac{u}{u_e}\left[\frac{T_w}{T_e}-1-\tfrac{1}{2}(\gamma-1)M_e^2\right]-\frac{\frac{1}{2}u^2}{u_e^2}(\gamma-1)M_e^2 \tag{3.98}$$

as anticipated in Section 2.11 [see equation (2.61)].

As already remarked the above transformation can be applied to the similar family of solutions corresponding to

$$U_e = \text{const.}\ X^m = K\,X^m, \quad \text{say} \tag{3.99}$$

and with S_w = const. This was done by Cohen and Reshotko[3.21] and by Li and Nagamatsu.[3.22]

In the incompressible flow we have as before

$$\psi = (u_e X \nu_0)^{1/2}\,[2/(m+1)]^{1/2}\,f(\xi)$$

with $\quad \xi = Z[(m+1)U_e/(2\nu_0 X)]^{1/2}$

leading to the ordinary differential equation derived from equation (3.93)

$$f''' + ff'' = \beta(f'^2 - 1 - S) \tag{3.100}$$

with $\beta = 2m/(m + 1)$. The energy equation (3.95) becomes

$$S'' + fS' = 0 \tag{3.101}$$

The boundary conditions are $\xi = 0, f = f' = 0, S = S_w$; $\xi = \infty, f' = 1$. We note that

$$f' = U/U_e = u/u_e \quad (3.102)$$

From equation (3.89) the relation between x and X is

$$dx = \frac{1}{C}(a_0/a_e)^{(3\gamma-1)/(\gamma-1)}\, dX$$

and since $a_0^2/a_e^2 = 1 + \frac{1}{2}(\gamma - 1)U_e^2/a_e^2$

$$dx = \frac{1}{C}\left[1 + \tfrac{1}{2}(\gamma - 1)\frac{K^2X^{2m}}{a_e^2}\right]^{(3\gamma-1)/(2\gamma-2)} dX \quad (3.103)$$

so that in general there is no simple power relation between u_e and x. However, as before, when $m > 0$ the flow is an accelerating one, i.e. u_e is an increasing function of x, and the converse holds when $m < 0$.

A number of solutions of equations (3.100) and (3.101) for ranges of positive and negative values of β and S_w have been numerically determined by Cohen and Reshotko and by Li and Nagamatsu. These show that the value of β corresponding to a separation profile becomes less negative as S_w increases. This is because heat transfer from the plate to the fluid reduces the fluid density near the surface and renders the boundary layer more susceptible to the adverse pressure gradient. It is to be noted that for $\beta < 0$ there are two possible solutions.

The sets of solutions to the above equations determined by the cited authors provide useful test cases against which to check the accuracy of approximate methods, and they are also valuable in providing empirical relations as inputs in the development of approximate methods. Such methods are the subject of the next chapter.

References

3.1 Blasius, H. (1908) 'Grenzschichten in Flüssigkeiten mit kleiner Reibung'. *Z. Math. Phys.*, **56**, p. 1. English Translation NACA TM 1256.

3.2 Cebeci, T. and Bradshaw, P. (1977) *Momentum Transfer in Boundary Layers*, Hemisphere Publishing Corporation.

3.3 Griffith, A.A. and Meredith, F.W. (1936) 'The possible improvement in aircraft performance due to the use of boundary layer suction'. ARC Rep. 2315.

3.4 Pohlhausen, E. (1921) 'Der Wärmeaustausch zwischen festen Körpern und Flüssigkeiten mit kleiner Reibung und kleiner Wärmerleitung'. *ZAMM* **1**, p. 115.

3.5 Busemann, A. (1935) 'Gasströmung mit laminarer Grenzschicht entlang einer Platte'. *ZAMM* **15**, p. 23.

3.6 Von Karman, T. and Tsien, H.S. (1938) 'Boundary layers in compressible fluids'. *J. Aero. Sc.*, **5**, p. 227.
3.7 Emmons, H.W. and Brainerd, J.G. (1941, 2) 'Temperature effects in a laminar compressible fluid boundary layer on a flat plate'. *J. App. Mech.*, **8**, p. 105, and **9**, p. 1.
3.8 Hantsche, W. and Wendt, H. (1940) Zur Kompressibilitätseinfluss bei der laminaren Grenzschicht der ebenen Platte. *Jb. Dt. Luftfahrtforschung*, **I**, p. 517; ibid. (1942), p. 40.
3.9 Crocco, L. (1941) 'Sullo strato limite laminare nei gas lungo una lamina piano'. *Rend. Mat. Univ. Roma*, **V**, 2, p. 738.
3.10 Cope, W.F. and Hartree, D.R. (1948) 'The laminar boundary layer in a compressible flow'. *Phil. Trans. Roy. Soc. A*, **241**, p. 1.
3.11 Young, A.D. (1949) 'Skin friction in the laminar boundary layer in high speed flow'. *Aero. Qu.*, **I**, p. 137.
3.12 Young, A.D. (1948) 'Note on the velocity and temperature distributions attained with suction on a flat plate of infinite extent in compressible flow'. *Q. J. Mech. Appl. Math.*, **1**, p. 70.
3.13 Falkner, V.M. and Skan S.W. (1931) 'Some approximate solutions of the boundary layer equations'. *Phil. Mag.*, **12**, p. 865.
3.14 Hartree, D.R. (1937) On an equation occurring in Falkner and Skan's approximate treatment of the equations of the boundary layer'. *Proc. Camb. Phil. Soc.*, **33**, II, p. 223.
3.15 Stewartson, K. (1954) 'Further solutions of the Falkner–Skan equation'. *Proc. Camb. Phil. Soc.*, **50**, p. 454.
3.16 Koh, J.C. and Hartnett, J.P. (1961) 'Skin friction and heat transfer for incompressible laminar flow over porous wedges with suction and variable wall temperature'. *Int. J. Heat & Mass Transfer*, **2**, 3.
3.17 Zamir, M. and Young, A.D. (1967) 'Similar and asymptotic solutions of the incompressible laminar boundary layer equations with suction'. *Aero. Qu.*, **XVIII**, p. 103.
3.18 Illingworth, C.R. (1946) 'The laminar boundary layer associated with retarded flow in a compressible fluid'. ARC RM. 2590.
3.19 Stewartson, K. (1949) 'Correlated compressible and incompressible boundary layers'. *Proc. Roy. Soc. A*, **200**, p. 84.
3.20 Illingworth, C.R. (1949) 'Steady flow in the laminar boundary layer of a gas'. *Proc. Roy. Soc. A*, **199**, p. 533.
3.21 Cohen, C.B. and Reshotko, E. (1956) 'The compressible laminar boundary layer with heat transfer and arbitrary pressure gradient'. NACA Rep. 1294.
3.22 Li, T.Y. and Nagamatsu, H.T. (1953) 'Similar solutions of compressible boundary layer equations'. *J. Aero. Sc.*, **20**, p. 653; ibid. (1955), **22**, p. 607.

Chapter 4

Some Approximate Methods of Solution for the Laminar Boundary Layer in Steady Two Dimensional Flow

4.1 Introduction

In Chapter 2, Section 2.11, we developed various integral equations and remarked that they present mean descriptions of the equations of motion or energy averaged over the boundary layer thickness and as such provide a basis for approximate methods of solution. In essence such methods are characterised by the use of assumptions about the forms of the velocity and temperature profiles which involve a small number of non-dimensional parameters, the values of which are determined by satisfying the relevant integral equations and some of their boundary conditions sometimes together with a number of auxiliary relations. We gave a simple example of the use of the momentum integral equation in Section 2.12 for the basic case of a steady laminar boundary layer on a flat plate at zero incidence in incompressible flow.

In this chapter we shall develop this approach in more detail and present some approximate methods of wide ranging applicability, which have been found to be of adequate accuracy for many engineering purposes yet are simple and economic in computing requirements. They are particularly useful in yielding reasonably reliable estimates of such overall quantities as skin friction, momentum and displacement thicknesses, wall heat transfer etc., but they are not primarily designed to yield distributions of local quantities e.g. velocity or shear stress; when they have been used for this purpose, using the assumed profile forms, the accuracy is often poor.

One of the first of such methods and a classic of the literature is Pohlhausen's method[4.1] for incompressible flow based on the momentum integral equation. In its earliest form it was developed with a quartic form for the velocity profile in terms of the non-dimensional distance normal to the surface with a parameter involving the local pressure gradient and the boundary layer thickness. It will be presented here only in very summary form because it has been superseded in practice by subsequent simpler and more accurate methods. It is described in some detail by Schlichting[2.2] together with an improved form of the method by Holstein and Bohlen[4.2].

4.2 Incompressible flow

4.2.1 Pohlhausen's method – quartic profile

The momentum integral equation [Equation (2.50)] is

$$\frac{d\theta}{dx} + \frac{\theta}{u_e}\frac{du_e}{dx}(H + 2) = \frac{\tau_w}{\rho u_e^2} \tag{4.1}$$

With an assumed finite boundary layer thickness δ, the boundary conditions for the velocity profile at any station can be written (see Section 2.10):

$$\left.\begin{aligned} z = 0,\ u = 0, \quad & \nu\left(\frac{\partial^2 u}{\partial z^2}\right)_w = -u_e\frac{du_e}{dx} \\ z = \delta,\ u = u_e, \quad & \frac{\partial u}{\partial z} = \frac{\partial^2 u}{\partial z^2} = \frac{\partial^3 u}{\partial z^3} = \ldots = 0 \end{aligned}\right\} \tag{4.2}$$

We assume that u/u_e can be expressed as a quartic in η, where $\eta = z/\delta$, i.e.

$$u/u_e = a\eta + b\eta^2 + c\eta^3 + d\eta^4 \tag{4.3}$$

Then we can satisfy four of the above boundary conditions in addition to $u = 0$ when $z = 0$, in order to determine the coefficients a, b, c, d. We choose to satisfy:

$$\eta = 0, \quad \nu\left(\frac{\partial^2 u}{\partial z^2}\right)_w = -u_e\frac{du_e}{dx}$$

$$\eta = 1, \quad u/u_e = 1, \quad \frac{\partial u}{\partial z} = \frac{\partial^2 u}{\partial z^2} = 0$$

and we find

$$a = 2 + \frac{\lambda}{6}, \quad b = \lambda/2, \quad c = -2 + \frac{\lambda}{2}, \quad d = 1 - \frac{\lambda}{6} \tag{4.4}$$

where $\quad \lambda = \dfrac{\delta^2}{\nu}\dfrac{du_e}{dx}$

Hence the velocity distribution can be written:

$$u/u_e = F(\eta) + \lambda G(\eta)$$

where $\quad F(\eta) = 2\eta - 2\eta^3 + \eta^4, \qquad G(\eta) = \tfrac{1}{6}\eta(1 - \eta)^3 \qquad (4.5)$

Thus, the velocity profiles are expressed as a uni-parametric family with λ as the parameter, this is readily supported by the results of more exact solutions for favourable (negative) pressure gradients but becomes less easy to justify for adverse (positive) pressure gradients. We see that

$$\frac{\tau_w}{\rho u_e^2} = \frac{\nu}{u_e\delta}[F'(0) + \lambda G'(0)] = \frac{\nu}{u_e\delta}\left(2 + \frac{\lambda}{6}\right) \tag{4.6}$$

and we can likewise derive expressions for θ and H as functions of λ. Hence the momentum integral equation (4.1) can be expressed as a differential equation for λ (or δ) which can be written after some algebra in the form:

$$\frac{dZ}{dx} = h(\lambda)\,\frac{d^2u_e}{dx^2}\,Z^2 + g(\lambda)/u_e \tag{4.7}$$

where $Z = \dfrac{\delta^2}{\nu} = \lambda \Big/ \left(\dfrac{du_e}{dx}\right)$

and $h(\lambda)$ and $g(\lambda)$ are functions which can be evaluated and tabulated. Equation (4.7) can be readily solved numerically by a step by step process to give Z as a function of x, but a starting value is needed for λ. At a leading edge either $\delta = 0$ (as for a flat plate at zero incidence), and then $\lambda = 0$ there, or $u_e = 0$ (as on a round nosed wing). For the latter case it follows from equation (4.7) that $g(\lambda) = 0$ there and this leads to a starting value of $\lambda = 7.052$. We note that since in the region of a stagnation point $u_e = \beta x$, where β is a constant dependent on the nose shape, then from equation (4.7) $\delta = (7.052\ \nu/\beta)^{1/2}$, i.e. the boundary layer thickness is finite and constant for small x there.

However, the method is clearly not restricted to the quartic form of velocity profile, other polynomials or transcendental functions can be used. In general, the results prove to be of acceptable accuracy in the presence of small or favourable pressure gradients but become less satisfactory when the pressure gradients are strongly adverse. Thus, it will be seen from equation (4.6) that the skin friction is zero and hence separation begins when $\lambda = -12$. This value is, however, appreciably larger in magnitude than the values of λ given by exact solutions for which values of -7 to -8 are usual. As already noted an important source of error appears to be the assumption that the velocity profiles are uni-parametric.

4.2.2 Thwaites' method[4.3]

Thwaites defined two parameters l and m, such that

$$l = \frac{\theta}{u_e}\left(\frac{\partial u}{\partial z}\right)_w = \frac{\theta\tau_w}{u_e\mu}, \qquad m = \frac{\theta^2}{u_e}\left(\frac{\partial^2 u}{\partial z^2}\right)_w \tag{4.8}$$

We note that since $\nu(\partial^2u/\partial z^2)_w = -u_e\, du_e/dx$, from equations (2.32) and (2.36),

$$m = -\frac{\theta^2}{\nu}\frac{du_e}{dx} \tag{4.9}$$

We see that l is directly related to the frictional stress at the wall whilst m is related to the external pressure or velocity gradient. The assumption is again made that the family of laminar boundary layer velocity profiles are uni-parametric and we take m as the parameter. It then follows that l and H can be regarded as functions of m only. From the above, equation (4.1) is then readily transformed to

$$u_e \, d\theta^2/dx = 2\nu[m(H + 2) + l] = \nu L(m), \quad \text{say} \tag{4.10}$$

where $$L(m) = 2[m(H + 2) + l] \tag{4.11}$$

Thwaites examined the values of $L(m)$ for a number of known exact and approximate solutions covering a wide range of pressure distributions and found that the results fitted quite closely the simple linear relation (see Fig. 4.1)

$$L(m) = 0.45 + 6m \tag{4.12}$$

Such a linear relation had in fact been suggested earlier by Walz[4.4] and by Tani,[4.5] but with slightly different coefficients, being based on the somewhat more restricted data then available. Equations (4.10) and (4.12) then yield

$$u_e \, d(\theta^2)/dx = 0.45\nu - 6\theta^2 \, du_e/dx$$

or $$\frac{d}{dx}(u_e^6\theta^2) = 0.45\nu u_e^5 \tag{4.13}$$

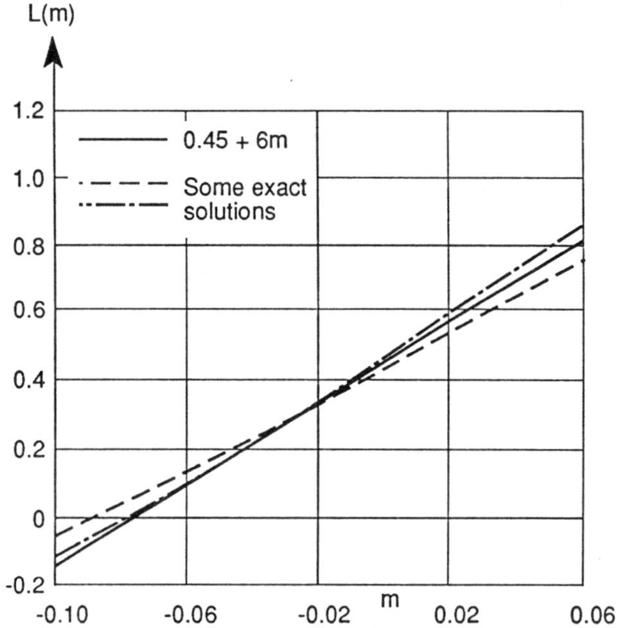

Fig. 4.1 The function $L(m)$.

Hence, if we integrate this equation from the leading edge, $x = 0$, where either $u_e = 0$ (stagnation point) or $\theta = 0$, to x_1, say we have

$$[\theta^2] \text{ at } x_1 = 0.45 \frac{\nu}{u_e^6} \int_0^{x_1} u_e^5 \, dx \tag{4.14}$$

Thus we have the remarkable result that θ at any station x can be determined by a simple integration of u_e^5 with respect to x starting at $x = 0$. The accuracy of this result is very good as it depends solely on the well substantiated empirical expression for $L(m)$ given by equation (4.12). A very similar relation but with slightly different constants was derived in 1940 by Young and Winterbottom[4.6] using a different approach. Their method was subsequently improved and adapted to compressible flow and will be discussed in that context later in Section 4.3.3.

We note that near the forward stagnation point on a round nosed aerofoil where $u_e = \beta x$, say, then equation (4.14) leads to

$$\theta^2 = 0.45\nu/6\beta \tag{4.15}$$

and hence θ is a constant independent of x in that region. This is consistent with δ being constant there, as already noted. Having θ^2 as a function of x we can determine the corresponding distribution of m from equation (4.9). Of major practical interest, however, are such quantities as the skin friction distribution, and therefore l, and the displacement thickness, and therefore H. Plots of l and H as functions of m derived from available exact solutions showed an encouraging fit to mean curves for favourable pressure gradients, i.e. for $m < 0$, but there was considerable scatter amongst the different solutions for adverse pressure gradients ($m > 0$). This scatter is a reflection of the breakdown of the assumption when $m > 0$ that the velocity profiles are uni-parametric. Thwaites therefore selected those exact solutions that seemed most relevant to the flow over an aerofoil, with particular emphasis on a solution for a linear adverse velocity distribution due to Howarth.[4.7] From these he derived mean values of l and H as unique functions of m which he recommended for general use in determining the overall boundary layer characteristics for aerofoils.

Curle and Skan[4.8] subsequently suggested some small changes to Thwaites' recommended values as a result of an analysis of exact solutions near separation using additional exact solutions not available to Thwaites. Their recommended values are listed in Table 4.1.

We see that according to this table the skin friction is zero ($l = 0$) and the flow is on the point of separating when $m = 0.090$ and $H = 3.55$. The corresponding earlier Thwaites values were $m = 0.082$ and $H = 3.7$.

4.2.2.1 *Examples of the use of Thwaites' method*

(1) Consider the case when $u_e = u_0(1 - bx)$, where b is a positive

Table 4.1

m	l	H
−0.25	0.500	2.00
−0.20	0.463	2.07
−0.14	0.404	2.18
−0.12	0.382	2.23
−0.10	0.359	2.28
−0.080	0.333	2.34
−0.064	0.313	2.39
−0.048	0.291	2.44
−0.032	0.268	2.49
−0.016	0.244	2.55
0	0.220	2.61
0.016	0.195	2.67
0.032	0.168	2.75
0.040	0.153	2.81
0.048	0.138	2.87
0.056	0.122	2.94
0.060	0.113	2.99
0.064	0.104	3.04
0.068	0.095	3.09
0.072	0.085	3.15
0.076	0.072	3.22
0.080	0.056	3.30
0.084	0.038	3.39
0.086	0.027	3.44
0.088	0.015	3.49
0.090	0	3.55

constant. Howarth's solution for this case of a linearly retarded main stream flow is a classic of the literature and although it was obtained long before digital computers were available, it has been shown to be reasonably accurate by comparison with subsequent calculations using modern computers.

From equation (4.12) we obtain

$$[\theta^2] \text{ at } x_1 = \frac{0.45\nu}{(u_e)^6}\int_0^{x_1} u_0^5(1 - bx)^5\,dx = \frac{0.45\nu}{6bu_0}[(1 - bx)^{-6} - 1]$$

Hence $$m = -\frac{du_e}{dx}\frac{\theta^2}{\nu} = \frac{0.45}{6}[(1 - bx)^{-6} - 1]$$

For flow about to separate ($l = 0$) we have from Table 4.1 that $m = 0.09$ and this occurs when

$$1 - bx_1 = (1/2.2)^{1/6} = 0.876, \qquad \text{or } bx_1 = 0.124$$

This is in very good agreement with Howarth's value of 0.120, but it is as well to note that Howarth's results were used to provide an important input in the derivation of Table 4.1.

We note that the pressure coefficient

$$\begin{aligned} c_p &= (p - p_0)/\tfrac{1}{2}\rho u_0^2 = 1 - (u_e/u_0)^2 \\ &= 0.233 \text{ at separation} \end{aligned}$$

This illustrates the relatively small pressure increase that can be sustained by a laminar boundary layer before separating.

(2) Given $u_e = u_0(x/L)^{-n}$, find n for which the boundary layer is just on the point of separating for all x.

From equation (4.14) we see that

$$(\theta^2) \text{ at } x_1 = 0.45\nu x_1^{1+n}/A(1 - 5n)$$

where $A = u_0 L^n$. Hence $m = 0.45n/(1 - 5n) = 0.090$ if the boundary layer is about to separate. Therefore, $n = 0.1$.

This again illustrates the fact that only a relatively small adverse pressure gradient can be sustained by the laminar boundary layer if separation is to be avoided.

4.2.3 Use of the kinetic energy integral equation (KIE)

The KIE for incompressible flow, equation (2.53), can be written

$$\frac{d}{dx}(u_e^3\delta_E/2) = \mathcal{D}/\rho$$

where the dissipation integral

$$\mathcal{D} = \int_0^h \mu\,(\partial u/\partial z)^2\,dz \tag{4.16}$$

Here, h is a length $> \delta$, the boundary layer thickness, and the reader is reminded that the kinetic energy thickness δ_E is given by

$$\delta_E = \int_0^h \frac{u}{u_e}\left[1 - \left(\frac{u}{u_e}\right)^2\right]dz$$

We write $\quad H_E = \delta_E/\theta, \qquad \beta = \mathcal{D}\theta/\mu u_e^2 \quad$ (4.17)

then if we multiply equation (4.16) by $2H_E u_e^3$ it can be written

$$\frac{d}{dx}(H_E^2 u_e^6 \theta^2) = 4H_E\beta\nu u_e^5$$

and hence $\quad (\theta^2)_{x_1} = \dfrac{4\nu}{(H_E^2 u_e^6)_{x_1}}\displaystyle\int_0^{x_1} H_E\beta u_e^5\,dx \quad$ (4.18)

We have here noted, as before, that at the leading edge, where $x = 0$ either $u_e = 0$ or $\theta = 0$.

Table 4.2

Flow	H_E	β
Stagnation	1.63	0.209
Blasius	1.57	0.173
Separation	1.52	0.157

The values of H_E and β listed in Table 4.2 are derived from exact solutions for the very different cases of stagnation point flow, zero pressure gradient flow (Blasius case) and a flow typical of incipient separation. It will be seen that H_E varies very little whilst the variation of β is somewhat greater but not large. Truckenbrodt[4.9] therefore suggested that in equation (4.18) both H_E and β should be taken as constant and given by their zero pressure gradient values, in which case equation (4.18) becomes

$$(\theta^2)_x = \frac{0.441\nu}{(u_e^6)_{x_1}} \int_0^{x_1} u_e^5 \, . \, dx \tag{4.19}$$

This is very remarkably close to Thwaites' result [equation (4.14)] derived from the MIE.

However, it is to be noted that both equations (4.10) and (4.18) involve no assumption additional to those of boundary layer theory. They can be combined to yield a general relation between l, m, H, H_E and β. Thus, from equation (4.10)

$$\frac{u_e}{2\nu}\frac{d\theta^2}{dx} = L(m)/2$$

and from (4.18)

$$\frac{u_e}{2\nu}\frac{d\theta^2}{dx} = \frac{2\beta}{H_E} + m\left(3 + \frac{H'_E u_e}{H_E u'_e}\right)$$

where the prime denotes differentiation with respect to x.

Hence $\quad L(m)/2 = m(H + 2) + l = \dfrac{2\beta}{H_E} + m\left(3 + \dfrac{H'_E u_e}{H_e u'_e}\right)$

and so $\quad l = -m(H - 1) + \dfrac{2\beta}{H_e} + m\dfrac{H'_E u_e}{H_E u'_e} \qquad (4.20)$

In view of the near constancy of H_E, it would seem acceptable to ignore the last term and then

$$l = -m(H - 1) + \frac{2\beta}{H_E} \tag{4.21}$$

From whatever forms are assumed for the velocity profiles (e.g. the Pohlhausen quartic, as was assumed by Tani,[4.10] or the Falkner–Skan set of similar profiles, as was assumed by Truckenbrodt) l, H, β and H_E can be obtained as functions of one of them, l say, and then from (4.21) l can be determined as a function of m, i.e. the skin friction distribution can be determined once θ^2 is determined from equation (4.14). Tani also suggested correcting for the initially neglected last term of equation (4.20) by an iterative process. This last term, although small, can be important near separation, for there l is small and tends to zero at separation.

These methods are somewhat more complex than Thwaites' method but are probably a little more accurate in the presence of adverse pressure gradients.

4.2.4 Two parameter methods: Head's method[4.11]

The manifest errors, particularly in the presence of adverse pressure gradients, arising from the assumption of a uni-parametric family of velocity profiles have led to the development of methods involving two parameter families. An early example was that of Wieghardt,[4.12] who assumed the profiles could be represented by an 11th order polynomial and he satisfied enough boundary conditions to leave two coefficients to be determined by satisfying both the MIE and the KIE. This was done by a simultaneous numerical solution of the two equations.

A more general form of Wieghardt's method was later developed by Head. He wrote the velocity profile in the form:

$$u/u_e = F(\eta) + \lambda_1 G_1(\eta) + \lambda_2 G_2(\eta)$$

where $\eta = z/\delta$, $F(\eta)$ is a close approximation to the Blasius profile, λ_1 and λ_2 are form parameters which can be l and m, and the functions G_1 and G_2 were determined from a study of known exact solutions. Hence, the overall quantities appearing in the MIE and KIE, such as H, H_E, β and L are expressible as functions of l and m, and so by the simultaneous numerical solution of the two equations these two parameters can be evaluated as functions of x. This method is regarded as the most accurate of the approximate methods and it has the advantage that the assumed form leads to fairly reliable velocity profiles. It can readily be extended to problems involving suction at the surface and, indeed, such problems provided much of the initial incentive for the development of the method.

4.2.5 Multi-layered approach

For situations where the boundary layer is subject to a rapidly developing or even sudden change of pressure or surface condition it seems reasonable to infer that the changes in the boundary layer response are largely manifest in an inner sub-layer in which viscosity effects associated with the imposed change are dominant. In the remaining (outer) part of the boundary layer the viscous effects can be assumed to be much the same as they would be in the absence of the change but the inner and outer regions must be presumed to merge smoothly. These ideas are the essence of what has been termed the multi-layer approach.

Such an approach was first pioneered by Karman and Millikan[4.13] who applied it to the prediction of separation due to the imposition of a strong adverse pressure gradient on a boundary layer on a flat plate downstream of a region of uniform pressure. Stratford[4.14] improved on their method postulating a more realistic velocity profile in the thin inner layer. His arguments led to the prediction that at separation:

$$\left[x^2 c_p \left(\frac{dc_p}{dx}\right)^2\right]_s = C \tag{4.22}$$

where the initial boundary layer just prior to the imposed adverse pressure gradient was assumed to be of the Blasius type with $p = p_0$, $u = u_0$, x was the distance from the start of the initial Blasius boundary layer, $c_p = (p - p_0)/\frac{1}{2}\rho u_0^2$, and C was a constant whose value depended on the assumed form of the inner region velocity profile. Given the latter as of the form:

$$u = az + bz^2 + cz^m$$

then for $m = 3$, 4 or 6 the corresponding values of C were found to be 0.010, 0.0065 and 0.0049, respectively. From an analysis of some exact solutions for cases leading to separation Curle and Skan[4.8] deduced that $C = 0.0104$ gave acceptable agreement with those solutions and suggested this value should be used in the above Stratford relation.

It will be noted that the method assumes an initial undisturbed region with uniform pressure. However, a typical pressure distribution on a wing will start with a stagnation point from which the pressure falls to some minimum value, which we shall denote by suffix m, followed by a region of pressure rise. To apply the criterion of equation (4.22) it is argued that the initial region of pressure fall can be regarded as equivalent to one of constant pressure p_m of extent x_{0e} resulting in the same momentum thickness at the start of the rise in pressure. Using Thwaites' formula, equation (4.14), it follows that

$$x_{0e} = \int_0^{x_m} (u_e/u_0)^5 \, . \, dx$$

Then in equation (4.22) x must be replaced by x_e measured from the start of the equivalent boundary layer, i.e. $x_e = x_{0e} + (x - x_0)$, whilst c_p and u_0 are to be taken as $(p - p_m)/\frac{1}{2}\rho u_m^2$ and u_m, respectively.

A later and more far reaching development is that of the 'triple-deck' concept for a suddenly perturbed boundary layer, due to Stewartson and Williams.[4.15] This concept has applications to the flow in the region of the trailing edge of a wing and to the region of interaction of a shock wave and boundary layer. It has been used in the context of solving simplified forms of the governing equations rather than integral equations. Here again the lowest layer or deck is one of rapid reaction to the perturbation with strong viscous effects, the middle or main deck comprises the remainder of the boundary layer in which viscous effects are relatively small and can be approximated in ways that depend on the problem considered, and the upper deck is one of inviscid flow but is a region of significant response to the perturbation. For a laminar boundary layer the thicknesses of these three decks are found to be of order $R^{-5/8}$, $R^{-1/2}$ and $R^{-3/8}$, respectively where R is the Reynolds number based on the distance L from the boundary layer origin, and the length of the strongly perturbed region is of order $R^{-3/8}$, assumed small compared with L. The analytical details cannot be given here, but the interested reader is referred to reference 4.15 and a review article by Stewartson.[4.16]

4.3 Compressible flow

4.3.1 Introduction

Over the years a variety of approximate methods have been developed for dealing with compressible flow boundary layers. These make use of the momentum integral equation and sometimes the energy integral equation and, as with the Stewartson–Illingworth approach (see Section 3.5), involve transformations that reduce these equations to forms close to if not exactly the same as the corresponding incompressible flow equations, so that their solution can follow well established lines. The degree of closeness to incompressible flow practice depends on the assumptions made with regard to the Prandtl number, σ, the viscosity–temperature relation and the presence or otherwise of heat transfer at the surface. A useful review of many of these methods is given by Curle.[4.17] In addition to the methods discussed below,[4.18–21] the methods of Curle,[4.22] Poots[4.23] and Lilley[4.24] are noteworthy.

However, the rapid development in speed, power and cheapness of digital computers has reduced the value of such methods as compared with the direct numerical solution of the basic boundary layer equations and only a moderate degree of complexity in those methods can be accepted if they are to retain a recognised advantage in terms of speed and cheapness of operation.

It is outside the scope of this book to discuss all these methods and we shall confine ourselves to presenting a summary of two of them. The first is that due to Cohen and Reshotko.[4.18] It is an extension to compressible flow of Thwaites' method and offers a similar appeal due to the simplicity of the underlying physical concepts. However, it involves the assumptions that $\sigma = 1$ and $\mu \propto T$, or $\omega = 1$. The second method discussed is that of

Luxton and Young.[4.19] This is not limited by such assumptions about σ and ω, beyond requiring them to be near unity, and it is relatively simple to apply. It must be noted that it does not yield estimates of the heat transfer as it does not involve the direct solution of the energy integral equation. Consideration of the latter along with the momentum integral equation is essential to arrive at reliable estimates of heat transfer.

4.3.2 The method of Cohen and Reshotko

In Section 3.5 we described the Stewartson–Illingworth transformation, which with the assumptions of $\sigma = 1$ and μ/T = constant (i.e. $\omega = 1$) and with a constant wall temperature or with zero heat transfer at the wall, yielded equations of motion and energy closely similar to the incompressible forms of these equations. We then summarised how Cohen and Reshotko, following Li and Nagamatsu, adapted the Falkner–Skan family of similar solutions for incompressible flow to provide related families of solutions for ranges of values of the parameters S_w and β determining the wall temperature [see equation (3.90)] and the pressure distribution [see equations (3.99) and (3.100)].

Cohen and Reshotko then introduced the following boundary layer thicknesses in terms of the transformed coordinates X and Z:

$$\begin{aligned}
\delta_t^* &= \int_0^\infty (1 + S - U/U_e)\mathrm{d}Z \\
\theta_t &= \int_0^\infty (U/U_e)(1 - U/U_e)\mathrm{d}Z \\
\delta_{st} &= \int_0^\infty (U\,S/U_e)\mathrm{d}Z
\end{aligned} \tag{4.23}$$

and they derived the transformed MIE and total energy integral equation (TEIE) in the forms:

$$\frac{\mathrm{d}\theta_t}{\mathrm{d}X} + \frac{1}{U_e}\frac{\mathrm{d}U_e}{\mathrm{d}X}(\delta_t^* + 2\theta_t) = \frac{\nu_0}{U_e^2}\left(\frac{\partial U}{\partial Z}\right)_w \tag{4.24}$$

[cf. equation (2.50)], and

$$\frac{\mathrm{d}\delta_{st}}{\mathrm{d}X} + \frac{1}{U_e}\frac{\mathrm{d}U_e}{\mathrm{d}X}\delta_{st} = -\frac{\nu_0}{U_e}\left(\frac{\partial S}{\partial Z}\right)_w \tag{4.25}$$

Following Thwaites they introduced the parameters l_t and m_t, where

$$\left.\begin{aligned}
l_t &= \frac{\theta_t}{U_e}\left(\frac{\partial U}{\partial Z}\right)_w \\
m_t &= \frac{\theta_t^2}{U_e(1 + S_w)}\left(\frac{\partial^2 U}{\partial Z^2}\right)_w = -\frac{(\mathrm{d}U_e/\mathrm{d}X)}{\nu_0}\theta_t^2
\end{aligned}\right\} \tag{4.26}$$

They then converted the MIE to the form [cf. equation (4.10)]

$$\frac{U_e}{\nu_0}\frac{d\theta_t^2}{dX} = 2[m_t(H_t + 2) + l_t] = L_t, \quad \text{say} \tag{4.27}$$

where $H_t = \delta_t^*/\theta_t$. An analysis of the various 'similar' solutions that they had derived suggested that L_t, H_t and l_t could be regarded as functions of m_t and S_w, only, and Cohen and Reshotko have provided useful plots of these functions for ranges of m_t and S_w. With constant S_w a linear relation between L_t and m_t, cf. equation (4.12), was found to be an acceptable approximation so that equation (4.27) could then be integrated analytically, leading to an expression for θ_t^2 featuring a simple integration with respect to X of a power of U_e as for incompressible flow [cf. equation (4.14)].

By differentiating the equation of motion in the transformed coordinates [equation (3.93)] with respect to Z one deduces that at the wall

$$U_e\frac{dU_e}{dX}\left(\frac{\partial S}{\partial Z}\right)_w = -\nu_0\left(\frac{\partial^3 U}{\partial Z^3}\right)_w$$

and Cohen and Reshotko introduced a third parameter r, related to the heat transfer at the wall, where

$$r = m_t\theta_t\left(\frac{\partial S}{\partial Z}\right)_w = \frac{\theta_t^3}{U_e}\left(\frac{\partial^3 U}{\partial Z^3}\right)_w \tag{4.28}$$

They further assumed in the light of their 'similar' solutions that r, like l_t, was a function of m_t and S_w, only, and likewise provided plots of this function. It follows from equation (4.28) that having solved for θ_t as a function of X we can estimate the corresponding distribution of r and therefore of the heat transfer. Note that this procedure does not necessarily imply that the energy equation, equation (4.25) is satisfied, and it seems likely that the latter is required for a reliable estimate of heat transfer.

From the distribution of θ_t as a function of X and the Cohen and Reshotko plots the corresponding distributions can be derived for m_t, H_t, δ_t^*, l_t and r.

Then the distributions of θ, δ^*, τ_w and heat transfer in the physical plane are given by

$$\left.\begin{aligned}
\theta &= \theta_t(T_0/T_e)^{(\gamma+1)/2(\gamma-1)} \\
\delta^* &= \theta[H_t + \tfrac{1}{2}(\gamma - 1)M_e^2(H_t + 1)] \\
\frac{\tau_w}{\mu} &= \left(\frac{\partial u}{\partial z}\right)_w = \frac{u_e}{\theta}\frac{T_e}{T_w}l_t \\
\left(\frac{\partial T}{\partial z}\right)_w &= \frac{T_e}{\theta}\frac{T_0}{T_w}\frac{r}{m_t}
\end{aligned}\right\} \tag{4.29}$$

Comparison of the results predicted by this method with the results of available exact solutions, additional to the 'similar' family on which the method is based, shows very good agreement in cases of favourable or small adverse pressure gradients but in regions of strong adverse pressure gradients approaching separation the agreement is less satisfactory.

A development of this method involving some further simplifying assumptions without any evident serious deterioration of accuracy was made by Monaghan.[4.20]

4.3.3 The Luxton–Young method

This method starts with a simple transformation of the ordinate z normal to the wall to Z where now

$$Z = \int_0^z (\mu_e/\mu)\mathrm{d}z \tag{4.30}$$

A Pohlhausen quartic form of velocity profile is assumed, viz.

$$u = aZ + bZ^2 + cZ^3 + dZ^4 \tag{4.31}$$

with the boundary conditions: at $Z = 0$, $u = 0$, $T = T_w$; and at the outer edge of the boundary layer $Z = \delta_1$, say, $u = u_e$, $\partial u/\partial z = \partial^2 u/\partial z^2 = 0$. Also from the first equation of motion [equation (2.32)] it follows that at the wall

$$\frac{\partial}{\partial z}(\mu\,\partial u/\partial z)_w = \frac{\mathrm{d}p}{\mathrm{d}x} = -\rho_e u_e \frac{\mathrm{d}u_e}{\mathrm{d}x}$$

These conditions enable a, b, c and d to be determined in terms of δ_1 and a pressure gradient parameter, Λ, namely,

$$a = \frac{u_e}{6\delta_1}(12 + \Lambda),\quad b = -\frac{u_e\Lambda}{2\delta_1^2},\quad c = -\frac{u_e(4-\Lambda)}{2\delta_1^3},\quad d = \frac{u_e(6-\Lambda)}{6\delta_1^4} \tag{4.32}$$

where $\quad \Lambda = \dfrac{\mathrm{d}u_e}{\mathrm{d}x}\,\delta_1^2\rho_e\,\dfrac{\mu_w}{\mu_e^2}$

Hence

$$\frac{\tau_w}{\rho_e u_e^2} = \mu_w\left(\frac{\partial u}{\partial z}\right)_w = \mu_e\left(\frac{\partial u}{\partial Z}\right)_w = \mu_e a = \frac{\mu_e(12+\Lambda)}{6\delta_1\rho_e u_e} \tag{4.33}$$

The method, in fact, makes no more use of the quartic velocity profile apart from accepting equation (4.33) relating the non-dimensional frictional stress and δ_1.

The MIE [equation (2.48)] is

$$\frac{\mathrm{d}\theta}{\mathrm{d}x} + \frac{1}{u_e}\frac{\mathrm{d}u_e}{\mathrm{d}x}\theta(H+2) + \frac{\theta}{\rho_e}\frac{\mathrm{d}\rho_e}{\mathrm{d}x} = \frac{\tau_w}{\rho_e u_e^2} \tag{4.34}$$

To solve this we need two more relations between θ, H and τ_w.

Write $\delta_1/\theta = f$, say, then, if we make use of equation (4.33), we have

$$\frac{\tau_w}{\rho_e u_e^2} = \frac{2\mu_e}{\rho_e u_e f\theta} + \frac{du_e}{dx}\frac{f\theta}{6u_e}\frac{\mu_w}{\mu_e}$$

and so equation (4.34) becomes

$$\frac{d}{dx}(\rho_e^2\theta^2) + 2\frac{\rho_e^2\theta^2}{u_e}\frac{du_e}{dx}\left[(H+2) - \frac{f}{6}\frac{\mu_w}{\mu_e}\right] = 4\frac{\mu_e\rho_e}{u_e f} \tag{4.35}$$

Luxton and Young then write

$$\left[(H_n + 2) - \frac{f_n\mu_w}{6\mu_e}\right] = g_n/2 \tag{4.36}$$

where H_n and f_n are the values of H and f at station x_n, say, and they are assumed to be constant over an interval of x from x_n to x_{n+1}.

Further $\quad \mu_w/\mu_e = (T_w/T_e)^\omega \qquad (4.37)$

and this ratio is also assumed to be constant over each such interval.

Then it follows that g_n is constant over that interval and is given by its value at x_n. Consequently, equation (4.35) can be integrated over the interval from x_n to x_{n+1} to yield

$$[\rho_e^2\theta^2 u_e^{g_n}]_{x_n}^{x_{n+1}} = 4\int_{x_n}^{x_{n+1}} \frac{\rho_e\mu_e u_e^{g_n-1}}{f_n}\,dx \tag{4.38}$$

[Compare equation (4.14).] It is to be noted that g_n occurs in this equation in such a manner that, provided the variation of u_e with x is reasonably small over the interval, the value of θ is insensitive to small variations or errors in g_n. Hence, approximations in its value can be accepted for the purpose of determining θ, and also the intervals can be large for regions where the variations in free stream Mach number (or velocity) are small.

To complete the solution we need to determine $f = \delta_1/\theta$ and H as functions of M_e, T_w/T_e (or S_w), ω, σ and Λ.

Now for zero pressure gradient, i.e. with uniform external flow, equation (4.38) yields:

$$[\rho_e^2\theta^2]_x = 4\mu_e x/f_0 u_e$$

where f_0 is the value of f for zero pressure gradient flow. Therefore

$$\theta\, R_x^{1/2}/2x = (1/\mathrm{f}_0)^{1/2}$$

and so $\quad c_f = \dfrac{2\tau_w}{\rho_e u_e^2} = 2\dfrac{d\theta}{dx} = \dfrac{2}{(f_0 R_x)^{1/2}}$

But we had, equation (3.55),

$$c_f R_x^{1/2} = 0.664\left[0.45 + 0.55\frac{T_w}{T_e} + 0.09(\gamma - 1)M_e^2\,\sigma^{1/2}\right]^{(\omega-1)/2}$$

and hence $$f_0 = 9.072\left[0.45 + 0.55\frac{T_w}{T_e} + 0.09(\gamma - 1)M_e^2\,\sigma^{1/2}\right]^{1-\omega} \qquad (4.39)$$

An analysis of the Cohen and Reshotko 'similar' results shows that the effect of pressure gradient on f can be approximated with acceptable accuracy by writing

$$f = f_0(1 + k_1\Lambda) \qquad (4.40)$$

where k_1 is shown as a function of S_w in Fig. 4.2.

Equations (4.39) and (4.40) enable us to determine f in terms of M_e, S_w, ω, σ and Λ. To determine H we recall first (Section 3.3.4) that with $\omega = 1$ (or $\rho\mu$ = const.) and zero pressure gradient then u/u_e is a unique function of $Z = z_1$, independent of Mach number. In that case we have

$$\theta = \int_0^\infty \frac{\rho u}{\rho_e u_e}\left(1 - \frac{u}{u_e}\right)dz = \int_0^\infty \frac{u}{u_e}\left(1 - \frac{u}{u_e}\right)dZ = \theta_{i,0} \qquad (4.41)$$

where the suffixes i, 0 denote the value in incompressible flow (or, more strictly, $M_e = 0$) and with zero pressure gradient. Also

$$\delta^* = \int_0^\infty \left(1 - \frac{\rho u}{\rho_e u_e}\right)dz = \int_0^\infty \left(\frac{\mu}{\mu_e} - \frac{u}{u_e}\right)dZ = \int_0^\infty \left(\frac{i}{i_e} - \frac{u}{u_e}\right)dZ$$

Now, with $\sigma = 1$, we have Crocco's relation [equation (2.61)]

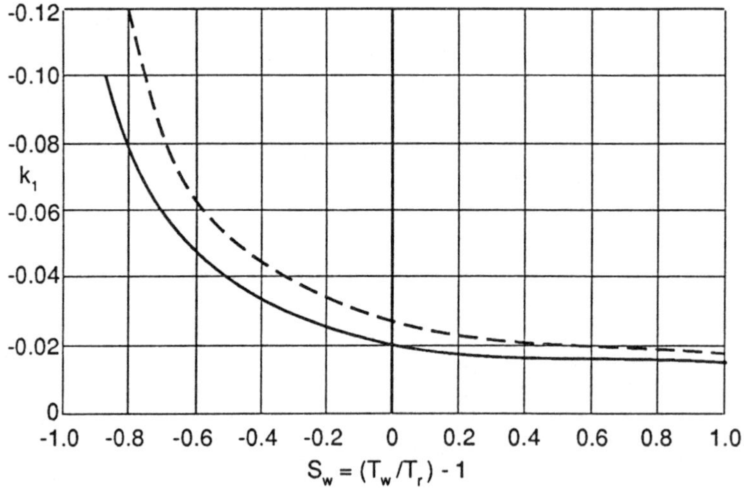

Fig. 4.2 Correction factor k_1,[4.19] see equation (4.40).
———— favourable pressure gradients
– – – – – adverse pressure gradients.

$$T_{\mathrm{H}} = T + \frac{u^2}{2c_{\mathrm{p}}} = k_1 u + k_2$$

With the boundary conditions $T = T_{\mathrm{w}}$ when $u = 0$, and $T = T_{\mathrm{e}}$ when $u = u_{\mathrm{e}}$, it follows that

$$\frac{T}{T_{\mathrm{e}}} = \frac{T_{\mathrm{w}}}{T_{\mathrm{e}}} + \left[\left(1 - \frac{T_{\mathrm{w}}}{T_{\mathrm{e}}}\right) + \frac{u_{\mathrm{e}}^2}{2c_{\mathrm{p}}T_{\mathrm{e}}}\right]\eta - \frac{u_{\mathrm{e}}^2}{2c_{\mathrm{p}}T_{\mathrm{e}}}\eta^2$$

where $\eta = u/u_{\mathrm{e}}$. Noting that $u_{\mathrm{e}}^2/2c_{\mathrm{p}}T_{\mathrm{e}} = \frac{1}{2}(\gamma - 1)M_{\mathrm{e}}^2$, we infer that

$$\frac{i}{i_{\mathrm{e}}} = \frac{i_{\mathrm{w}}}{i_{\mathrm{e}}} + \left[\left(1 - \frac{i_{\mathrm{w}}}{i_{\mathrm{e}}}\right) + \tfrac{1}{2}(\gamma - 1)M_{\mathrm{e}}^2(1 - \eta)\right]\eta$$

and hence $$\delta^* = \frac{i_{\mathrm{w}}}{i_{\mathrm{e}}} \,.\, \delta^*_{\mathrm{i},0} + \tfrac{1}{2}(\gamma - 1)M_{\mathrm{e}}^2\theta_{\mathrm{i},0}$$

Therefore $$H = \delta^*/\theta = \frac{i_{\mathrm{w}}}{i_{\mathrm{e}}} H_{\mathrm{i},0} + \tfrac{1}{2}(\gamma - 1)M_{\mathrm{e}}^2 \qquad (4.42)$$

From the Blasius solution for incompressible flow and zero pressure gradient we have that $H_{\mathrm{i},0} = 2.59$.

We defined $S = (T_{\mathrm{H}}/T_{\mathrm{H0}}) - 1$, but now T_{H0} is the total temperature when the air is brought to rest and this equals T_{r}, the recovery temperature. Hence $S_{\mathrm{w}} = (T_{\mathrm{w}}/T_{\mathrm{r}}) - 1$. Therefore, equation (4.42) can be written

$$H = (1 + S_{\mathrm{w}})\frac{T_{\mathrm{r}}}{T_{\mathrm{e}}} H_{\mathrm{i},0} + \tfrac{1}{2}(\gamma - 1)M_{\mathrm{e}}^2 \qquad (4.43)$$

It is now assumed that this relation can be generalised to flows with non-zero pressure gradient and values of σ and ω near to but not necessarily equal to unity by

$$H = [(1 + S_{\mathrm{w}})H_{\mathrm{i},0} + \phi(\Lambda, S_{\mathrm{w}})](T_{\mathrm{r}}/T_{\mathrm{e}}) + \tfrac{1}{2}(\gamma - 1)M_{\mathrm{e}}^2 \qquad (4.44)$$

where now [equation (3.43)]

$$T_{\mathrm{r}} = 1 + \tfrac{1}{2}(\gamma - 1)M_{\mathrm{e}}^2\,\sigma^{1/2}$$

and $\phi(\Lambda, S_{\mathrm{w}})$ is a function representing the effects of a non-zero pressure gradient and a non-zero surface heat transfer. It was determined from an analysis of the Cohen and Reshotko 'similar' solutions and is shown in Fig. 4.3 as a function of Λ for different values of S_{w}. With the aid of equations (4.39), (4.40) and (4.44) plus Figs. 4.2 and 4.3 we can determine the value of g_n in terms of the local values at x_n of M_{e}, T_{w}, Λ, S_{w}, ω and σ for use in equation (4.38). Therefore given the external flow over an aerofoil, say, we can solve for θ as a function of x by a numerical step by step process. It appears that the steps can be quite large, steps in external Mach number of the order of 0.3 to 0.5 give acceptable accuracy.

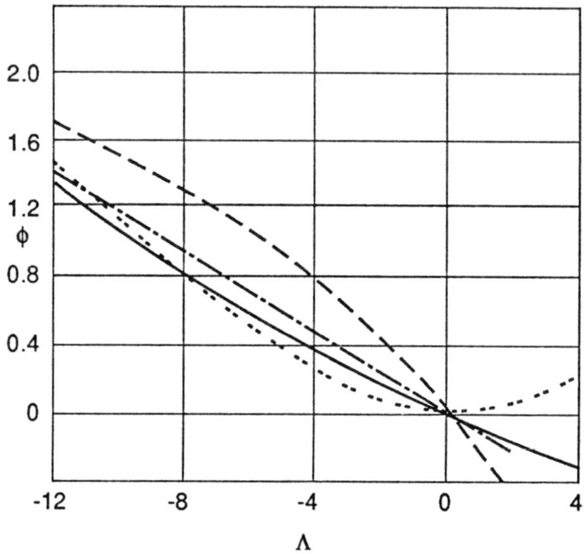

Fig. 4.3 Correction factor $\phi(\Lambda, S_w)$,[4.19] see equation (4.44),
——— $S_w = 0$ – – – – – $S_w = -0.8$
–·–·– $S_w = -0.4$ ·· ·· ·· ·· $S_w = 0.4$.

The solution yields also the distributions of H, f and Λ and hence the corresponding distributions of δ^* and τ_w [see equation (4.33)] can be determined.

Results obtained by the method show excellent agreement with the results of 'similar' solutions for both strongly favourable and adverse pressure distributions and wide ranging Mach numbers. However, it has to be noted that the method involves some empirical relations based on the 'similar' solution family.

A series of cases calculated by the method show that in the presence of a favourable pressure gradient cooling of the surface decreases the skin friction, because the increased density of the fluid near the surface due to the cooling causes the flow to be less readily accelerated there and the frictional stress is therefore reduced. The same argument explains the converse result that the skin friction is increased by wall cooling in the presence of an adverse pressure gradient and then separation occurs further downstream. Wall heating obviously has opposite effects to wall cooling in the presence of pressure gradients.

The method was also used by Young [4.21] to examine the effects of small changes in σ and ω on skin friction. This investigation showed that the effects for a given wall temperature can be predicted with adequate accuracy by applying the variations predicted by the zero pressure gradient relation [equation (3.55)] for the local free stream conditions.

References

4.1 Pohlhausen, K. (1921) 'Zur näherungsweisen Integration der Differentialgleichung der laminaren Reibungsschicht'. *ZAMM*, **I**, p. 252.

4.2 Holstein, H. and Bohlen, T. (1940) 'Ein einfaches Verfahren zur Berechnung laminarer Reibungsschichten die den Näherungsverfahren von K. Pohlhausen genügen'. *Lilienthal Bericht S.*, **10**, p. 5.

4.3 Thwaites B. (1949) 'Approximate calculation of the laminar boundary layer'. *Aero. Qu*, **I**, p. 245.

4.4 Walz, A. (1941) 'Ein neuer Ansatz für das Geschwindigskeitprofil der laminaren Reibungsschicht'. *Lilienthal Bericht*, **141**, p. 8.

4.5 Tani, I. (1949) 'A simple method for determining the laminar separation point'. *Rep. Aero. Res. Inst.*, Tokyo, No. 199.

4.6 Young, A.D. and Winterbottom, N. (1940) 'Note on the effect of compressibility on the profile drag of an aerofoil'. ARC RM 2400.

4.7 Howarth, L. (1938) 'On the solution of the laminar boundary layer equations'. *Proc. Roy. Soc. A*, **164**, p. 547.

4.8 Curle, N. and Skan, S.W. (1957) 'Approximate methods for predicting separation properties of laminar boundary layers'. *Aero. Qu.*, **8**, p. 257.

4.9 Truckenbrodt, E. (1952) 'Ein Quadraturverfahren zur Berechnung der laminaren und turbulenten Reibungsschichten bei ebener und rotations symmetrischer Strömung'. *Ing. Arch.*, **20**, p. 211.

4.10 Tani, I. (1954) 'On the approximate solution of the laminar boundary layer equations'. *J. Aero. Sc.*, **31**, p. 487.

4.11 Head, M.R. (1957) 'An approximate method of calculating the laminar boundary layer in two dimensional incompressible flow'. ARC RM 3123.

4.12 Wieghardt, K. (1946) 'On an energy equation for the calculation of laminar boundary layers'. Rep. ARC 9825.

4.13 v. Karman, T. and Millikan, C.B. (1934) 'On the theory of laminar boundary layers involving separation'. NACA Rep. 504.

4.14 Stratford B.S. (1954) 'Flow in the laminar layer near separation'. ARC RM 3002.

4.15 Stewartson, K. (1969) 'Self-induced separation'. *Pt. 2. Mathematica*, **20**, p. 98.

4.16 Stewartson K. (1974) 'Multi-structured boundary layers on flat plates and related bodies'. (Ed. C.S. Yih) *Adv. Appl. Mech.*, **14**, p. 145.

4.17 Curle, N. (1962) 'The Laminar Boundary Layer Equations'. *Oxford Mathematical Monographs*. OUP.

4.18 Cohen, C.B. and Reshotko, E. (1956) 'The compressible laminar boundary layer with heat transfer and arbitrary pressure gradient'. NACA Rep. 1294.

4.19 Luxton, R.E. and Young, A.D. (1962) 'Generalised methods for the calculation of the laminar compressible boundary layer characteristics with heat transfer and non-uniform pressure distribution'. ARC RM 3233.

4.20 Monaghan, R.J. (1961) Effects of heat transfer on laminar boundary layer development under pressure gradients in compressible flow'. ARC RM 3218.

4.21 Young, A.D. (1964) 'The effects of small changes of Prandtl number and the viscosity temperature index on skin friction'. *Aero. Qu.*, **XV**, p. 393.

4.22 Curle, N. (1962) 'Heat transfer through a compressible laminar boundary layer'. *Aero. Qu.*, **13**, p. 255.
4.23 Poots, G. (1960) 'A solution of the compressible laminar boundary layer equations with heat transfer and adverse pressure gradient'. *Qu. J. Mech. App. Maths.* **13**, p. 57.
4.24 Lilley, G.M. (1959) 'A simplified theory of skin friction and heat transfer for a compressible laminar boundary layer'. Coll. Aeronautics, Cranfield, Note No. 93.

Chapter 5

Transition

5.1 Introductory remarks

The transition process from laminar to turbulent flow in a shear layer has been a subject of study for over a century, but it still remains a challenging and important area for research. It was first noted by Reynolds in 1883[5.1] when he showed in a classic experiment that a thin filament of dye in water flowing along a transparent glass tube would remain well defined and straight if the Reynolds number of the flow, R_d, was less than about 2500, but for higher Reynolds numbers the filament rapidly became convoluted, thickened and broke up, the dye dispersing throughout the tube. Here $R_d = u_m d/\nu$, where u_m is the mean velocity in the tube and d is the diameter of the tube. Evidently, the laminar flow developed some form of instability above that Reynolds number, an instability that culminated in what we now call turbulent flow, characterised by a much larger scale of irregular movements of fluid elements and more intense mixing than the random molecular movements and associated viscous mixing of laminar flow. This considerably more powerful mixing mechanism of the turbulent flow resulted in a more nearly uniform velocity distribution over much of the pipe cross section, but with a consequent higher rate of shear and frictional stress very close to the pipe surface.

Later workers found that the so-called critical Reynolds number above which transition was observed was dependent on the conditions of the experiment; by carefully avoiding sources of vibration and making sure that the inlet flow was initially still and undisturbed workers found that the critical Reynolds number could be raised to as high as 4×10^4. It could therefore be inferred that instability of the laminar flow increased with Reynolds number above a critical value but the latter depended on the nature and scale of any disturbances present. Such disturbances could be in the form of initial unsteadiness, eddying introduced by sharp entry edges, acoustic in origin, or due to excrescences or waviness of the pipe surface.

Closely similar observations were made of the flow in boundary layers which likewise showed that the process of transition was dependent on the Reynolds number (defined either in terms of a boundary layer thickness or distance from the leading edge) the nature and intensity of disturbances in the external flow and the smoothness of the surface. To illustrate the effect of the Reynolds number it may be noted that with zero pressure gradient and a smooth surface it is difficult to achieve transition on a flat plate if $R_\theta = u_e \theta/\nu$ is less than about 200, but for $R_\theta > 5 \times 10^3$ it is difficult to avoid it. Further, it readily became evident that the streamwise pressure gradient was a very important factor, a favourable pressure gradient ($dp/dx < 0$) helped to delay transition whilst an adverse one ($dp/dx > 0$) hastened it.

It must be emphasised that the process of transition in a boundary layer is not instantaneous but occurs over a finite distance, which we call a transition region. At low Reynolds numbers the extent of this region can be a large part of the wing chord; however, at the Reynolds numbers characteristic of typical wings or bodies in moderate to large wind tunnels (of the order of 5×10^6 in terms of distance from the leading edge) the transition region is small and in two dimensional flow may be conveniently referred to as a point but not necessarily steady.

A feature of fully turbulent flows is the irregular spiky nature of the records shown by hot wire anemometer measurements as the resistance of the wire responds practically instantaneously to the cooling effects of the random fluctuations in velocity due to the turbulence (see Fig. 5.1). However, in a transition region the records show periods of smooth laminar flow alternating with periods of turbulent flow and the fraction of the time that the latter occupy increases from zero at the beginning of the transition region to unity at the end for a probe in the lower half of the boundary layer. This fraction is sometimes referred to as the *intermittency*.

Once it was realised that turbulence in a boundary layer can in some way be initiated by the growth of perturbations in conditions in which the boundary layer was unstable, much effort, with the aim of understanding and predicting transition, was devoted to the analysis of the stability of laminar flows, in the first place in the presence of small perturbations for which a linearised theory was applicable. Later the analysis was extended to the effects of large perturbations where non-linear contributions can be important.

We shall discuss these analytical developments in more detail later, but the following simple argument (Sections 5.2 and 5.3) is instructive in demonstrating how a perturbation can extract energy from the mean flow and therefore grow, and the vital role of viscosity (or Reynolds number) in that process. However, it must be emphasised that the essentially three dimensional nature of turbulence cannot in the final analysis be ignored, it plays a key role even when the mean flow is two dimensional. It is also worth anticipating what will be evident from the remainder of this

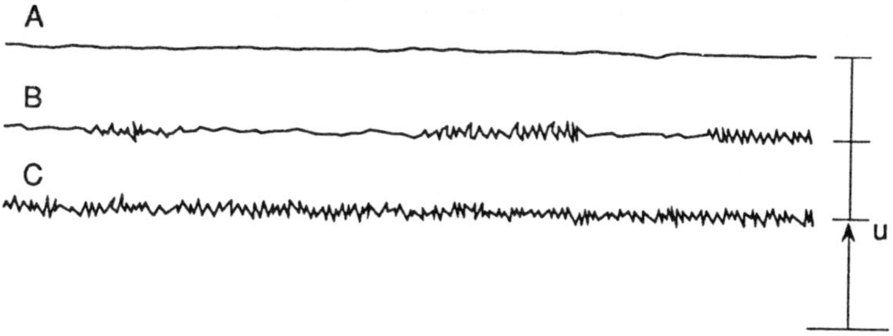

Fig. 5.1 Illustration of hot wire records in a boundary layer.
A – laminar B – transitional C – turbulent.

chapter, namely that the physics of the transition process is still far from fully understood, and major questions remain to be resolved.

5.2 The Reynolds stresses[5.2]

Any form of timewise varying perturbation or unsteadiness present in a flow results in rates of transport of momenta of the fluid particles additional to those of the unperturbed flow. The values of these perturbation momenta fluxes, which can be defined relative to a convenient system of axes, can be averaged over a suitable time interval, long compared with a typical perturbation period, and such averages may be interpreted as stresses which are additional to the viscous stresses of laminar flow. (We may note that the latter similarly result from the much smaller scale random molecular movements.) Strictly, the use of time averages implies that the perturbations at a particular point have statistical characteristics that are invariant with time, i.e. that they are stochastic, and for transitional and fully turbulent flows that are steady in the mean this is generally an acceptable assumption. It is still acceptable if the mean flow varies in time but with a time scale that is long compared with a typical turbulence fluctuation.

To be more specific we define the mean of any quantity q, say, at a point $O(x, y, z)$ at time t_1, as

$$\overline{q} = \frac{1}{2T}\int_{t_1-T}^{t_1+T} q \,.\, dt \qquad (5.1)$$

where T is a time interval assumed long enough compared with a typical perturbation period for $\overline{q}$ to be independent of T. We then denote the instantaneous velocity components relative to the (x, y, z) axes as

$$U = u + u', \qquad V = v + v', \qquad W = w + w' \qquad (5.2)$$

here $\quad (u, v, w) = (\bar{U}, \bar{V}, \bar{W})$ denotes the mean flow $\quad$ (5.3)

and (u', v', w') denotes the perturbation components, with

$$\overline{u'} = \overline{v'} = \overline{w'} = 0 \tag{5.4}$$

Across a small area δA in the yz plane embracing O the flow of mass in time δt is

$$(\rho U \, \delta A)\delta t$$

and this is associated with a momentum transport across δA having components

$$(\rho U^2 \, \delta A)\delta t, \qquad (\rho UV \, \delta A)\delta t, \qquad (\rho UW \, \delta A)\delta t$$

The mean values of these components per unit time per unit area are

$$\overline{(\rho+\rho')(u+u')^2}, \quad \overline{(\rho+\rho')(u+u')(v+v')}, \quad \overline{(\rho+\rho')(u+u')(w+w')}$$

or

$$\left.\begin{aligned} &\rho u^2 + \rho\overline{u'^2} + 2u\overline{\rho' u'} + \overline{\rho' u'^2} \\ &\rho uv + \rho\overline{u' v'} + u\overline{\rho' v'} + v\overline{\rho' u'} + \overline{\rho' u' v'} \\ &\rho uw + \rho\overline{u' w'} + u\overline{\rho' w'} + w\overline{\rho' u'} + \overline{\rho' u' w'} \end{aligned}\right\} \tag{5.5}$$

We see that in addition to the rates of momenta flux per unit area due to the mean velocity components there are components arising from the perturbations.

If we confine ourselves at this stage to incompressible flow then ρ is constant and $\rho' = 0$. The additional components are then

$$\rho\overline{u'^2}, \quad \rho\overline{u' v'}, \quad \rho\overline{u' w'}$$

These components of momentum flux per unit area due to the perturbations can be regarded as equivalent to three stress components acting in the opposite directions to the momentum flux components on the elementary area δA at O, and we can write these components as

$$\tau_{xx} = -\rho\overline{u'^2}, \qquad \tau_{xy} = -\rho\overline{u' v'}, \qquad \tau_{xz} = -\rho\overline{u' w'} \tag{5.6}$$

Here we adopt the convention for the stress components $\tau_{\alpha\beta}$ that the first suffix denotes the normal to the surface element considered and the second denotes the direction along which the component is taken. We see that the normal stress component τ_{xx} is not zero; and the other two shear stress components are also non-zero as long as u' is not completely independent of v' and w', i.e. as long as there is some measure of correlation between u' and the other two perturbation velocity components. Such a correlation is likely as long as the perturbations have some kind of structure of finite scale.

Similarly, if we had taken elements of area at O in the zx and xy planes we would infer the existence of stress components

$$\left.\begin{array}{lll} \tau_{yx} = -\rho\overline{u'v'}, & \tau_{yy} = -\rho\overline{v'^2}, & \tau_{yz} = -\rho\overline{v'w'} \\ \tau_{zx} = -\rho\overline{w'u'}, & \tau_{zy} = -\rho\overline{w'v'}, & \tau_{zz} = -\rho\overline{w'^2} \end{array}\right\} \qquad (5.7)$$

Thus, we have nine perturbation stress components forming a tensor, which is symmetric since $\tau_{xy} = \tau_{yx}$, etc.

These components are generally known as *Reynolds Stresses* as Reynolds first drew attention to them,[5.2] they are also sometimes called *eddy stresses*. They play a most important part in the turbulent boundary layer where they are usually much greater than the corresponding viscous stresses, but it will be evident that they can also arise whenever any time varying perturbation is present in the laminar boundary layer.

In a two dimensional turbulent boundary layer flow the dominant stress is

$$\tau_{xz} = -\rho\overline{u'w'}$$

where z is taken normal to the surface. In this instance the simple argument can be postulated (as was done by Prandtl) that fluid particles moving towards the surface (negative w') come from strata where the streamwise velocity was greater than in this area of mixing (postive u'), and vice versa, and so $\tau_{xz} > 0$.

In perturbed laminar flow the magnitude and sign of τ_{xz} depend on the nature of the perturbation and the mean flow. However, if we think of the perturbation as a train of simple harmonic waves we find that the effects of viscosity adjacent the surface in enforcing the 'no-slip' condition there produce a phase-shift between u' and w' such that τ_{xz} is positive in that region, given a mean velocity distribution of the normal laminar flow type. In addition, we find that viscous effects become important in the region where the wave train velocity (or phase) equals the local mean flow velocity. Here the transverse gradient of the perturbation velocity becomes relatively large and theory indicates that the viscous effects become important there and again a phase change between u' and w' results to produce local non-zero values of τ_{xz}.

5.3 Energy transfer between perturbations and mean flow

Lorentz[5.3] produced the following simple argument to illustrate the dependence on Reynolds number of a possible transfer of energy from the mean flow to the perturbations – a necessary condition for instability.

We consider a two dimensional laminar boundary layer on a flat plate at zero incidence and we approximate to it by regarding the mean flow as locally parallel so that we ignore its rate of change with x, i.e. we write

$$U = u(z) + u'(x, z, t), \qquad W = w'(x, z, t)$$

where x is in the undisturbed stream direction parallel to the plate and z is normal to it. The mean kinetic energy of the disturbances per unit mass is

$$\overline{e'} = \tfrac{1}{2}(\overline{u'^2} + \overline{w'^2})$$

The disturbance energy equation for a fluid particle can be derived from the Navier–Stokes equation and the continuity equation to yield the result that the rate of change of e' is equal to the rate of work done by the Reynolds stresses and by the pressure fluctuations minus the rate at which the disturbance energy is dissipated by viscosity. If the disturbance energy equation is then integrated over a volume of fluid in a domain D encompassing the boundary layer over an extent of x such that the disturbances can be assumed identical for the two end values of x (e.g. over a wavelength if the disturbance is sinusoidal) then the pressure fluctuation terms vanish. We are then left with the equation

$$\frac{\overline{DE'}}{Dt} = -\rho \iint \overline{u'w'} \frac{\partial u}{\partial z} \, . \, dx\,dz - \mu \iint \overline{\eta'^2} dx\,dz \tag{5.8}$$

where $\overline{E'}$ is the integrated mean disturbance kinetic energy over the domain D, and

$$\eta' = \frac{\partial w'}{\partial x} - \frac{\partial u'}{\partial z}$$

i.e. the disturbance vorticity. The first term on the RHS of equation (5.8) can be interpreted as the integrated rate of work done by the Reynolds stress $-\rho\overline{u'w'}$ on the mean motion and the second term can be shown to be the rate of dissipation of the mean disturbance kinetic energy due to viscosity. Thus, the RHS of equation (5.8) can be written as

$$\rho(M - \nu N)$$

where N is always positive but M can be of either sign. If we now non-dimensionalise M and N by writing

$$M' = M/(U_0^3 h), \qquad N' = N/U_0^2$$

where U_0 and h are a characteristic velocity and length, respectively, then

$$\rho(M - \nu N) = \rho U_0^2[hU_0 M' - \nu N'] = \rho U_0^2 \nu[RM' - N'] \tag{5.9}$$

where $R = U_0 h/\nu$.

We infer that the disturbance motion is stable or unstable according as R is less than or greater than N'/M', respectively. It follows that for a specified form of disturbance and a given mean flow there will be a critical value of R, namely N'/M', below which the flow will be stable. This argument by itself does not imply that for a given mean flow there is a critical Reynolds number that applies for all forms of disturbance. However, it will be evident that as the Reynolds number increases instability becomes more probable, since it is always possible to conceive of a disturbance for which $M' > 0$.

To pursue this argument in more detail we must consider disturbances that are hydrodynamically possible, i.e. that are consistent with the equations of motion. This consideration leads us to the important classical work of Orr,[5.4] Sommerfeld,[5.5] Tollmien[5.6] and Schlichting[5.7] on the stability of a laminar boundary layer to infinitesimal harmonic disturbances. Comprehensive accounts of this subject will be found in the books by Schlichting and Rosenhead.[2.2,2.3] Here we confine ourselves to a brief account in the next section.

5.4 Classical small perturbation analysis for two dimensional incompressible flow

For a detailed account of the history and development of this topic see Schlichting.[2.2] The assumption of locally parallel mean flow for which u is independent of x is accepted. The perturbation flow is assumed to be a wave-like function of x, z and t of small amplitude and its decay or growth is a function of t only. Such perturbations are sometimes referred to as *temporal*. An alternative approach would be to treat the perturbation amplitude as a function of x, only; in this case the perturbations are referred to as *spatial*. The latter approach is more realistic for boundary layers, but for the present we will develop the analysis for temporal disturbances.

The equation of continuity is

$$\frac{\partial}{\partial x}(u + u') + \frac{\partial w'}{\partial z} = 0$$

and hence $$\frac{\partial u'}{\partial x} + \frac{\partial w'}{\partial z} = 0$$

implying a perturbation stream function ψ such that

$$u' = -\frac{\partial \psi}{\partial z}, \qquad w' = \frac{\partial \psi}{\partial x} \tag{5.10}$$

ψ is then assumed to be of the form

$$\left.\begin{aligned} \psi &= F(z) \,.\, \exp\,[i(\alpha x - \beta t)] \\ &= F(z) \,.\, \exp\,[i\alpha(x - ct) \end{aligned}\right\} \tag{5.11}$$

We see that if the wavelength of the disturbance is λ, say, then

$$\lambda = 2\pi/\alpha \tag{5.12}$$

α is sometimes called the wave number and for temporal perturbations it is real, whilst $c = \beta/\alpha$ is in general complex $= c_r + ic_i$, say, or

$$\beta = \beta_r + i\beta_i \tag{5.13}$$

We note that

$$c_r = \beta_r/\alpha = \text{wave (or phase) velocity}$$

whilst the rate of growth (or damping) of the disturbance is given by $\exp(\beta_i t)$ or $\exp(\alpha c_i t)$. Thus if $c_i > 0$ the disturbance grows with time, and conversely if $c_i < 0$ it damps out with time. We can regard β_i or c_i as coefficients of amplification. We can express quantities non-dimensionally in terms of U_0, a representative velocity of the mean flow, and δ^*, the displacement thickness of the mean flow, with the non-dimensional time $= tU_0/\delta^*$. Thus, we write

$$F(z) = U_0\delta^*\phi(\zeta), \qquad \text{where} \quad \zeta = z/\delta^* \tag{5.14}$$

Then after linearising the Navier–Stokes equation by retaining only first order perturbation terms we eventually derive the equation

$$(u - c)(\phi'' - \alpha^2\phi) - u''\phi = -i(\phi'''' - 2\alpha^2\phi'' + \alpha^4\phi)/\alpha R_{\delta^*} \tag{5.15}$$

where $R_{\delta^*} = U_0\delta^*/\nu$, the primes denote differentiation with respect to ζ, and the terms have been non-dimensionalised as described above. The boundary conditions are

$$\zeta = 0, \phi = \phi' = 0, \qquad \zeta = \infty, \phi = \phi' = 0$$

This equation is known as the Orr–Sommerfeld equation after the two scientists who independently derived it in the early years of this century. It will be noted that the terms on the RHS represent the viscous effects and they are small for large R_{δ^*}. One is therefore tempted to consider solutions neglecting them – the so-called inviscid flow (or infinite Reynolds number) solutions. Although such solutions provide results of significant physical interest it has to be borne in mind that the RHS terms involve the highest order derivatives of ϕ and in regions of large rate of change of ϕ with ζ, such as close to the surface where the no-slip condition must hold, these terms cannot be neglected. Further in the region of the flow where $(u - c_r)$ is small the LHS of (5.15) becomes small and the viscous terms again cannot be neglected there.

The mathematical details of the various available methods of solution of equation (5.15) are too complicated to be presented here and we can only offer a brief discussion of the important physical features of the results. However, it is to be noted that the advent of high speed digital computers has considerably eased the problem of obtaining reliable numerical solutions. Not surprisingly, the earliest work was directed at the simpler problem of inviscid flow instability, neglecting the terms on the RHS of equation (5.15). The resulting equation is referred to as the Rayleigh equation. A key theorem emerged, namely:

> A boundary layer for which the mean velocity profile has a point of inflection (i.e. $u'' = 0$) is unstable. This can be expanded to saying that if for a given disturbance phase velocity c_r the gradient of the mean flow vorticity with respect to z is zero or positive [i.e. $(\partial/\partial z)(\partial u/\partial z) \geqslant 0$] where $u = c_r$, then the boundary layer is unstable to that disturbance.

When the viscous terms are taken into account this criterion is found to hold for finite R_{δ^*} but with decreasing effect as R_{δ^*} is reduced. We recall that the direct viscous damping effect is to reduce the instability until, for low enough Reynolds number, the flow is stable. However, the importance of the criterion lies in the fact that in the presence of an adverse pressure gradient the velocity profile always has a point of inflection (see Section 2.10). We may therefore expect the boundary layer to show then a marked readiness to become unstable except at very low Reynolds numbers. Conversely, a favourable pressure gradient results in a profile that has no point of inflection and hence will be stable for infinite R_{δ^*}. Nevertheless, for finite Reynolds numbers we recall the possibility of some destabilisation produced by viscosity insofar as the no-slip requirement at the wall and the avoidance of physically unacceptable rates of shear at the critical layer where $u = c_r$ result in a relative phase change between u' and w' such that the Reynolds stress $-\rho\overline{u'w'} > 0$, and the disturbance can then extract energy from the mean flow. We note, therefore, that viscous effects can be stabilising insofar as they involve damping of the disturbances, but they can be destabilising insofar as they can result in phase changes between the disturbance velocity components so as to augment the disturbance energy extracted from the mean flow.

Since equation (5.15) is homogeneous, as are its boundary conditions, it constitutes a characteristic or eigen-value problem; i.e. for any given mean velocity distribution there exists a functional relation between the quantities α, R_{δ^*} and c which the solution must satisfy. This relation being complex, can be resolved into two relations of the form

$$c_r = c_r(\alpha, R_{\delta^*}), \qquad c_i = c_i(\alpha, R_{\delta^*}) \tag{5.16}$$

Hence using α and R_{δ^*} as ordinates we can plot a series of curves for each of which c_r is a constant and we can similarly plot a series for which c_i has constant values.

The particular curve for $c_i = 0$ corresponds to neutral stability and it separates areas of boundary layer stability to the small disturbances considered ($c_i < 0$) from areas of instability ($c_i > 0$). Likewise for given values of the parameters α and R_{δ^*} we can determine the phase velocity c_r. The neutral stability curve for the boundary layer on a flat plate at zero incidence[5.8] is illustrated in Fig. 5.2. Within the loop area encompassed by the two branches of the curve the boundary layer is unstable to a range of disturbance frequencies, as indicated by the lines of constant c_i. It will be noted that there is a critical value of R_{δ^*} ($\simeq$ 500) below which the boundary layer is stable to small disturbances of all frequencies, whilst at high values of R_{δ^*} the range of frequencies to which the boundary layer is unstable decreases and tends to zero as the Reynolds number tends to infinity. It will also be noted that there is some point on the α, R_{δ^*} plot within the neutral curve for which the amplification factor is a maximum.

The strong effects of pressure gradient on the stability were demonstrated by early calculations of Schlichting and Ulrich[5.9] using a Pohlhausen type family of profiles with $\lambda = (\mathrm{d}u_e/\mathrm{d}x)\delta^2/\nu$ as parameter, and similar calculations were made by Pretsch[5.10] using the Falkner–Skan family, ($u_e = u_0 x^m$, see Section 3.4), with $\beta = 2m/(m + 1)$ as parameter. Curves

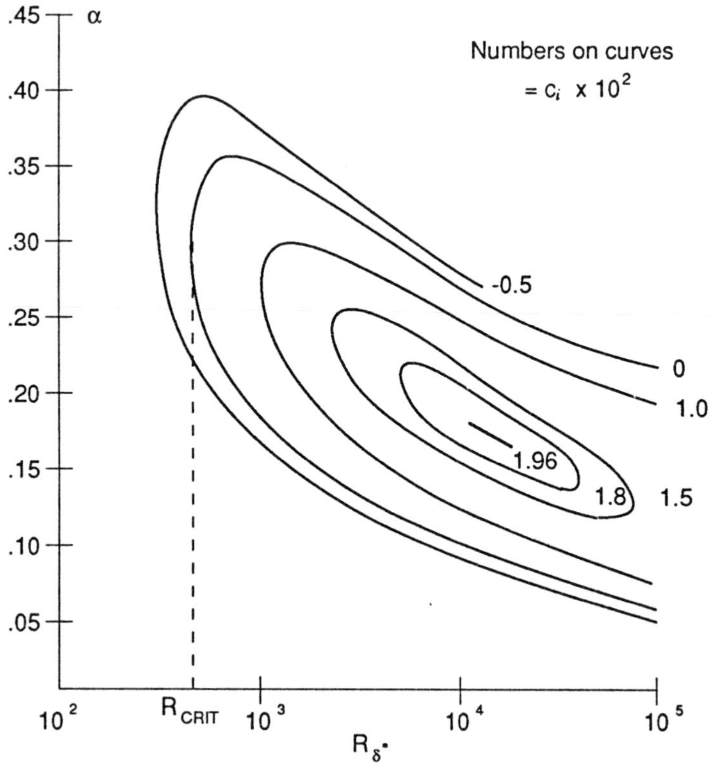

Fig. 5.2 Curves of constant temporal amplification rate for the boundary layer on a flat plate at zero incidence.[5.8,5.51]

of neutral stability obtained by Schlichting and Ulrich for different values of λ are shown in Fig. 5.3. It will be seen that with λ decreasing below zero (adverse pressure gradient) the critical Reynolds number decreases and the range of frequencies, or wavelengths, to which the boundary layer is unstable increases. Further, we note that with the associated point of inflection in the velocity profile there is a range of α, or c_r, for which the boundary layer is unstable for $R_{\delta^*} \to \infty$. Likewise, as λ increases above zero (favourable pressure gradient) the critical Reynolds number increases and the range of frequencies for which the boundary layer is unstable decreases. It is of interest to note that Pretsch's calculations lead to a relation between the critical Reynolds number and the form parameter $H = \delta^*/\theta$, illustrated in Fig. 5.4, which appears to have wide applicability including profiles obtained in the presence of wall suction (see Section 5.10). The important effect of pressure gradient on stability is further manifest in a rapid increase of the amplification factor as the pressure gradient passes from negative to positive.

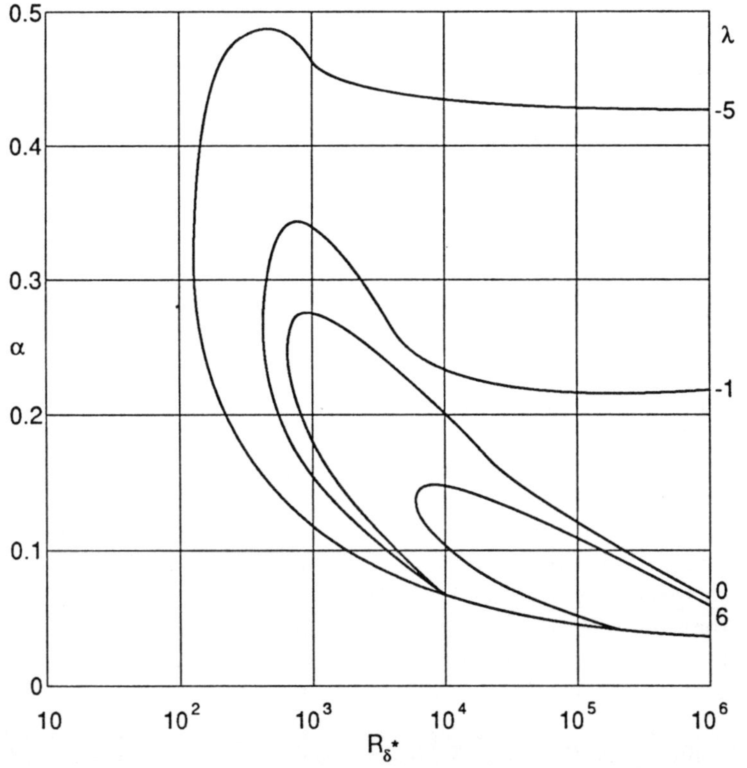

Fig. 5.3 Neutral stability curves for velocity profiles with parameter $\lambda = (\delta^2/\nu)(du_e/dx)$.[2.2,5.9]

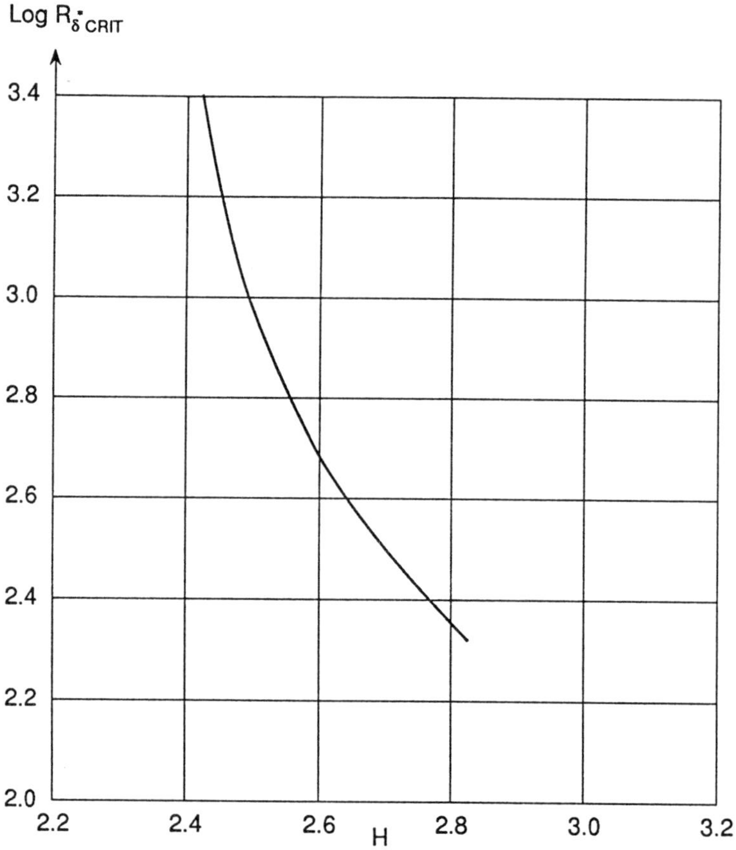

Fig. 5.4 Critical Reynolds number as a function of shape parameter.[5.8, 5.10]

At an early stage doubts were raised about the relevance of such calculations to boundary layers on wings or bodies mainly because of concern about the assumption of a parallel mean flow and the consequent neglect of the spatial rate of change of the mean flow and of the disturbance flow. However, a remarkable series of experiments by Schubauer and Skramstad[5.11] provided solid support for the theory. In these experiments, using a specifically designed low turbulence wind tunnel, small sinusoidal disturbances were generated in the boundary layer on a flat plate by means of a vibrating ribbon held parallel to the plate and normal to the stream. The results showed that in the absence of strong external turbulence and surface imperfections instabilities developed as predicted by the theory as *Tollmien–Schlichting waves*, as these small disturbance waves are now called; the neutral curve, critical Reynolds number and amplification rates were encouragingly close to the theoretical results.

However, the possibility of treating the perturbation more realistically as spatially varying was first raised by Gaster.[5.12] His approach was to regard β as real and α as complex in equation (5.11) so that

$$\psi = F(z) \exp [i(\alpha_r x - \beta t)] \, . \, \exp (-\alpha_i x) \tag{5.17}$$

The disturbance then damps out or grows with x depending on whether $\alpha_i >$ or < 0, respectively. The spatial and temporal approaches both lead to the same neutral curve, since both α and β are then real, but they differ somewhat in the amplification rates. Gaster argued that the group velocity of the waves, $c_g = \partial\beta/\partial\alpha_r$ for constant α_i, was the correct velocity to use to relate an element of time in the history of a temporal wave and the corresponding distance travelled by an element of a spatial wave, and was thus able to compare the corresponding amplification factors. The spatial approach has been increasingly adopted in recent years because of its greater realism.

Further improvements in the small perturbation theory were subsequently made by allowing for the non-parallel features of the mean flow, viz. the existence of a mean velocity component normal to the surface and of the rates of change of the velocity components with x. A comparison by Saric and Nayfeh[5.13] of the theoretical results for the neutral curve for the Blasius boundary layer for parallel and non-parallel flows with corresponding experimental results is shown in Fig. 5.5. It will be seen that the experiments agree more closely with the non-parallel flow theory and the latter predicts a slightly smaller critical Reynolds number than the parallel flow theory.

To sum up, the classical small disturbance stability theory, so brilliantly developed by Tollmien and Schlichting has been well authenticated experimentally in circumstances appropriate to the theory, i.e. a low level of external disturbance and a smooth surface free from waviness. The theory provides valuable insight into the effects of Reynolds number and pressure gradient. That it can be improved by taking account of spatial variations and non-parallel flow effects is evident, but these factors do not undermine its basic validity.

5.5 Effects of curvature

As noted in Section 2.8, surface curvature with the accompanying curvature of the mean streamlines in the boundary layer results in centrifugal forces acting normal to the streamlines, and for steady mean flow these forces are essentially balanced by pressure gradients along the normals. Thus for a two dimensional flow along a curved surface we have

$$\partial p/\partial z = \rho u^2/r + \text{small viscous terms}$$

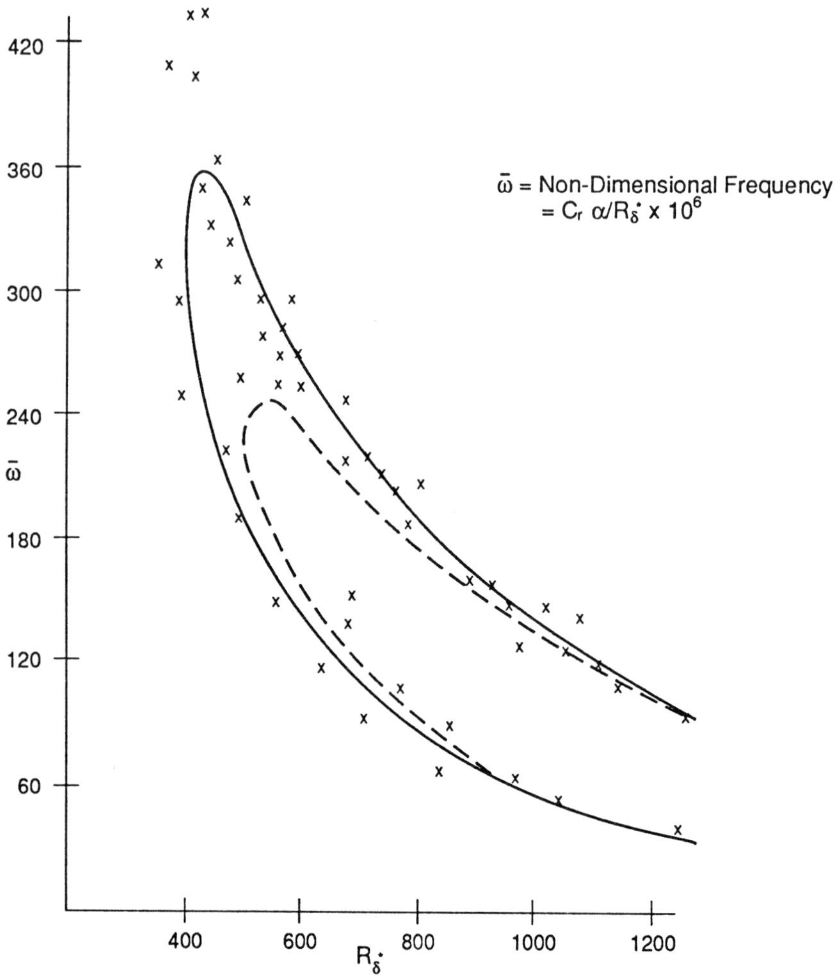

Fig. 5.5 Neutral stability curves[5.13] for the Blasius boundary layer.
- - - - - parallel flow
——— non-parallel flow
× experimental results.

where r is the radius of curvature of the streamline and is positive if the streamline is convex upwards. For attached flow r would normally differ little from r_0, the local radius of curvature of the surface.

Consider the flow over a concave surface and suppose that an element of fluid is displaced by some small disturbance from z to $z + \Delta z$, say, where Δz is > 0. Assume that the angular momentum ur about the centre of curvature is conserved. Then the particle could find itself with a smaller value of $|ur|$ than the surrounding fluid, and it can be readily shown that then the local outwards pressure gradient $-\partial p/\partial z$ would be greater than that needed to keep the particle following the local streamline, and it would tend to move further away from the surface causing its displacement from its original position to increase, and so on. Likewise if it were displaced initially towards the surface it would experience a change of pressure gradient tending to move it closer to the surface. The flow is therefore unstable except at low enough Reynolds numbers for the viscous damping effects to be significant. The resulting disturbance pattern tends to take the form of streamwise vortex pairs as illustrated in Fig. 5.6. This form of instability was first brilliantly demonstrated in a classical investigation by Taylor on the flow between concentric cylinders.[5.14] With the outer one at rest and the inner one rotating, the flow close to the surface of the outer one is effectively that over a concave wall.

Instability then manifested itself in the form of ring-like vortices which appeared when the so-called Taylor Number

$$T_a = \frac{u_1 d}{\nu}\sqrt{\frac{d}{r}} > 41, \quad \text{approx.}$$

where d = gap width between the cylinders, r_1 = radius of the inner cylinder, and u_1 = peripheral velocity of the inner cylinder. The regular pattern of laminar vortices arising from the initial instability is found to persist for values of T_a up to about 4000 but for higher values of T_a the flow becomes turbulent and loses its regularity of pattern. We note that a factor in T_a is the Reynolds number $u_1 d/\nu$.

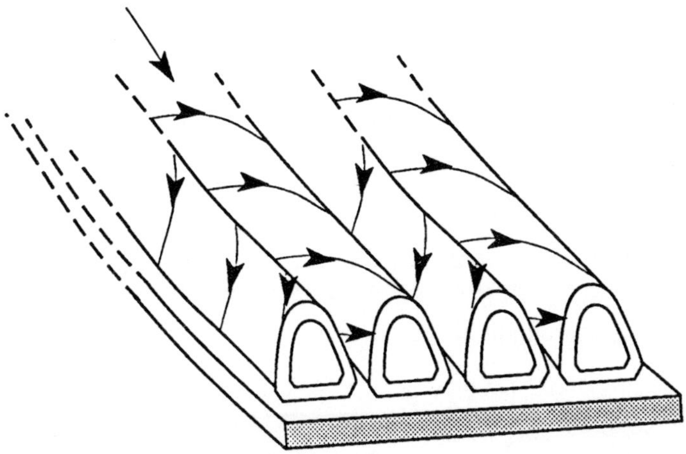

Fig. 5.6 Formation of Görtler vortices in the boundary layer on a concave wall.

Subsequently, Görtler[5.15] demonstrated that a similar pattern of streamwise vortices developed in the boundary layer on a concave plate, as in Fig. 5.6, and such vortices are often referred to as Görtler vortices or Taylor–Görtler vortices.

Görtler's theory showed that such vortices will develop if

$$\frac{u_e\theta}{\nu}\sqrt{\frac{\theta}{|r_0|}} > 0.58$$

where r_0 is the radius of curvature of the plate, and θ is the boundary layer momentum thickness. It is of interest to note that in experiments made by Liepmann[5.16] transition to turbulent flow did not appear until

$$\frac{u_e\theta}{\nu}\sqrt{\frac{\theta}{|r_0|}} > 7.3$$

We see that just as with Tollmien–Schlichting waves, the appearance of curvature induced instabilities in the form of streamwise vortices is a forerunner to the subsequent development of turbulent flow but the process of development can be lengthy and complex.

For the boundary layer flow over a convex surface we can likewise deduce that the curvature is stabilising. For a detailed and comprehensive discussion of the flows near rotating bodies the reader is referred to Wimmer.[5.17]

5.6 Effects of heat transfer and compressibility

If there is heat transfer at the surface the viscosity will vary in response to the temperature variation in the fluid and so will the density if the flow is such that compressibility effects are significant. Both these variations can result in changes in the flow stability.

Consider first incompressible flow in two dimensions. Then from the boundary layer equation of motion (2.32) we deduce that at the surface

$$\frac{\partial}{\partial z}\left(\mu\frac{\partial u}{\partial z}\right)_w = \frac{dp}{dx} = -\rho u_e\frac{du_e}{dx}$$

or $$\left(\frac{\partial^2 u}{\partial z^2}\right)_w = -\frac{1}{\mu_w}\left(\frac{\partial\mu}{\partial z}\right)_w\left(\frac{\partial u}{\partial z}\right)_w - \frac{1}{\mu_w}\frac{dp}{dx}$$

If the pressure gradient is zero then

$$\left(\frac{\partial^2 u}{\partial z^2}\right)_w = -\frac{1}{\mu_w}\left(\frac{\partial\mu}{\partial z}\right)_w\left(\frac{\partial u}{\partial z}\right)_w \tag{5.18}$$

If the surface is hotter than the fluid, and there is therefore transfer of heat from the surface to the fluid, then it follows that for air $(\partial\mu/\partial z)_w < 0$, since $\mu \propto T^{\omega}$ with $\omega = 0.8 - 0.9$ (see Section 1.3). If the boundary layer is unseparated then $\partial u/\partial z > 0$ at the surface and hence from equation (5.18) $(\partial^2 u/\partial z^2)_w > 0$. However, near the edge of the boundary layer $\partial^2 u/\partial z^2 < 0$ and hence there must be a point of inflection in the velocity profile where $\partial^2 u/\partial z^2 = 0$. We can infer, therefore, that for a gaseous fluid heat transfer from the wall to the fluid will help to promote instability to small disturbances and so hasten transition. Conversely with a cooled wall with heat transfer from the fluid to the wall, the flow will become more stable to small disturbances and transition will be delayed. Therefore wall cooling is a feasible process for achieving extensive regions of laminar flow.

It is interesting to note that for a liquid μ falls as the temperature is increased and the effects of heat transfer on the stability are the opposite of those with a gas.

We come now to the effects of compressibility. A key point to note is that the condition for instability for infinite Reynolds number in incompressible flow, viz. the existence of a point of inflection in the velocity profile, or

$$\partial^2 u/\partial z^2 = 0, \text{ at some value of } u = c_r$$

is generalised for compressible flow to

$$\frac{\partial}{\partial z}(\rho \, \partial u/\partial z) = 0 \tag{5.19}$$

at some point in the boundary layer for which $u > u_e - a_e$, where a_e is the speed of sound just outside the boundary layer (see Lees and Lin[5.18]). This last condition implies that the critical disturbance of phase velocity $c_r = u$ travels subsonically relative to the external flow. We see, therefore, that

$$c_r/u_e > 1 - (1/M_e)$$

where M_e is the external flow Mach number. Such disturbances are referred to as subsonic. The requirement for instability embodied by equation (5.19) implies that at the critical layer the angular momentum reaches a maximum. The analysis for the boundary layer response to subsonic disturbances follows very closely that for incompressible flow and they can be regarded as essentially the same as Tollmien–Schlichting waves.

For the flow over a flat plate at zero incidence with zero heat transfer the surface temperature (the recovery temperature) increases as the Mach number increases (see Section 3.3.5) and it can be readily shown that the critical condition of equation (5.19) then occurs within the boundary layer at a distance from the surface that increases with Mach number. Consequently we may expect some corresponding increase of instability which generally takes the form of a decrease of the critical Reynolds number and a broadening of the unstable range of wavelengths (or frequencies) for higher Reynolds numbers. However, the maximum amplification factor (β_{imax}) varies inversely with respect to a measure of the boundary layer thickness, e.g. δ^*, and since the latter increases with M_e the net effect can be a decrease of the amplification with Mach number.

It will be seen from the above that as M_e increases the region of the boundary layer where subsonic disturbances are possible involves a decreasing portion of the outer part of the boundary layer where the condition for instability, equation (5.19) is decreasingly likely to be met. This suggests that the stability characteristics for subsonic disturbances will at some Mach number begin to improve with further increase of Mach number. However, we must then consider the possibility of disturbance waves laterally inclined to the main stream direction at an angle ψ, say. For such waves the effective external Mach number is $M_e \cos \psi$, and so the region in which equation (5.19) can be met is increased as ψ increases. Thus, we may expect with increase of M_e an increasing range of inclined disturbance waves for which the flow is unstable. Such waves are still of the Tollmien–Schlichting type but for incompressible flow the most unstable waves are non-inclined, as was demonstrated by Squire.[5.19] Non-inclined waves are referred to as two dimensional waves of the first mode, whilst inclined waves are called three dimensional waves of the first mode – the first mode being of the Tollmien–Schlichting type.

To complicate the picture further we must now take note of the fact that for flow regions where $u_e - c_r > a_e$ an infinite number of unstable disturbances are possible, they are referred to as supersonic disturbances and they do not depend on the velocity profile satisfying equation (5.19) at some point. They are sometimes called second mode disturbances and the most unstable ones are two dimensional. The results of some calculations by Mack[5.20] of the effects of Mach number on the maximum amplification factors for an insulated flat plate for the two and three dimensional first modes and the second mode disturbances are illustrated in Fig. 5.7. It will be seen that as generators of instability the three dimensional first mode waves are important for $1 < M_e < 4$ whilst the second mode waves are dominant for $M_e > 4$. The sparse available experimental data on transition at high Mach numbers appears to be in rough accord with these calculations.

However, transition measurements in wind tunnels at such speeds present serious problems in interpretation, because of the difficulty of assessing in quantitative terms the effects of tunnel noise, free stream turbulence, model surface finish and scale effects. These difficulties have not been effectively resolved as yet, since the frequency spectrum of the noise and turbulence inputs determine the receptivity of the laminar boundary layer in ways that are not yet properly understood.

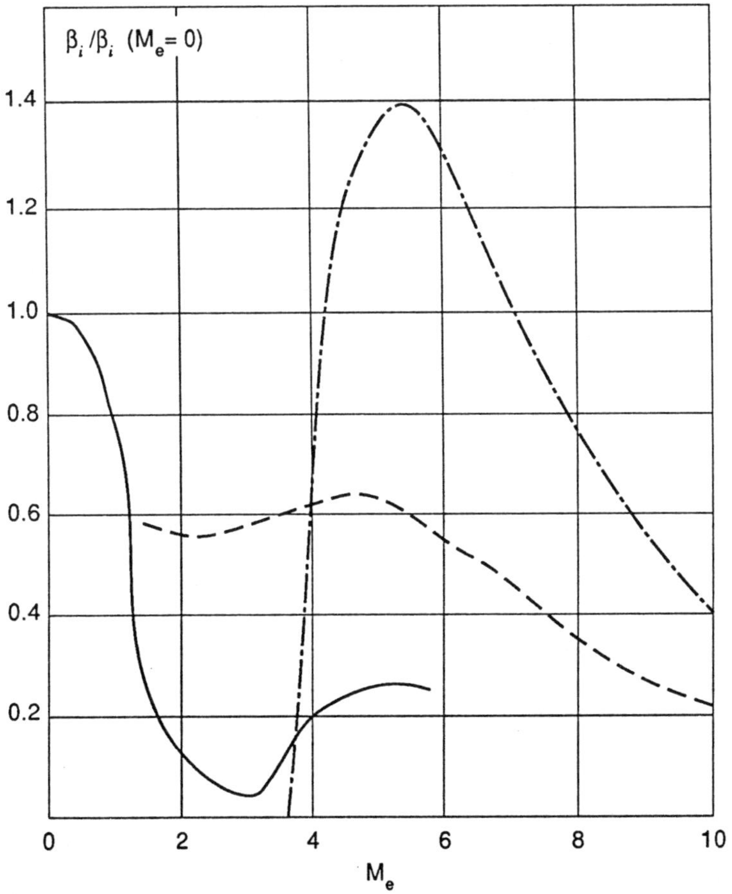

Fig. 5.7 Effect of Mach number on maximum amplification factors for small disturbances.[2.2,5.20]

——— 2D waves of 1st mode
– – – – – three dimensional (inclined) waves of first mode with ψ in range 45°–60°
—·—·—·· second mode waves.

5.7 Effects of surface excrescences and of external turbulence

A fairly detailed account of the effects of excrescences on transition is given in Section 2.2 of reference 5.21. Here we will confine ourselves to a brief review of the main points.

If we consider first an isolated excrescence on the surface of a body we find that depending on its shape, location on the body, the fluid viscosity and the external velocity distribution there is a critical excrescence height, which we will call k_{crit1}, below which there is no effect on transition. As the size of the excrescence is increased above this value transition moves upstream in the wake of the excrescence from its initial smooth surface position until at a second critical value, k_{crit2}, transition occurs just aft of the excrescence. For an excrescence in the form of a roughness element with breadth and height of the same order, a turbulent wedge then results, trailing downstream from the excrescence at its apex and of semi-angle of the order of about 10°. With a two dimensional excrescence, e.g. a skin joint or a transition trip wire, turbulent flow occurs over the whole span immediately aft of the excrescence for $k > k_{crit2}$.

These two critical heights are of considerable practical importance: k_{crit1} defines the stage at which the roughness begins to produce a noticeable effect on drag due to the forward transition movement (for smaller roughnesses in air the surface is referred to as aerodynamically smooth), whilst k_{crit2} is a useful guide to the height of a trip required to bring the transition to a specified location. The latter is often used to help to reduce the uncertainties of scale effect prediction in the use of wind tunnel models to simulate full scale behaviour in flight.

It has long been known that for the flow past a bluff shape in steady motion there is a critical Reynolds number above which the wake behind the body becomes unsteady and vortices are generated at the body and move downstream in the form of a vortex street. For a circular cylinder this Reynolds number is about 150 in terms of its diameter. It was therefore argued by Schiller[5.22] that there was similarly a roughness critical Reynolds number which would determine k_{crit1}, such that for a Reynolds number less than the critical no eddies were shed by the roughness and hence there would be no downstream effect on the transition position. Thus we can expect to find that:

$$R_{k_{crit1}} = (u_k\, k/\nu)_{crit1} = \text{const.} = C, \quad \text{say} \tag{5.20}$$

where, however, C will be a function of the roughness geometry. Here, u_k is the velocity in the boundary layer at the height k above the surface in the absence of the excrescences. If $k/\delta \ll 1$, then

$$u_k \simeq k(\partial u/\partial z)_w = \frac{k}{2\nu}\, c_f u_e^2 = \frac{0.664}{2\nu\, R_{xk}^{1/2}}\, k u_e^2 \tag{5.21}$$

for a flat plate at zero incidence in incompressible flow (see Section 3.2.1). Here $R_{xk} = u_e x_k/\nu$, where x_k = distance of the roughness position aft of the leading edge. Hence

$$R_k = u_k k/\nu = \frac{0.664k^2}{2\nu^2\, R_{xk}^{1/2}}\, u_e^2 = 0.332(k/x_k)^2\, R_{xk}^{3/2} \tag{5.22}$$

and $\quad u_e k/\nu = R_k u_e/u_k = (R_k/0.332)^{1/2}\, R_{xk}^{1/4} \qquad (5.23)$

Thus, given the values of $R_{k_{crit1}}$ and x_k we can determine the corresponding value of k_{crit1}.

If k/δ is not small compared with 1 then we must assume or calculate the velocity distribution in the boundary layer to determine u_k. If we assume for zero pressure gradient

$$u/u_e = \sin(\pi z/2\delta)$$

and use $\delta/x \simeq 5/R_x^{1/2}$ (see Section 2.12), then we readily deduce that

$$R_k = u_k k/\nu = \frac{u_e k}{\nu} \sin\left(\frac{\pi}{10} \frac{u_e k}{\nu} R_{xk}^{-1/2}\right) \qquad (5.24)$$

A plot of this relation is shown in Fig. 5.8. However, it must be noted that in the presence of a non-uniform pressure distribution a more accurate result for the value of u_k requires the solution of the boundary layer equations.

For information on $R_{k_{\text{critl}}}$ we must appeal to experimental data. Braslow[5.23] has analysed a wide range of data and his results are most conveniently presented in the form shown in Fig. 5.9 where $(R_{k_{\text{critl}}})^{1/2}$ is plotted as a function of d/k, d is the spanwise dimension of a typical roughness. It will be seen that his results fall into a band reflecting significant scatter (masked somewhat by the scale used). Thus, for a hemi-spherical roughness ($d/k = 2$) $(R_{k_{\text{critl}}})^{1/2} \simeq 23 \pm 6$, whilst for much larger d/k as for a spanwise ridge or trip $(R_{k_{\text{critl}}})^{1/2} \simeq 11 \pm 4$. The scatter is in part due to the fact that d/k is only one of the possible parameters to describe the excrescence shape. For tests in tunnels of low turbulence it

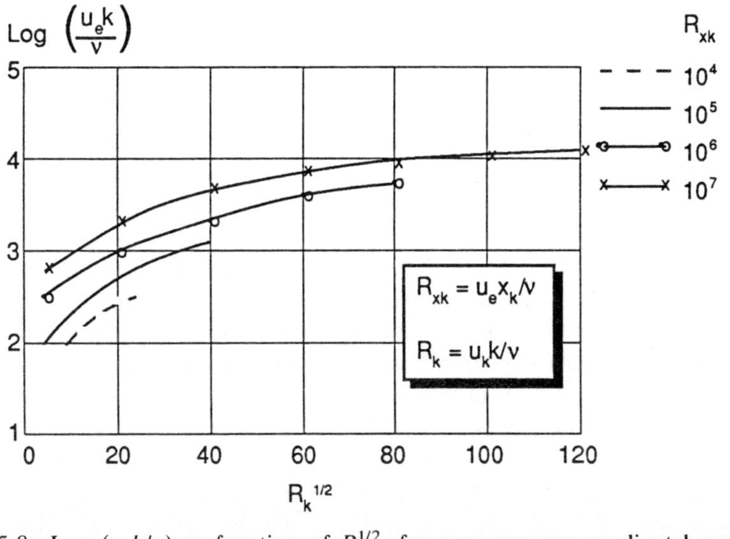

Fig. 5.8 Log $(u_e k/\nu)$ as function of $R_k^{1/2}$, for zero pressure gradient boundary layer. k_{crit} can be determined given $R_{k_{\text{crit}}}$ and R_{xk}.

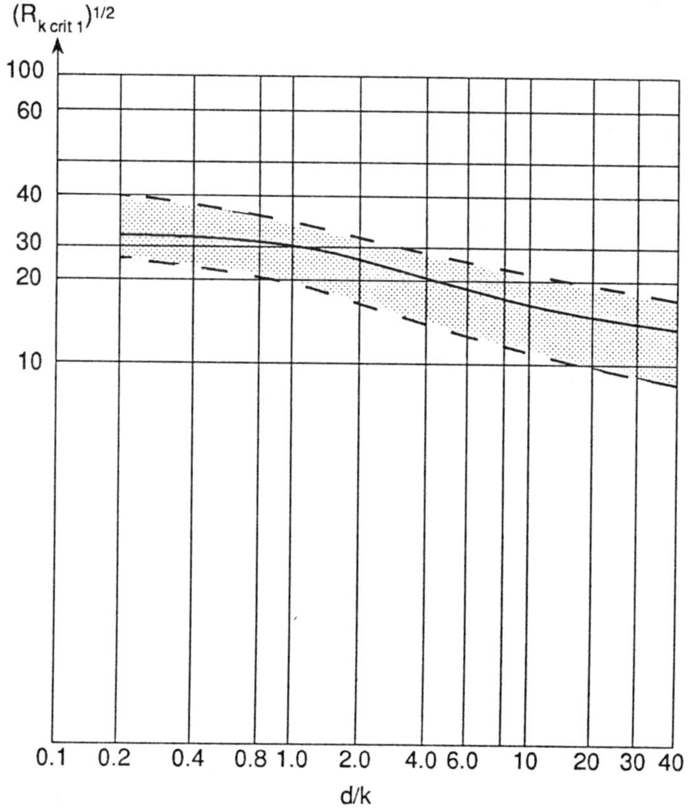

Fig. 5.9 $R_{k_{crit1}}^{1/2}$ as function of roughness shape parameter d/k.[5.23]

seems likely that the value of $R_{k_{crit1}}$ will be found in the upper part of the band, but for a conservative estimate of the permissible roughness the lower region of the band should be used. It is to be noted that Smith and Clutter[5.24], who did measurements on a variety of excrescences, detected no consistent effect of pressure gradient on $R_{k_{crit1}}$, presumably because of countervailing effects on the velocity at the roughness height and on stability, whilst Braslow found no effect of compressibility on $R_{k_{crit1}}$ for Mach numbers less than about 3.

If we now consider k_{crit2} we find that it is of the order of twice k_{crit1}. In incompressible flow $(R_{k_{crit2}})$ is about 400 for a transition trip in the form of a wire on the surface and about 600 for a band of small roughnesses (e.g. ballotini). An empirical formula derived by Van Driest and Blumer,[5.25] based on tests of bands of spherical roughnesses over a range of subsonic and supersonic Mach numbers but with zero pressure gradient, is

$$\frac{u_e k_{crit2}}{\nu} = 42.6 R_{xk}^{1/4}[1 + \tfrac{1}{2}(\gamma - 1)M_e^2] \qquad (5.25)$$

In addition to surface excrescences disturbances can exist in the external flow in the form of turbulence and noise. Regrettably, we are only a little nearer achieving a comprehensive understanding of their effects on transition than we were when studies of them first started. Depending on the environment, i.e. wind tunnel or flight, such disturbances can cover wide ranges of frequency, wavelength, direction and velocity of propagation and these affect, in ways that are at present far from fully understood, how the disturbances are received by the boundary layer and initiate Tollmien–Schlichting waves or higher order modes of instability. We can only briefly summarise a couple of simple empirical correlations and refer the reader to more detailed discussions.[5.26–28]

The results of an analysis by Dryden[5.29] of some experiments on a flat plate at zero incidence are reproduced in Fig. 5.10 where the Reynolds number (R_{eT}) based on the free stream velocity u_e and the distance from the leading edge to transition are plotted against T_u, a simple measure of turbulence intensity in the free stream

$$= \left(\frac{\overline{u'^2} + \overline{v'^2} + \overline{w'^2}}{3}\right)^{1/2} \Big/ u_e$$

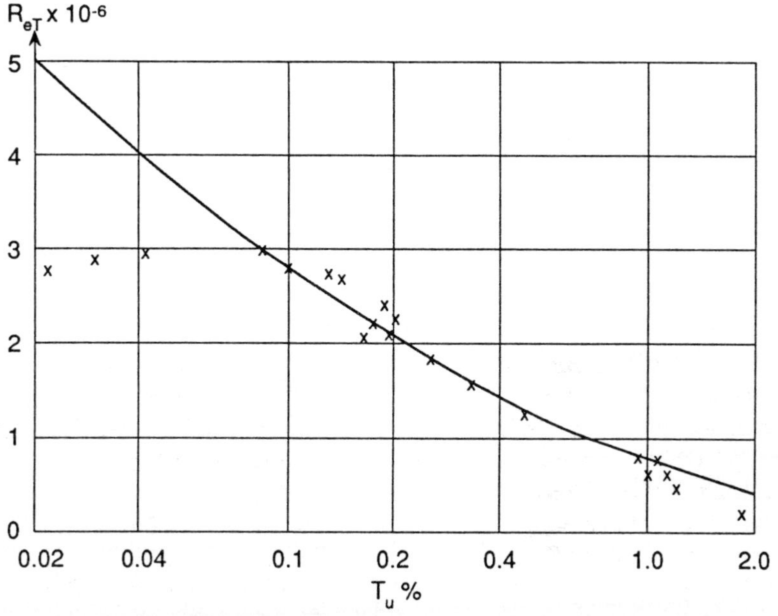

Fig. 5.10 Transition Reynolds number as a function of free stream turbulence.
× experimental results;[5.29]
——— e^n prediction [equation (5.32),[5.46]].

It will be noted that for T_u less than about 0.1% there appears to be no significant effect of external turbulence on transition, but with further increase of T_u there is a continuous decrease of the transition Reynolds number. An analysis of the transition Reynolds number in flow in a channel by Wells[5.30] gave a similar result but with higher values of R_{eT} for a given value of T_u.

An analysis of a considerable array of wind tunnel experiments on 5° cones at zero incidence by Whitfield and Dougherty[5.28] for Mach numbers from 0.3 to 4.5 and with zero heat transfer led to the empirical relation:

$$R_{eT} = 3.7 \times 10^6(\sqrt{\overline{p'^2}}/50\rho u_e^2)^{-1/4} \tag{5.26}$$

with a scatter of about 20%. Here $\sqrt{\overline{p'^2}}$ is the time-averaged root mean square of the fluctuating part of the static pressure as measured on the wind tunnel wall by a flush mounted microphone, frequency integrated over a band width from about 10 Hz to about 30 kHz. This is used rather than T_u as it is easier to measure and is in accordance with the idea that a major source of disturbance in wind tunnels, particularly at high speeds, derives from the turbulent boundary layers on the tunnel walls.

5.8 The transition process

One can infer from the foregoing and the available experimental data that the onset of turbulent flow in the boundary layer requires the presence of sufficiently strong disturbances, but that the required strength of the disturbances depends on their form and the Reynolds number. In the absence of such sources of finite disturbance as surface excrescences or waviness and with a low level of turbulence in the external flow (T_u less than about 0.1%) the process begins with the amplification of unsteady Tollmien–Schlichting waves and/or the appearance of streamwise Taylor–Görtler vortices associated with surface or streamline curvature. These in turn depend on the Reynolds number and the pressure distribution. However, there may then be a large streamwise distance between the positions at which these waves or vortices first appear and the onset of turbulence, and our knowledge of the physics of the process that develops within that distance is far from complete.

An important line of attack that has been pursued for many years was to extend the classical small perturbation stability theory, described in Section 5.4, to disturbances of finite amplitude by including the non-linear terms in the equations of the perturbed flow.[5.31–33] This brought into play harmonics of the basic sinusoidal disturbance considered and demonstrated that even for conditions of stability to linear modes the instantaneous velocity profiles resulting from the combined mean and disturbance velocities could readily be sufficiently inflected to be destabilised over a part of the disturbance wave with an associated strong amplification rate. The wavelengths of the disturbances that are destabilised tend to be large relative to the boundary layer thickness ($\simeq 10\delta$) so that the inflected profiles can be present over a significant extent of the surface and for time intervals of the order of $10\delta/u_e$. These regions of inflected profiles can introduce the possibility of wider ranges of frequency for which unstable disturbances can develop in addition to those of linear theory, and these in turn can lead to others.

Thus, in principle, one can speculate on the possibility of a chain of instabilities developing over a wide spectrum leading to a flow that could be described as turbulent with a mean flow differing markedly from the original laminar flow. It should be noted that non-linearities can also result in stabilising effects which lead to new equilibrium conditions at large enough amplitudes. For reviews of recent developments in non-linear theory see Stuart[5.34] and Herbert.[5.35]

However, the most readily accepted ideas on the development of the transition process from the appearance of Tollmien–Schlichting waves to turbulence have emerged from the work of Emmons and Bryson,[5.36] Schubauer and Klebanoff[5.37] and the subsequent work of Narasimha.[5.38] For a useful valuable review see reference 5.39. From observations of the flow on a water table, Emmons formulated the view that turbulence was not initiated along a continuous front but started in the form of randomly distributed 'spots' of turbulence each of which expanded as it moved downstream. Fully developed turbulent flow then began when these expanding regions of turbulence merged across the whole span. The transition region can therefore be said to begin at the streamwise position where the spots first appear and to end where they have merged to form a continuous front. Emmons first suggested that the probability of the generation of spots was constant with distance x downstream from the beginning of the transition region but later modified this by assuming the probability was proportional to x^n where $n \geqslant 1$.

The fraction of time over which the flow at a point in the transition region is turbulent, when recorded over a long enough period with, say, a hot wire anemometer, is referred to as the *intermittency*, denoted by γ; it will be evident that $\gamma = 0$ at the onset of transition and $\gamma = 1$ at the end when the turbulent flow is fully developed. For a flat plate at zero incidence it is found that γ can be described as a function of ξ, where

$$\xi = (x - x_t)/\lambda$$

and $$\lambda = x_{\gamma=0.75} - x_{\gamma=0.25} \qquad (5.27)$$

x_t marks the beginning of transition.

However, Narasimha has shown that Emmon's hypothesis for the generation of spots over the transition region leads to a distribution of γ that does not fit the experimental data, he then postulated that spots form at a preferred streamwise position but randomly in time and spanwise location. This postulate led to the relation for a flat plate at zero incidence:

$$\gamma = 1 - \exp(-0.412\xi^2) \tag{5.28}$$

which is in better agreement with the experimental data. Destabilising factors such as adverse pressure gradients, turbulence in the external flow, noise and increase of Reynolds number increase the rate of generation of the spots.

From a small surface excrescence a continuous stream of spots are generated which expand as they move downstream to form the familiar turbulence wedge of semi-apex angle $\simeq 10°$. Such wedges have long been observed in wind tunnels and in flight on wings and bodies with the aid of suitable flow visualisation techniques (see Fig. 5.11).

Details of the geometry of the spots and their growth with movement downstream were demonstrated in experiments by Schubauer and Klebanoff.[5.37] A spot was introduced into a laminar boundary layer on a flat plate in a wind tunnel by means of an electrical spark discharged normal to the surface through the boundary layer. The spot's subsequent development into a roughly triangular region of turbulence is illustrated

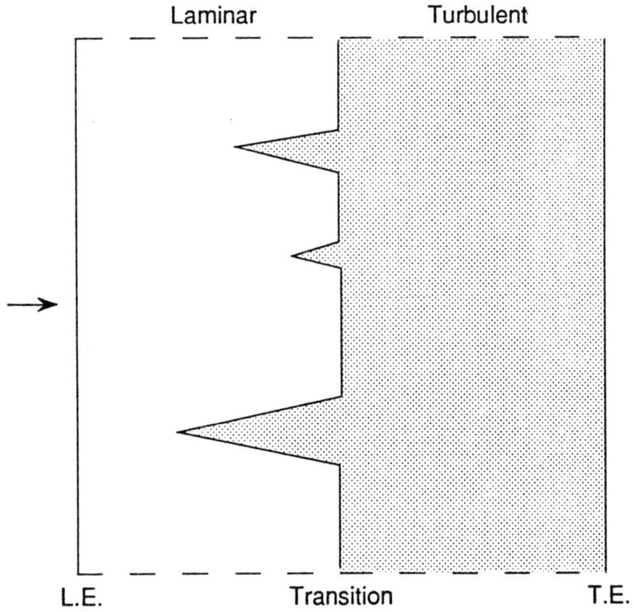

Fig. 5.11 Typical turbulence 'wedges' from isolated protuberances.

in Fig. 5.12, where it will be seen that its forward apex moved downstream at close to the main stream velocity, but its rear edge convected downstream at about half that speed. The spot expanded outwards so that its tips defined the sides of a wedge of semi-angle $\simeq 11°$ and apex at the spot origin. Hot wire records within the growing 'triangular' region were much the same as those of fully developed turbulent flow but upstream of this region at any instant the flow was smooth and laminar. Consistent with this, it was found that the instantaneous velocity profiles in the transition region were for part of the time characteristic of turbulent flow, and at other times they were those of a laminar boundary layer – hence the concept of intermittency.

The obvious question then is – how do the spots arise? A partial answer to this question was provided by Klebanoff *et al.*[5.40] and by Stuart.[5.41] It was shown that even in nominally two dimensional flow the Tollmien–Schlichting waves that develop in regions of instability do not remain rectilinear but show evidence of spanwise warping or bending which becomes increasingly marked with movement downstream. The warping appears to arise from initially small irregularities in the main flow such as might be caused in a wind tunnel by small variations in a smoothing screen. Hence, the associated disturbance vorticity develops streamwise components which grow as the warping increases, and these in turn induce secondary flows to and from the surface which help to

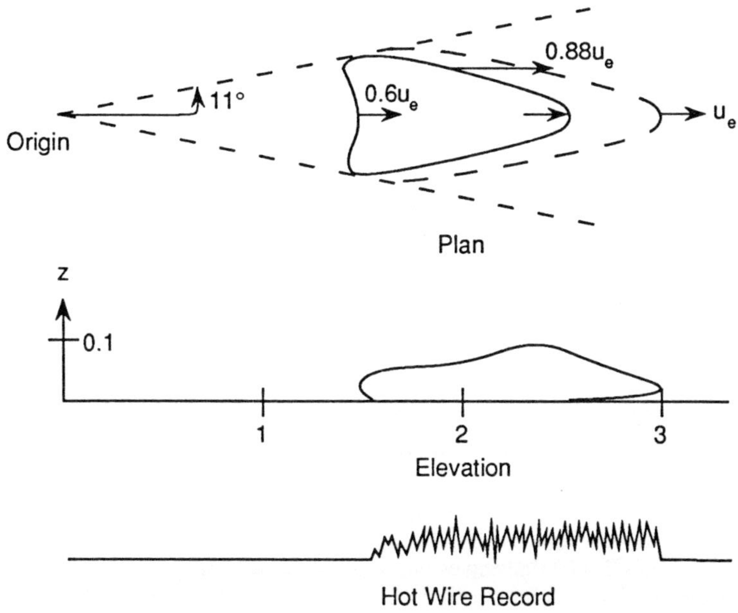

Fig. 5.12 Sketch of developing turbulence 'spot' initiated by spark discharge.[5.37]

intensify the warping. The disturbance vorticity lines thus form hairpin-like bends and as the bends grow the vorticity lines stretch and strengthen along the inclined sides of the bends. This results in 'peaks' and 'hollows' of high and low disturbance intensity with higher amplification rates at the 'peaks', signifiantly higher than predicted by linearised theory. Energy is transferred from the valleys to the peaks by the growing spanwise component of the disturbance velocity, v'. The bends or loops of vorticity induce regions of high streamwise shear above them and also a jet-like flow between the limbs of each loop (see Wortmann[5.42]). These introduce additional regions of inflected velocity profiles and therefore additional sources of instability and disturbance growth. At some stage the hot wire records at a peak show 'spikes' of a kind that can be identified as due to turbulence and we can infer that a spot is born. The sequence postulated in the above sketchy description is illustrated in Fig. 5.13.

It seems that surface excrescences and, to a lesser extent, external turbulence and noise can in some measure bypass the initial Tollmien–Schlichting wave stage of the above process, since they can introduce directly disturbances of large enough amplitude associated with highly unstable three dimensional flow patterns which can readily generate 'spots'.

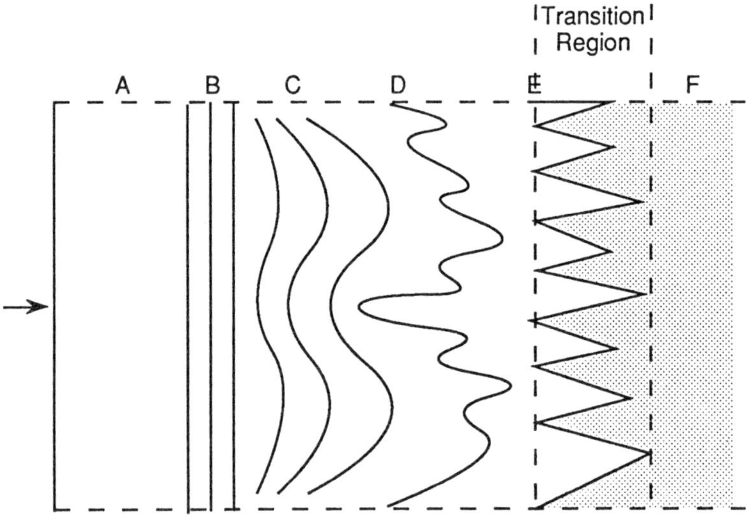

Fig. 5.13 Sketch of suggested stages in transition process.

A – stable laminar flow
B – Tollmien–Schlichting waves
C – warping of waves, formation of vortex loops
D – development of non-linear instabilities
E – formation and growth of turbulent 'spots'
F – fully developed turbulent flow.

5.9 Transition prediction

It will not be surprising in the light of the foregoing to learn that we are still far from being able to develop a general and reliable prediction method reflecting adequately all the physical processes involved in transition. We must therefore appeal to simple empirically based methods derived from data culled from a sufficiently varied range of cases to engender a degree of confidence for applications where the circumstances are not too far removed from those of the cases analysed.

The most successful and widely used of such methods is the so-called e^n method suggested at about the same time (1956) by Smith and Gamberoni[5.43] and by Van Ingen.[5.44] This method is based on the idea that the position of transition can be correlated with the position where the overall maximum amplification of a disturbance by Tollmien–Schlichting waves has reached a level of the form e^n where n can be reliably predicted from an analysis of experimental results.

For the temporal mode, we have from equation (5.11) that if A is the amplitude of the disturbance stream function ψ at any time t, then

$$\frac{1}{A}\frac{\mathrm{d}A}{\mathrm{d}t} = \beta_i, \qquad \text{and hence } \frac{A}{A_0} = \exp \int_0^t \beta_i \,\mathrm{d}t \tag{5.29}$$

Here, suffix 0 defines the initial state where the Tollmien–Schlichting waves are neutrally stable. For the spatial mode we have similarly from equation (5.17) that

$$\frac{1}{A}\frac{\mathrm{d}A}{\mathrm{d}x} = -\alpha_i, \qquad \text{and hence } \frac{A}{A_0} = \exp \int_0^x -\alpha_i \,\mathrm{d}x \tag{5.30}$$

Following Gaster[5.12] it is generally accepted that to compare results in the temporal and spatial modes we can write $-\alpha_i = \beta_i/c_g$, where $c_g = (\partial \beta_r/\partial \alpha_r)\alpha_i$ is the group velocity. The spatial mode permits a simpler development of the argument and is easier to check against experiment. It follows from equation (5.30) that

$$A/A_0 = \exp\left[R \int_{X_0}^{X} \frac{(-\alpha_i \delta^*)}{R_{\delta^*}} \mathrm{d}X\right] = e^n, \quad \text{say} \tag{5.31}$$

where $X = x/c$, $R = u_\infty c/\nu$, $R_{\delta^*} = u_\infty \delta^*/\nu$, u_∞ is a reference velocity and c is a reference length.

We may assume that we have available $\alpha_i \delta^*$ as a function of X and $\overline{\omega}$ ($= \beta_r\, \nu/u_c^2$) for a given boundary layer from the solution of the Orr–Sommerfeld equations (see Section 5.4) and hence for a given $\overline{\omega}$ we can calculate A/A_0 as a function of $R_x = u_\infty x/\nu$ as illustrated in Fig. 5.14. The envelope of the resulting curves for different values of $\overline{\omega}$ represents the maximum value of A/A_0 as a function of R_x. From the known experimental values of R_x at the beginning or end of the transition region we can determine the corresponding values of n. The applicability of the method lies in the fact that in a wide range of cases the values of n are found to vary relatively little, for the beginning of transition n usually falls in the range 8–9 and for the end of transition n usually lies in the range 9–11. If a high Reynolds number flow is considered then the transition region may be assumed small and a value of $n = 10$ is often assumed. Cases do arise where values of n outside these ranges have been inferred, these seem to be associated with significant surface curvature, and Malik and Poll[5.45] have made some allowance for the effects of curvature in deriving modified Orr–Sommerfeld equations and have then recovered values of n consistent with those quoted above.

To cope with the effects of external turbulence Mack[5.46] has suggested from an analysis of Dryden's measurements that for $T_u > 0.1\%$ (see Fig. 5.10)

$$n = -8.43 - 2.4 \ln T_u \tag{5.32}$$

Van Ingen[5.47] has analysed a wide range of data distinguishing between the start and end of the transition region and has suggested

$$\left.\begin{array}{ll} n = 2.13 - 6.18 \log T_u & \text{at the start} \\ n = 5 - 6.18 \log T_u & \text{at the end of the transition region} \end{array}\right\} \tag{5.33}$$

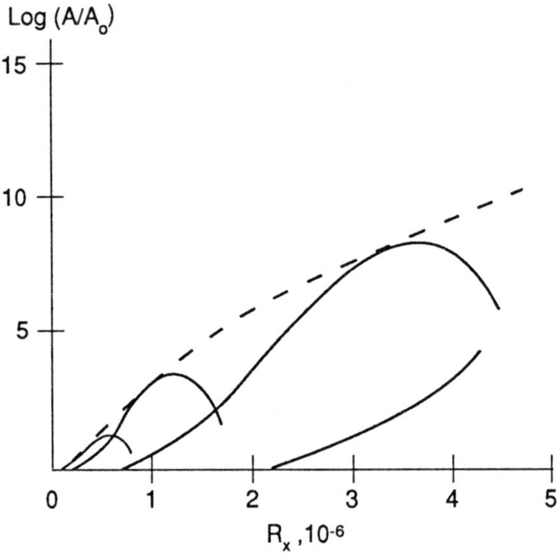

Fig. 5.14 Illustration of the amplification factor as function of R_x for a boundary layer, according to linearised disturbance theory. Solid lines are for constant values of $\overline{\omega} = \beta_r \nu / u_e^2$: the dashed line is the envelope of the solid lines (locus of n).

For extensions of the method to cases involving suction and separated flows the reader is referred to reference 5.47. For a detailed critique of the method and its extension to compressible flow reference should be made to the cited paper by Mack.[5.46]

5.10 Effect of suction

We have already seen that suction at the wall both reduces the boundary layer thickness and makes the velocity profile fuller. On both counts it helps to stabilise the boundary layer. The small perturbation stability analysis for the simple flat plate asymptotic suction profile (see Section 2.2), viz.

$$u/u_e = 1 - \exp(w_w z/\nu)$$

leads to a critical Reynolds number ($u_e \delta^*/\nu$) of about 70 000, as compared with about 500 for a Blasius profile. Now, for the asymptotic suction profile

$$\delta^* = -\nu/w_w$$

Hence, since

$$c_Q = Q/Su_e = -w_w/u_e = \nu/u_e\delta^*$$

where S is the surface area sucked and Q is the rate of volume flow sucked, it follows that

$$c_Q > 1.4 \times 10^{-5}$$

to ensure stability to small disturbances. This is a remarkably small value and the consequent power expended in suction need be no more than a small fraction of the reduction in power required in flight with a laminar rather than a turbulent boundary layer. It is also to be noted that the maximum value of the amplification factor β_i is about a tenth of that for the Blasius profile, and for Reynolds numbers greater than critical the range of frequencies to which the boundary layer is unstable is considerably reduced.

For the more realistic case of the flat plate with constant suction for which the profile changes from the Blasius one at the leading edge to the asymptotic form far downstream, calculations by Iglisch[5.48] lead to the result that for stability

$$c_Q > 1.2 \times 10^{-4}$$

This is still a relatively small suction coefficient from a practical point of view as it would pose very modest power demands. It leads to a small increase of skin friction drag over that of an unsucked laminar boundary layer but the skin friction drag is no more than about one tenth of that for a turbulent boundary layer at a Reynolds number of 10^8.

A considerable amount of work was done both in flight and in wind tunnels some twenty or more years ago to demonstrate conclusively that substantial reductions of drag were possible up to high subsonic cruising Mach numbers by the use of modest amounts of suction through porous or slotted surfaces resulting in extensive regions of laminar flow over wings and bodies.[5.49, 5.50] However, it was clear that at Reynolds numbers representative of large transport aircraft great care was needed to maintain the surfaces free from waves and imperfections under operational conditions; rain and impacting insects were a problem and the required holes or slits were difficult to drill or cut with the necessary regularity and smoothness of finish to ensure that they did not themselves introduce disturbances having an adverse transition effect. The relatively low price of oil at that time did not result in a strong enough economic spur for the requisite research and development work to be undertaken to solve these engineering and operational problems, and such work ceased until relatively recently.

However, in recent years a number of factors have arisen to cause a major renewal of interest in the development of 'laminar flow' aircraft. These include the ever increasing demand of airlines for operational economies and the consequent sharpening of competition between designers, the current and likely future cost of fuel as a large fraction of direct operating costs, the advent of new materials (e.g. composites) that more readily offer the needed surface integrity and smoothness than traditional materials, and new computer controlled manufacturing techniques for achieving relatively easily the holes or slits to the requisite accuracy. The results of this recent work are very promising.

The development of aircraft with extensive regions of laminar flow is likely to occur in three successive stages of increasing design complexity.

(1) *Natural laminar flow (NLF).* This involves the use of wing sections with extensive regions of favourable pressure gradient which would be effective for small executive type of aircraft operating at relatively low Reynolds numbers.
(2) *'Hybrid' boundary layer control.* Here it is envisaged that suction will be applied over the leading edge region of the wing where destabilising influences are most likely to originate and grow (particularly with leading edge sweepback, see Section 10.2) followed by extensive regions of favourable pressure gradient to maintain laminar flow. This is seen as applicable to a wide range of small to moderate sized aircraft.
(3) *Full boundary layer control.* Here the suction will be applied over much of the surface of the wing, particularly the upper surface, to

ensure extensive regions of laminar flow and low skin friction drag for large aircraft cruising at high Reynolds numbers.

References

5.1 Reynolds, O. (1883) 'An experimental investigation of the circumstances which determine whether the motion of water should be direct or sinuous, and of the law of resistance in channels'. *Phil. Trans. Roy. Soc.*, **174**, p. 935.
5.2 Reynolds, O. (1895) 'On the dynamical theory of incompressible viscous fluids and the determination of the criterion'. *Phil. Trans. Roy. Soc. A*, **186**, p. 123.
5.3 Lorentz, H.A. (1907) *Abhandlung über theoretische Physik*, **1**, p. 43.
5.4 Orr, W.M.F. (1907) 'The stability or instability of the steady motions of a perfect liquid and of a viscous liquid'. *Proc. Roy. Irish Acad.*, **27**, p. 9, p. 69.
5.5 Sommerfeld, A. (1908) 'Ein Beitrag zur hydrodynamischen Erklärung der turbulenten Flüssigkeitsbewegungen'. Atti del 4 Congr. Internat. dei Mat. III, Roma, p. 116.
5.6 Tollmien, W. (1929) 'Über die Enstehung der Turbulenz'. *1* Mitt. Nachr. Ges. Wiss., Göttingen, *Math. Phys. Kl.*, **21**. English translation NACA TM 609 (1931).
5.7 Schlichting, H. (1933) 'Zur Entstehung der Turbulenz bei der Plattenströmung'. Nachr. Ges. Wiss. Göttingen, *Math. Phys. Kl.*, p. 182; also *ZAMM* (1933) **13**, p. 171.
5.8 Obremski, H.T., Morkovin, M.V. and Landahl, M.T. (1969) 'A portfolio of stability characteristics of incompressible boundary layers'. AGARDograph 134.
5.9 Schlichting, H. and Ulrich A. (1942) 'Zur Berechnung des Umschlages laminar-turbulent'. *Jb. dt. Luftfahrtforschung*, **1**, p. 8.
5.10 Pretsch, J. (1941) 'Über die Stabilität einer ebenen Laminarströmung bei Druckgefälle und Druckanstieg'. *Jb. dt. Luftfahrtforschung*, **1**, p. 58.
5.11 Schubauer, G.B. and Skramstad, H.K. (1947) 'Laminar boundary layer oscillations and stability of laminar flow'. *JAS*, **14**, p. 69; see also NACA Rep. (1947) 909.
5.12 Gaster, M. (1962) 'A note on the relation between temporally-increasing and spatially-increasing disturbances in hydrodynamic stability'. *JFM*, **14**, p. 222.
5.13 Saric, W.G. and Nayfeh, A.W. (1977) 'Non-parallel stability of boundary layers with pressure gradients and suction'. AGARD CP No **224**, Paper 6.
5.14 Taylor, G.I. (1923) 'Stability of a viscous fluid contained between rotating cylinders'. *Phil. Trans. A* **223**, p. 289; see also *Proc. Roy. Soc. A*, **151**, p. 494 (1935); **157**, p. 546–565 (1936).
5.15 Görtler, H. (1949) 'Über ein dreidimensionale Instabilität laminarer Grenzschichten an konkaven Wänden'. Nachr. Ges. Wiss., Göttingen, *Math. Phys. Kl.* **1**, p. 1; see also NACA TM 1375.
5.16 Liepmann, H.W. (1945) 'Investigation of boundary layer transition on concave walls'. NACA Wartime Rep., W–87.

5.17 Wimmer, M. (1988) 'Viscous flow instabilities near rotating bodies'. *Prog. Aero. Sc.*, **25**, 1, p. 45.
5.18 Lees, L. and Lin, C.C. (1946) 'Investigation of the stability of a laminar boundary layer in a compressible fluid'. NACA TN 1115.
5.19 Squire, H.B. (1933) 'On the stability for three dimensional disturbances of viscous fluid flow between parallel walls. *Proc. Roy. Soc. A*, **142**, p. 621.
5.20 Mack, L.M. (1969) 'Boundary layer stability theory'. JPL Pasadena, Cal., Rep. 900–277.
5.21 Young, A.D. and Paterson, J.H. (1981) 'Aircraft excrescence drag'. AGARDograph 264.
5.22 Schiller, L. (1932) *Handbuch der Exp. Physik.*, Leipzig, **IV**, 4, p. 189.
5.23 Braslow, A.L. (1960) 'Review of the effect of distributed surface roughness on boundary layer transition'. AGARD Rep 254.
5.24 Smith, A.M.O. and Clutter, D.W. (1959) 'The smallest height of roughness capable of affecting boundary layer transition'. *JAS* Apr., p. 229.
5.25 Van Driest, E.R. and Blumer C.B. (1960) 'Effect of roughness on transition in supersonic flow'. AGARD Rep. 255.
5.26 Rogler, H.L. (1971) 'Laminar-turbulent transition'. AGARD CP 224, paper 16.
5.27 Cousteix, J., Houdeville, R. and Desopper, A. (1971) 'Laminar-Turbulent Transition'. AGARD CP 224, paper 17.
5.28 Whitfield, J.D. and Dougherty, N.S. (1971) 'Laminar-turbulent transition'. AGARD CP 224, paper 25.
5.29 Dryden, H.L. (1959) 'Transition from laminar to turbulent flow'. *Turbulent Flows and Heat Transfer*, (ed. C.C. Lin), Princeton Univ. Press, N.J., p. 1.
5.30 Wells, C.S. (1967) 'Effects of free stream turbulence on boundary layer transition'. *AIAA J*, **5**, p. 172.
5.31 Stuart, J.T. (1960) 'On the non-linear mechanics of wave disturbance in stable and unstable parallel flow', Pt. 1. *JFM* **9**, p. 355.
5.32 Watson, J. (1960) 'On the non-linear mechanics of wave disturbance in stable and unstable parallel flow', Pt 2. *JFM*, **9**, p. 371.
5.33 Reynolds, W.C. and Potter, M.C. (1967) 'Finite amplitude instability of parallel shear flows'. *JFM*, **27**, p. 465.
5.34 Stuart, J.T. (1986) 'Instability of flows and their transition to turbulence'. *ZFW*, **10**, p. 379.
5.35 Herbert, T. (1984) 'Non-linear effects in hydrodynamic stability'. AGARD Rep. 709, paper 6.
5.36 Emmons, H.W. and Bryson, A.E. (1951) 'The laminar-turbulent transition in a boundary layer', Pt. I, *JAS* **18**, p. 490; Pt. II, Proc. 1st. US Nat. Congr. Appl. Mech. (1952) p. 859.
5.37 Schubauer, G.B. and Klebanoff, P.S. (1955) 'Contributions on the mechanics of transition'. NACA TN 3489; also IUTAM Symp. (1958) p. 84.
5.38 Narasimha, R. (1957) 'On the distribution of intermittency in the transition region of a boundary layer'. *JAS*, **24**, p. 711.
5.39 Narasimha, R. (1985) 'The laminar-turbulent transition zone in the boundary layer'. *Prog. Aero. Sc.*, **22**, 1, p. 24.
5.40 Klebanoff, P.S., Tidstrom, K.D. and Sargent, L.R. (1962) 'The three dimensional nature of boundary layer instability'. *JFM*, **12**, p. 1.

5.41 Stuart, J.T. (1965) 'The prediction of intense shear layers by vortex stretching and convection'. AGARD Rep. 514.
5.42 Wortmann, F.X. (1977) 'The incompressible fluid motion downstream of two dimensional Tollmien–Schlichting waves'. AGARD CP 224, paper 12.
5.43 Smith, A.M.O. and Gamberoni, N. (1956) 'Transition, pressure gradient and stability theory'. *Proc. Int. Congr. App. Mech. 9*, **4**, p. 234.
5.44 Van Ingen, J.L. (1956) 'A suggested semi-empirical method for the calculation of the boundary layer transition region'. VTH 74, Dept. of Aero. Eng. Tech. Univ. Delft.
5.45 Malik, M.R. and Poll, D.I.A. (1985) 'Effect of curvature on cross flow instability'. IUTAM Symp., Novosibirsk, *Laminar-Turbulent Transition*, Springer Verlag, Berlin.
5.46 Mack, L.M. (1977) 'Transition prediction and linear stability theory'. AGARD CP 224, paper 1.
5.47 Van Ingen, J.L. (1977) 'Transition, pressure gradient, suction separation and stability theory'. AGARD CP 224, paper 20.
5.48 Iglisch, R. (1944) 'Exacte Berechnung der laminaren Reibungsschicht an der langsangeströmten ebenen Platte mit homogener Absaugung'. *Schrift. dt. Akad. d. Luftfahrtforschung*, **8B**, p. 1; see also NACA RM 1205 (1949).
5.49 Pfenninger, W. (1977) 'Laminar flow control'. AGARD Rep. 654, paper 3.
5.50 Edwards, B. (1977) 'Laminar flow control – concepts, experiences, speculations'. AGARD Rep. 654, paper 4.
5.51 Wazzan, A.R. *et al.* (1968) 'Spatial and temporal stability charts for the Falkner–Skan boundary layer profiles'. Douglas Aircraft Co. Rep DAC-67086.

Chapter 6

Turbulence and the Structure of Attached Turbulent Boundary Layers: Some Basic Empiricisms

6.1 Introduction

Turbulence is a dynamic, ever changing flow process which at present can only be described quantitatively in statistical terms that reflect little of its physical detail or transient flow patterns. We have seen (Section 5.2) that in the presence of mean flow velocity gradients turbulent motion is associated with rates of transport of momenta which can be interpreted as stresses – the Reynolds or eddy stresses. We can infer that a major factor in the rate of growth or decay of the turbulent motion is the rate of work done by these eddy stresses on the mean flow. This work results in an exchange of energy between the turbulence and the mean flow.

Another major factor is the rate of work done by the viscous stresses associated with the small scale turbulence (where the instantaneous velocity gradients and hence these stresses are large) and this appears as a cause of dissipation of the turbulence. Other factors arise from the turbulence interacting with itself, viz. the work done by the pressure fluctuations on the turbulent rates of strain and the rate of advection of turbulent energy by the turbulent velocity components.

In spite of the underlying complexity of turbulent motion it is a fortunate fact that for two dimensional attached turbulent boundary layers in a steady mean flow, relatively simple empirical relations have been developed that, backed by dimensional analysis, enable important time-averaged characteristics (e.g. skin friction, velocity distribution, momentum and displacement thicknesses, mean turbulence parameters) to be predicted with an accuracy that is generally adequate for most engineering purposes.

In more complex cases, e.g. separated boundary layers, three dimensional flows, or flows where boundary layers interact with each other or with wakes from upstream surfaces, these empirical relations and prediction methods become less satisfactory. Our inadequate knowledge of the detailed physics of turbulence is then evident as a source of weakness in our efforts to develop predictive methods of acceptable accuracy.

In this chapter we shall mainly confine ourselves to a discussion of the basic dimensional arguments and empirical concepts which, as we have noted, have provided a useful foundation from which the time-averaged characteristics of simple attached two dimensional boundary layers can be readily predicted with an accuracy that is adequate for many practical purposes. We shall here consider in the main incompressible flow over a flat plate at zero incidence. In later chapters we shall deal with more general cases and the associated prediction methods.

In recent years increasing attention has been focussed on the time dependent transient turbulence flow patterns that have become evident with the use of modern forms of anemometry and digital sampling techniques. These so-called coherent structures will not be discussed at this stage, but a brief discussion is presented in Chapter 10.

6.2 Turbulence characteristics

Useful measures of the magnitudes of the turbulent velocity components are their root mean squares:

$$\sqrt{\overline{u'^2}}, \quad \sqrt{\overline{v'^2}}, \quad \sqrt{\overline{w'^2}}$$

The overall mean intensity of the turbulence kinetic energy per unit mass can likewise be represented by:

$$\overline{e'} = \tfrac{1}{2}(\overline{u'^2} + \overline{v'^2} + \overline{w'^2}) \tag{6.1}$$

This quantity is frequently denoted by k, but to avoid confusion with other uses of the symbol k we shall for the present adopt the symbol $\overline{e'}$. Typical distributions of these three components in a turbulent boundary layer on a flat plate at zero incidence[6.1] are shown in Fig. 6.1.

For any two points denoted 1 and 2, say, within a turbulent eddy one can expect some correlation between the turbulent velocity components measured at those points. Such correlation can be represented by a number of coefficients such as

$$\mathcal{R}_s = \frac{\overline{u'_1 u'_2}}{\sqrt{\overline{u'^2_1}}\sqrt{\overline{u'^2_2}}} \tag{6.2}$$

which is the correlation coefficient between the u' components at points 1 and 2. We can infer that if $\mathcal{R}_s$ is near unity then $u'_1 \propto u'_2$ and the two points are well within a typical eddy, but if $\mathcal{R}_s$ is zero then the turbulent patterns at the two points are independent. Thus a measure of the size of a typical eddy at point 1 in a given direction s, say, is

$$L = \int_0^{s_0} \mathcal{R}_s(s)\,\mathrm{d}s \tag{6.3}$$

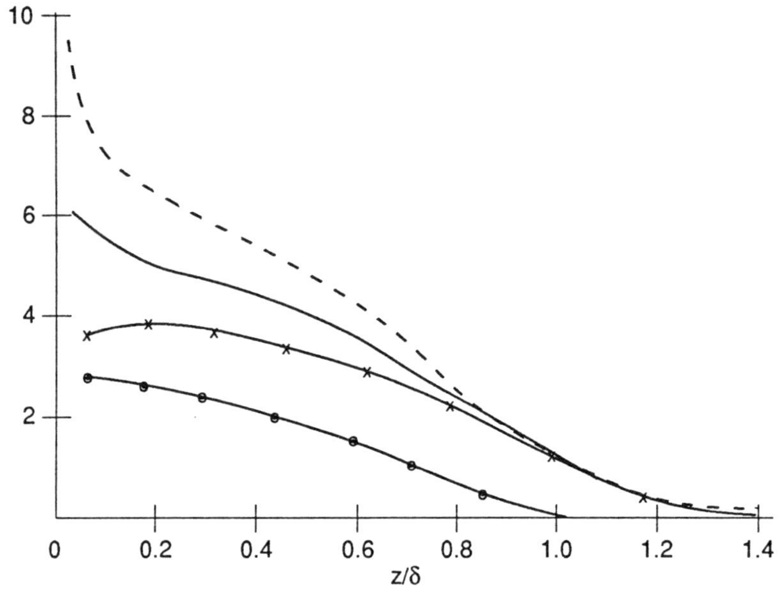

Fig. 6.1 Mean turbulence components and eddy stress distribution in the boundary layer on a flat plate at zero incidence. From top to bottom, the curves show:

- - - - - $(\sqrt{\overline{u'^2}}/u_e) \times 10^2$ —— $(\sqrt{\overline{v'^2}}/u_e) \times 10^2$

×——× $(\sqrt{\overline{w'^2}}/u_e) \times 10^2$ ⊙——⊙ $(-\rho\overline{u'w'}/\frac{1}{2}\rho u_e^2) \times 10^3$

where s is measured from point 1, and s_0 is the distance to the point where $\mathcal{R}_s = 0$. Likewise, the time of passage of a typical eddy past a fixed point can be represented by

$$T = \int_0^{t_0} \mathcal{R}_t(t)\mathrm{d}t \tag{6.4}$$

where $$\mathcal{R}_t = \frac{\overline{u'(0)\,u'(t)}}{\sqrt{\overline{u'(0)^2}}\,\sqrt{\overline{u'(t)^2}}}$$

t is the time interval from instant 0, say, and t_0 is time at which $\mathcal{R}_t = 0$. $\mathcal{R}_t$ is sometimes called an auto-correlation coefficient. It is evident that a number of such spatial and temporal correlation coefficients can be generated by ringing the changes on u', v' and w'. It will also be evident that they are closely related to the Reynolds stresses defined in Section 5.2.

However, as we shall see, there are other ways of defining length scales for turbulence. We must bear in mind that a turbulent flow comprises eddies covering a wide spectrum of sizes and the use of a single length scale can only be regarded as a very crude approximation to a complex reality. Thus we can adopt the energy spectrum concept and regard each turbulence velocity component as comprising harmonic oscillations covering ranges of frequency and amplitude. We can then denote the proportion of the total disturbance energy lying within a frequency band from n to $n + \delta n$, say, as $f(n).\delta n$, where $f(n)$ is called the energy spectrum function of the turbulence. A typical distribution of $f(n)$ for a boundary layer[6.1] is shown in Fig. 6.2.

It will be noted that the turbulence energy is mainly associated with the eddies of low frequency. A similar result is found for channel flow.[6.2] It is inferred that there is a cascading process of energy transfer from the larger eddies to the smaller ones, and energy is finally dissipated within the smallest eddies by the action of viscosity.

6.3 The mean velocity distribution

An examination of the structure of an attached turbulent boundary layer on a smooth surface suggests that it can conveniently be regarded as

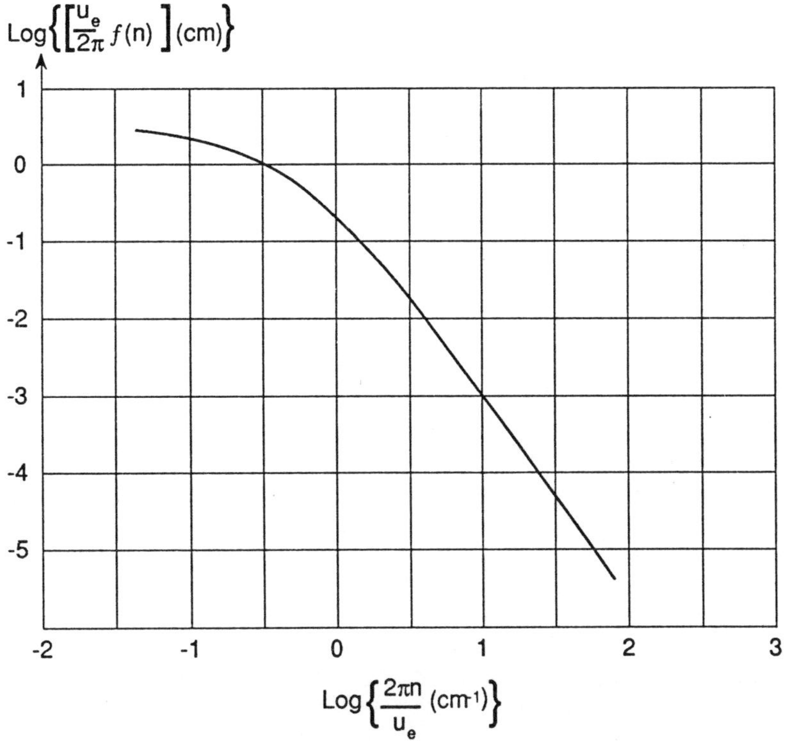

Fig. 6.2 Typical spectral distribution for $\overline{u'^2}$ in a boundary layer on a flat plate.[6.1]

comprising three regions.[6.3] Immediately adjacent the surface there is a so-called *viscous sub-layer* in which the turbulent fluctuations are relatively small and the dominant shear stress is the viscous one, namely

$$\tau_l = \mu \, \partial u / \partial z$$

The mean velocity is practically linear with z in the sub-layer and the stress is usually assumed constant there and equal to its value at the surface τ_w. The thickness of the sub-layer is very small, of the order of one-hundredth, or less, of the boundary layer thickness, δ. If we denote it as z_l then the non-dimensional quantity

$$z_l^+ = z_l \, u_\tau / \nu = O(10) \tag{6.5}$$

where u_τ, the friction velocity, $= (\tau_w/\rho)^{1/2}$. With $\tau_l = \tau_w$, it follows that

$$u/u_\tau = \tau_w \, z/(\mu \, u_\tau) = z \, u_\tau/\nu = z^+ \tag{6.6}$$

where z^+ denotes $z \, u_\tau/\nu$, (sometimes referred to as 'wall units'). Above the sub-layer there is an inner region of the boundary layer about 0.4δ thick in which the turbulence intensity is high and the dominant shear stress is the eddy stress

$$\tau_t = -\rho \overline{u' w'}$$

In this region there is a wide spectrum of eddy sizes and frequencies, and instantaneous velocity measurements (as with a hot wire anemometer) shows no intermittency of turbulence as illustrated in Fig. 6.3.

Above this region there is an outer region characterised by large eddies and the instantaneous edge of the boundary layer is strongly convoluted (see Fig. 6.3). A hot wire anemometer recording the velocities in this region shows the flow to be intermittently turbulent. This intermittency is not unlike that of a transition region (see Section 5.8) and a measure of it is the proportion of a velocity record of sufficient duration, taken, say, with a hot wire anemometer, for which the flow is recorded as turbulent. If we denote this intermittency as γ_0 then we find that it is unity for z less than about 0.4δ (except in the viscous sub-layer) and as z increases above 0.4δ then γ_0 decreases to 0 at about 1.2δ (see Fig. 6.4).[6.1] A good fit to the intermittency distribution for zero pressure gradient as measured by Klebanoff is given by

$$\gamma_0 = \tfrac{1}{2}\left[1 - \operatorname{erf} 5\left(\frac{z}{\delta} - 0.78\right)\right] \tag{6.7}$$

where erf denotes the error function, viz.

$$\operatorname{erf} \sigma = \frac{1}{\sqrt{2\pi}} \int_{-\sigma}^{\sigma} e^{-\sigma^2/2} \, d\sigma$$

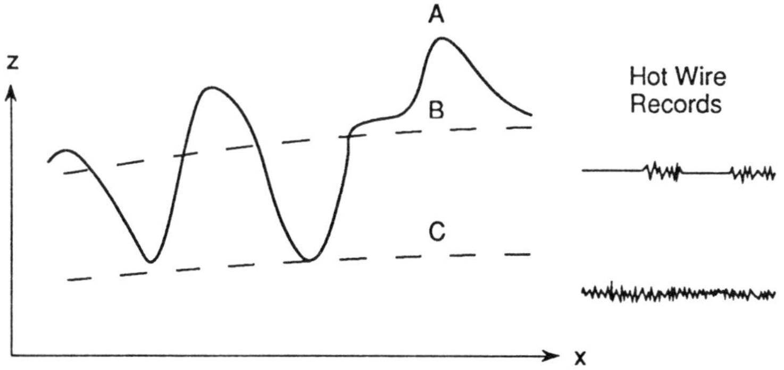

Fig. 6.3 Sketch illustrating the intermittent character of turbulence in the outer region of the boundary layer.
A – instantaneous edge of boundary layer; B – mean edge of boundary layer; C – edge of inner region.

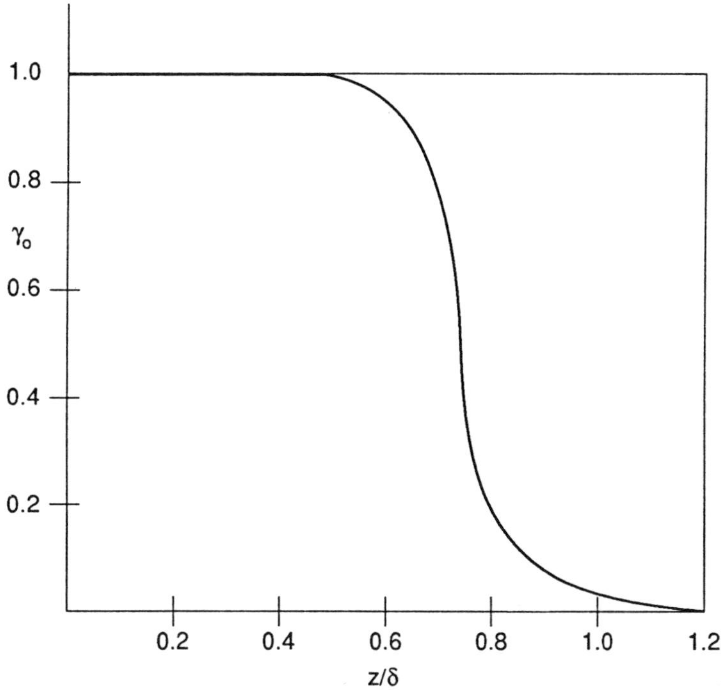

Fig. 6.4 Intermittency factor γ_0 in a flat plate boundary layer.[6.1]

We recall that δ is a time mean of the boundary layer thickness, whereas the outer eddies, and hence the convolutions, can instantaneously extend to about 1.2δ.

In the outer region the dominant stress is again the eddy or Reynolds stress τ_t and the turbulent mixing process is vigorous. Hence the velocity profile is more nearly uniform over much of a turbulent boundary layer, as compared with a laminar boundary layer, and for that reason the rate of strain adjacent the surface (and therefore the wall frictional stress) is much greater. It must, however, be emphasised that the three regions described above are not sharply distinct, they merge into each other, and their relative thicknesses can change with changes in the external pressure distribution and, to some degree, with changes in Reynolds number.

A complementary way of looking at the turbulent boundary layer proceeds from the argument that close to the wall, the shear stress there, τ_w, and the distance from the wall, z, are very important. If it is assumed that these factors, as well as ρ and ν are the only important ones to determine the velocity distribution close to the wall, then dimensional analysis readily leads to a relation of the form

$$u/u_\tau = f(z\,u_\tau/\nu) = f(z^+) \tag{6.8}$$

where f denotes a function to be determined. Such a relation is called *the law of the wall*. We note that the external flow conditions play no direct part in determining the flow in the 'wall region' in which this law holds, and therefore the latter should be independent of any external pressure distribution. We also note that in the viscous sub-layer, which is part of the wall region, the law of the wall takes the particularly simple form of equation (6.6). In practice, it is found that this argument is well borne out, the independence of the wall region flow to external flow conditions, apart from the dependence of τ_w on u_e, is evident over a wide range of conditions. The wall region is, however, thin – of the order of one tenth of the boundary layer thickness.

Above the wall region we have the remaining nine tenths of the boundary layer in which, it is argued, viscosity plays no direct part and the velocity defect relative to the external velocity u_e is solely a function of u_τ, z, δ and some non-dimensional pressure gradient parameter, which for the present we shall denote as P. Dimensional analysis then leads to the so-called *velocity-defect relation*:

$$(u_e - u)/u_\tau = F(z/\delta,\, P) \tag{6.9}$$

where F denotes a function to be determined.

We note that this relation neglects 'history effects', i.e. effects due to the functional dependence on x of u_τ, δ and P. This neglect amounts to assuming that the boundary layer is 'self-preserving' or in an 'equilibrium' condition. We shall say more about self-preserving turbulent boundary layers in Section 6.5, it may suffice here to note that they show a similarity of form for different x in much the same sense as the similar families of velocity profiles for the laminar boundary layer (see Section 3.4).

We shall now discuss more fully the forms the law of the wall and the velocity defect relation are found to take and the limitations on their validity.

6.4 Eddy viscosity and mixing length concepts; the law of the wall

A simple concept, first introduced by Boussinesq in 1877,[6.4,6.5] assumes the eddy shear stress $-\rho\overline{u'w'}$ and the mean rate of strain $\partial u/\partial z$ are related linearly, as in laminar flow, viz.

$$\tau_t = -\rho\overline{u'w'} = \mu_t\,\partial u/\partial z = \rho\nu_t\,\partial u/\partial z \tag{6.10}$$

where μ_t can be regarded as an *eddy viscosity coefficient* but which, unlike the normal molecular viscosity coefficient, is not a physical characteristic of the fluid but is a function of the local flow conditions. Likewise, ν_t can be regarded as an *eddy kinematic viscosity coefficient*. The dependence on the local flow conditions must be determined and this is generally done by a combination of dimensional reasoning plus the analysis of experimental data leading to useful empirical relations, and/or approximate solutions of the equations governing the transport of selected turbulence quantities (the latter process comes under the heading of 'turbulence modelling' and will be discussed in Chapter 9).

One notes that ν_t has the dimensions of velocity × length. In the law of the wall region, but outside the viscous sub-layer, one can therefore write

$$\nu_t = \text{const.}\ u_\tau z = K\,u_\tau z, \quad \text{say} \tag{6.11}$$

or more strictly

$$= K\,u_\tau(z - z_0) \tag{6.12}$$

where z_0 is related to the viscous sub-layer thickness, since it defines the lower boundary of the region in which the eddy stress is dominant. On dimensional grounds Squire[6.6] inferred that

$$z_0 = \text{const.}\ \nu/u_\tau = C\,\nu/u_\tau, \quad \text{say}$$

The total shear stress (eddy + viscous) is

$$\tau = \tau_t + \tau_l = (\mu_t + \mu_l)\partial u/\partial z$$

Experimental evidence (see, for example, Fig. 6.1) indicates that τ is nearly constant in the wall region, except when the external pressure gradient is strongly adverse. If we take τ as constant then we have

$$\tau = \tau_w = \rho(\nu_c + \nu)\partial u/\partial z = \rho[K\, u_\tau(z - z_0) + \nu]\partial u/\partial z$$

or $$u_\tau^2/\nu = \left[K \frac{u_\tau}{\nu}(z - z_0) + 1\right]\frac{\partial u}{\partial z}$$

From this and equation (6.6) we readily infer that

$$\begin{aligned} u/u_\tau &= \frac{1}{K}\ln\left[\frac{Ku_\tau}{\nu}(z - z_0) + 1\right] + C \\ &= \frac{1}{K}\ln\left[\frac{u_\tau}{\nu}(z - z_0) + \frac{1}{K}\right] + C + \frac{1}{K}\ln K \\ &= \frac{1}{K}\ln\left[\frac{u_\tau}{\nu}(z - z_0) + \frac{1}{K}\right] + B, \quad \text{say} \end{aligned} \tag{6.13}$$

where C and B are constants. K is known as the *Von Karman constant*[6.7] and is usually quoted as 0.4 or 0.41. We see that for $z \gg z_0$, i.e. z^+ greater than about 30,

$$u/u_\tau = \frac{1}{K}\ln z^+ + B \tag{6.14}$$

Equation (6.14) is a well known form of the law of the wall and is often referred to as the *logarithmic law*. However, since equation (6.13) is based on an allowance for the viscous as well as the eddy stresses, it provides a better overall fit to the experimental results,[6.8] particularly in the region between the viscous sub-layer and the logarithmic law region, i.e. for z^+ between about 10 and 30 (see Fig. 6.5). Given the values of the constants K and B the value of z_0 can be determined to achieve continuity between equation (6.13) and equation (6.6) for the viscous sub-layer.

In spite of the assumptions involved in the derivation of equations (6.13) and (6.14), in particular the neglect of the effects of variations in the external flow, such as pressure gradients, and the assumed constancy of τ in the law of the wall region, it is a fact that they provide a good fit to experimental results for that region even for markedly varying external flows. Thus, plots such as Fig. 6.5 of u/u_τ versus $\log z^+$ fit a linear relation between a value of z^+ of about 30 and a value in the range of 10^2 to 10^3 depending on the Reynolds number and the external pressure distribution. The range increases with increase of Reynolds number and decreases with increasingly adverse (positive) pressure gradients, for a boundary layer approaching separation it is sometimes difficult to identify a linear portion of the plot. For the velocity defect law region the plot, not surprisingly, becomes dependent on the external flow.

However, experimenters differ slightly in their derived values of the constants K, z_0^+ and B; favoured values are $K = 0.4$, $z_0^+ = 7.8$, $B = 5.5$, and $K = 0.41$, $z_0^+ = 7.2$, $B = 5.0$. Perhaps the latter set of values is the

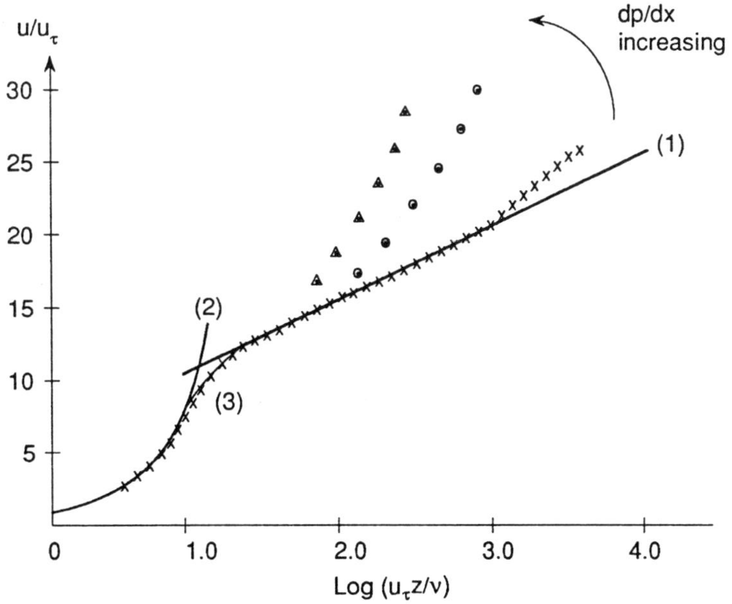

Fig. 6.5 Typical distributions of u/u_τ against log $(u_\tau z/\nu)$ in the 'law of the wall' region of the boundary layer on a flat plate.
(1) Equation (6.14) (2) equation (6.6) (3) equation (6.13)
× experimental results for zero pressure gradient
⊙ △ experimental results for adverse pressure gradient ($dp/dx > 0$).

more widely accepted. These differing experimental values in part reflect experimental errors, but there may be some small dependence of K and B on Reynolds number and strong pressure gradients.

The *mixing length concept*, due to Prandtl,[6.9] was an early attempt at what we now call 'turbulence modelling'. He proposed that there is a rough analogy between the relatively large scale random movements of fluid elements in turbulent motion and the much smaller scale random movements of molecules in a gas. The concept of the mean free path, that had been used in the description of the latter seemed a promising tool, therefore, for the representation of turbulent motion. Thus, a mixing length was postulated as the average distance normal to the wall travelled by a fluid particle over which its initial momentum or vorticity was conserved, but at the end of which it mixed with its surroundings. In Prandtl's formulation momentum was conserved over the mixing length, Taylor developed a parallel analysis based on the conservation of vorticity. We shall here discuss briefly the former approach, for the latter see reference 6.10.

It is assumed that the mixing length l_1 is small compared with the boundary layer thickness δ. Then (see Fig. 6.6) fluid particles at some instant at location z and mean velocity $u(z)$ will on the average have come from levels $z \pm l_1$ retaining their initial velocities $u \pm l_1\ \partial u/\partial z$, respectively, before mixing. Hence, we can identify the overall spread of instantaneous turbulent velocities at location z with these velocities, so that we can write approximately

$$\sqrt{\overline{u'^2}} = l_1\ \partial u/\partial z \tag{6.15}$$

We note that a strong negative correlation between u' and w' is implied by this concept, since particles of fluid from above ($w' < 0$) arrive with the higher instantaneous velocities ($u' > 0$) and conversely. Now if we also assume that u' and w' are of the same order of magnitude we can write

$$\tau_t = -\rho\overline{u'w'} = const.\ \rho\ l_1^2(\partial u/\partial z)\ |\partial u/\partial z| \tag{6.16}$$

Hence we can identify ν_t in equation (6.10) as equal to const. $l_1^2\ |\partial u/\partial z|$. We therefore write

$$\nu_t = l^2\ |\partial u/\partial z| \tag{6.17}$$

where l is a length proportional to l_1. The length l is presumed to be a function of z only and close to the wall it is assumed that

$$l = Kz \tag{6.18}$$

Thus, with $\partial u/\partial z > 0$, as is the case for attached boundary layers, equation (6.16) can be written

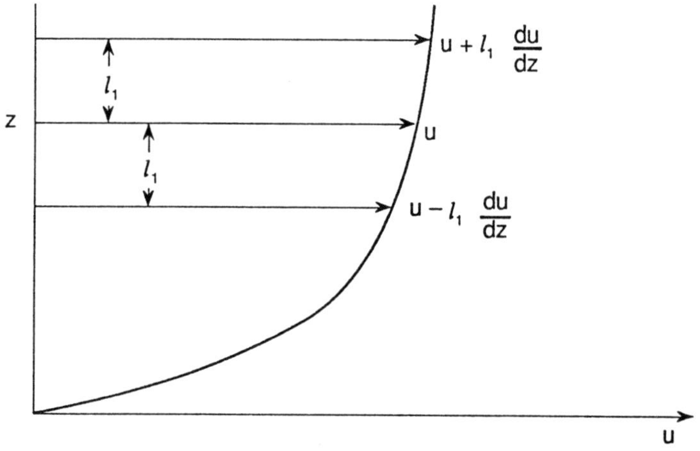

Fig. 6.6 Sketch illustrating the mixing length concept; l_1 = mixing length.

$$\tau_t = \rho K^2 z^2 (\partial u/\partial z)^2$$

and if we assume that close to the wall $\tau = \text{const.} = \tau_w$, then we infer that

$$\partial u/\partial z = u_\tau / Kz$$

Hence, $$u = \frac{u_\tau}{K} \ln z + \text{const.}$$

If we assume that this relation holds down to $z = z_l$ at the edge of the viscous sub-layer, then

$$u - u(z_l) = \frac{u_\tau}{K} \ln (z/z_l)$$

On dimensional grounds we can expect, as before, that z_l is a function of ν and u_τ, only, and so

$$z_l^+ = z_l\, u_\tau/\nu = \text{const.} = \beta, \quad \text{say}$$

We have earlier noted that z_l^+ is of the order of 10 [see equation (6.5)]. From equation (6.6)

$$u(z_l) = z_l\, u_\tau^2/\nu = \beta\, u_\tau$$

and hence $$u - \beta u_\tau = \frac{u_\tau}{K} \ln (z^+/\beta)$$

or $$u/u_\tau = \frac{1}{K} \ln z^+ + B \tag{6.19}$$

where $$B = \beta - \frac{1}{K} \ln \beta$$

We see that we have recovered the logarithmic law of the wall and it is evident that the latter follows essentially from the dimensional reasoning underlying the argument rather than the assumption of a mixing length *per se.* Indeed, this assumption does not reflect the complex physical processes that determine the dynamics of turbulence and the interactions of the turbulence and the mean flow. Further, the assumption that the mixing length is small compared with the boundary layer thickness is not always borne out by experiments in which the mixing length is deduced from measurements of the shear stress τ and the mean velocity gradient $\partial u/\partial z$. Nevertheless, the concept is of historical and didactic interest and the relations deduced from it can be regarded as providing a useful framework for empirical relations.

We have noted that the law of the wall region and in particular the merging between the viscous sub-layer and the main turbulent inner region was reasonably well fitted by Squire's analysis leading to equation (6.13). An alternative procedure whereby the viscous sub-layer is included in the mixing length approach is due to Van Driest[6.11] who suggested that the relation $l = Kz$ (equation 6.18) should be replaced by

$$l = Kz[1 - \exp(-z^+/A_0)] \tag{6.20}$$

The factor $[1 - \exp(-z^+/A_0)]$ is intended to account for the increasing damping effect on the turbulence as the wall is approached. The constant A_0 was found empirically by Van Driest to be 26, but some other authors favour the value 25. The resulting expression for u/u_τ is somewhat more complex than (6.13) but tends to (6.14) for z^+ greater than about 100 and to the viscous sub-layer relation (6.6) for small z^+.

If we multiply both sides of (6.14) by u_τ/u_e we get

$$u/u_e = \frac{u_\tau}{Ku_e}\ln\left(\frac{u_e z}{\nu}\right) + \frac{Bu_\tau}{u_e} - \frac{u_\tau}{Ku_e}\ln\left(\frac{u_e}{u_\tau}\right)$$

Thus, writing the local skin friction coefficient

$$c_{fe} = 2\tau_w/\rho u_e^2 = 2(u_\tau/u_e)^2$$

we have

$$\frac{u}{u_e} = \frac{1}{K}\sqrt{\tfrac{1}{2}c_{fe}}\ln\left(\frac{u_e z}{\nu}\right) + \sqrt{\tfrac{1}{2}c_{fe}}\left(B + \frac{1}{K}\ln\sqrt{\tfrac{1}{2}c_{fe}}\right) \tag{6.21}$$

Hence, given K and B, a chart can be prepared on which is plotted a family of curves of u/u_e against $\ln(u_e z/\nu)$ for different specified values of c_{fe}. From measured values of u/u_e and $u_e z/\nu$ in the law of the wall region we can infer from the chart the corresponding value of the skin friction coefficient. This method of determining c_{fe} is referred to as the *Clauser plot method* since Clauser first suggested it.[6.12] The independence of the flow in the law of the wall region from the external flow implies that this method can be generally applied provided that the boundary layer is attached and not too close to separation. Note that its validity depends on the assumed universality of the law of the wall and of the constants K and B.

A related method of equally general applicability is that due to Preston.[6.13] He argued that if a pitot tube were placed in contact with the surface and wholly immersed in the law of the wall region then the difference between the measured pitot pressure P and the measured static pressure p must depend only on ρ, ν, τ_w and d (the outer diameter of the tube). Hence, on the basis of dimensional reasoning we can expect a relation to hold of the form

$$\frac{(P - p)d^2}{\rho\nu^2} = F\left(\frac{\tau_w d^2}{\rho\nu^2}\right)$$

Preston did a range of experiments in pipes with different sized pitot tubes and established that such a relation exists and determined the function F. Once determined this relation can then be used to determine the local skin friction in a boundary layer from measured values of $(P - p)$ and d. Subsequent workers showed that the relation for a boundary layer on a plate was slightly different from that in pipe flow. Thus, the latter was found by Preston to be

$$\log_{10}\left[\frac{\tau_w d^2}{4\rho\nu^2}\right] = -1.396 + 0.875 \log_{10}\left[\frac{(P-p)d^2}{4\rho\nu^2}\right] \quad (6.22)$$

whilst in reference 6.14 the former was determined as

$$\log_{10}\left[\frac{\tau_w d^2}{4\rho\nu^2}\right] = -1.366 + 0.877 \log_{10}\left[\frac{(P-p)d^2}{4\rho\nu^2}\right] \quad (6.23)$$

Because of the small uncertainties in the values of the constants involved the method cannot be relied upon to be in error by less than 5% (as is the case with the Clauser plot method) but it has been found to have wide applicability and its simplicity is a major asset.

Given the distributions of τ_t and the mean velocity (e.g. from hot wire measurements) we can use equation (6.10) to determine the corresponding distribution of ν_t across the boundary layer, including the outer region. It is found that

$$\frac{\nu_t}{u_e\delta^*} = g(z/\delta), \quad \text{say} \quad (6.24)$$

where the function g is found to be practically the same for a wide range of self-preserving boundary layers. The function increases with z/δ up to a maximum at about $z/\delta \simeq 0.3$ and then it falls slowly. In the logarithmic law of the wall region it is closely given by the linear relation [equation (6.11)]

$$\frac{\nu_t}{u_e\delta^*} = K\left(\frac{\delta u_\tau}{\delta^* u_e}\right)\frac{z}{\delta} \quad (6.25)$$

Outside the law of the wall region it is found[6.15] that $g \simeq$ const. γ_0 so that

$$\frac{\nu_t}{u_e\delta^*} \simeq 0.0168\,\gamma_0 \quad (6.26)$$

Here γ_0 is the intermittency function [see equation (6.7)]. Likewise we can use equations (6.16) and (6.17) to determine the distribution of l across the boundary layer. We find that within the logarithmic law region

$$l/\delta = Kz/\delta$$

as expected, but outside that region l/δ tends to a nearly constant value of 0.08 to 0.09 for self-preserving boundary layers. A good empirical fit is

$$l/\delta = 0.085 \tanh\left(\frac{Kz/\delta}{0.085}\right) \tag{6.27}$$

6.5 Velocity defect relations: self preserving boundary layers; law of the wake

In the absence of a streamwise pressure gradient, we have from equation (6.9) that in the outer region

$$(u_e - u)/u_\tau = f(z/\delta), \quad \text{say}$$

We may note that similarly for fully developed flow in a pipe of circular section (i.e. sufficiently downstream of the pipe entry for the boundary layer to extend to the pipe axis)

$$(u_c - u)/u_\tau = f_p(z/a), \quad \text{say}$$

Here f and f_p are functions to be determined by experiment, u_c is the velocity on the pipe axis and a is the pipe radius of cross section. Strictly we should include in f_p a pressure gradient parameter (u_τ/u_e) but its effect is found to be very small.

However, although the logarithmic law of the wall holds strictly in a thin region close to the wall it is found to be not a bad approximation for some distance outside that region in the absence of a pressure gradient. Indeed, it has sometimes been used as an acceptable description of the flow across a zero pressure gradient boundary layer for the purpose of determining integral quantities, but of course it does not satisfy the boundary conditions at the wall or at the outer edge of the boundary layer. These considerations led Karman[6.16] to suggest that the mean velocity profile outside the viscous sub-layer can be written for zero pressure gradient flows:

$$u/u_\tau = \frac{1}{K}\ln\,(zu_\tau/\nu) + \phi(z/\delta) \tag{6.28}$$

Here the function ϕ is to be determined from experiment but it takes the constant value B in the law of the wall region. Hence.

$$u_e/u_\tau = \frac{1}{K}\ln\,(\delta u_\tau/\nu) + \phi(1)$$

and so, in accordance with equation (6.9),

$$(u_e - u)/u_\tau = -\frac{1}{K}\ln\,(z/\delta) + \phi(1) - \phi(z/\delta) = f(z/\delta), \quad \text{say} \tag{6.29}$$

The available experimental data fit such a relation, although again there are small differences between the results obtained by different experimenters and we must note that the so-called constants involved may be slowly varying functions of the Reynolds number. Good empirical fits for the functions ϕ and f are:

$$\left.\begin{aligned} \phi(z/\delta) &= B + 2.8 \sin^3 (\pi z/2\delta) \\ \text{and} \quad f(z/\delta) &= 2.8[1 - \sin^3 (\pi z/2\delta)] - (1/0.41) \ln (z/\delta) \end{aligned}\right\} \tag{6.30}$$

It should be noted that this expression for f does not meet the requirement that $\partial u/\partial z = 0$ for $z = \delta$. However, an alternative expression for ϕ which does meet this requirement due to Gaudet[6.28] is given in Section 6.11, equation (6.80).

Consider now the more general case of an external non-uniform pressure, then a *self preserving*, or *equilibrium*, turbulent boundary layer can be defined as one for which the velocity defect ratio $(u_e - u)/u_\tau$ is the same function of z/δ for all x. From the above discussion it can be inferred that the zero pressure gradient case meets this requirement, although strictly it does so more closely as the Reynolds number increases.

There are various related parameters that are functions of the pressure distribution and their constancy can be shown to ensure self-preserving boundary layers. These include:[6.17]

$$\beta_p = (\delta^*/\tau_w)\mathrm{d}p/\mathrm{d}x \tag{6.31}$$

and the Clauser parameter[6.12]

$$G = \int_0^\infty \left(\frac{u_e - u}{u_\tau}\right)^2 \mathrm{d}z \bigg/ \int_0^\infty \left(\frac{u_e - u}{u_\tau}\right) \mathrm{d}z \tag{6.32}$$

It can be readily shown that

$$G = \frac{H - 1}{H} u_e/u_\tau \tag{6.33}$$

where H is the form parameter δ^*/θ.

Coles[6.18] analysed a considerable body of experimental data and concluded that a good fit to the mean velocity distributions was given by

$$u^+ = \frac{u}{u_\tau} = \frac{1}{K} \ln (z^+) + B + \frac{\Pi(x)}{K} W(z/\delta) \tag{6.34}$$

The function $W(z/\delta)$ is such that $W(0) = 0$ and $W(1) = 2$ and it is very close to the normalised velocity distribution in a half wake. It is often referred to as the wake function and equation (6.34) is called the *law of the wake*. This equation will therefore be seen to describe the boundary layer mean velocity distribution as a combination of the logarithmic law of the wall with a wake component which is a product of the wake function and the function $\Pi(x)$ which depends on the pressure distribution. A good fit to the function $W(z/\delta)$ is:

$$W(z/\delta) = 2 \sin^2 (\pi z/2\delta) = 1 - \cos (\pi z/\delta) \tag{6.35}$$

If we put $z = \delta$, it follows from equation (6.34) that

$$u_e/u_\tau = \frac{1}{K}\ln\delta^+ + B + 2\,\frac{\Pi(x)}{K}$$

so that
$$\Pi(x) = \frac{K}{2}\left[u_e/u_\tau - \frac{\ln\delta^+}{K} - B\right] \qquad (6.36)$$

where $\delta^+ = \delta u_\tau/\nu$. Hence, $\Pi(x)$ can be determined at any station x given u_τ and δ there. It can be shown that $\Pi(x)$ is a function of the Clauser parameter G so that if it is constant the boundary layer is self-preserving. For the zero pressure gradient, boundary layer $\Pi \simeq 0.55$ for values of $R_\theta > 5000$.

From equations (6.34) and (6.35) the corresponding velocity defect relation is

$$(u_e - u)/u_\tau = -\frac{1}{K}\ln(z/\delta) + \frac{\Pi}{K}\,[2 - W(z/\delta)]$$

A comparison of the defect velocity relation as predicted by equations (6.29) and (6.30) and by Coles' relation for the zero pressure gradient case is shown in Fig. 6.7. Some additional discussion of Coles' relation will be found in Section 6.7.

A simple uni-parametric, approximate relation suggested by Spence[6.19] from his analysis of some experimental data is

$$u/u_e = (z/\delta)^{(H-1)/2}$$

This relation has a family resemblance to the power law relations discussed in the next section, but here H is a function of the pressure gradient.

6.6 Power laws and skin friction relations for boundary layers in zero pressure gradient

From measurements in pipe flows it was noted[6.20] that reasonable overall fits to the mean velocity distributions were given by formulae of the type

$$u/u_c = (z/a)^{1/n}$$

where u_c was the velocity on the pipe axis, a the pipe radius and z the distance from the pipe wall. The index n that gave the best fit depended on the pipe Reynolds number, $u_c a/\nu$. The latter ranged from 2×10^3 to 1.6×10^6 and the corresponding values of n for best fit ranged from about 6 to 10, increasing with Reynolds number. For a pipe Reynolds number of about 10^6, $n = 7$ was the preferred value.

This led to examinations of the applicability of such power laws to the mean velocity profiles in boundary layers on flat plates in zero pressure

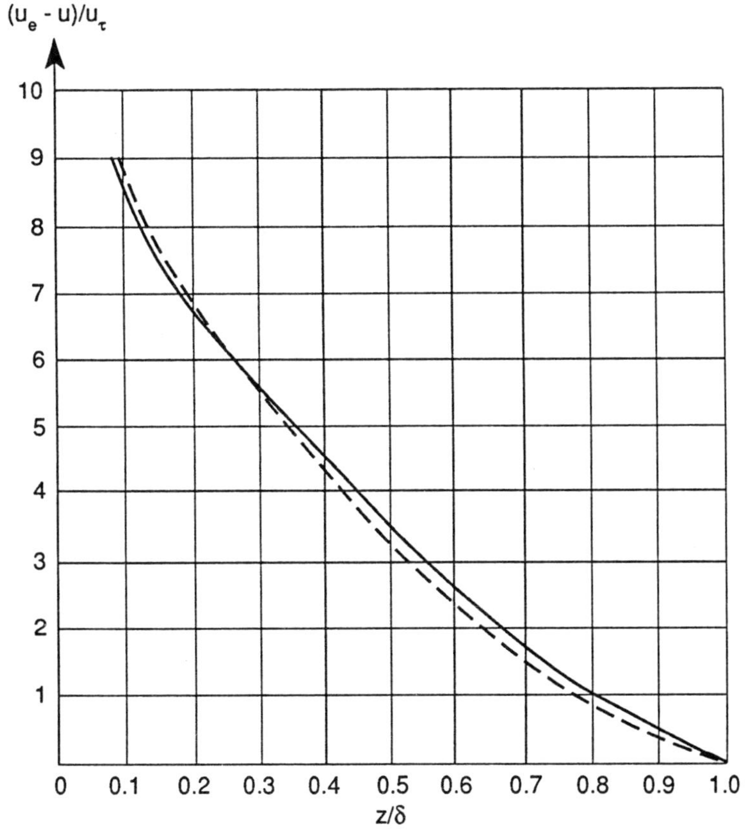

Fig. 6.7 Velocity defect relations for the boundary layer on a flat plate at zero incidence.
——— equations (6.29) and (6.30)
- - - - - Coles' relation, equation (6.34)

gradient. It was indeed found that the agreement was very acceptable, but again the best choice of n depended on the local Reynolds number, now defined by $u_e x/\nu$, where x is the streamwise distance from the leading edge. Thus, when this Reynolds number was between 5×10^5 and 10^7 a value of $n = 7$ gave a good fit, whilst for Reynolds numbers between 10^6 and 10^8 a value of $n = 9$ was acceptable. For a boundary layer a power law of the above type is consistent with the law of the wall in the form:

$$u/u_\tau = C_1(zu_\tau/\nu)^{1/n} \tag{6.37}$$

where the constant C_1 depends on n and therefore on the Reynolds number range of interest.

The pipe flow results indicated that for $n = 7$ the best value of C_1 was 8.74 whilst for $n = 9$ the best value was 10.6. These values also provide reasonable agreement with boundary layer measurements but small empirical adjustments of them give even closer agreement. However, again one must note that $\partial u/\partial z$ is incorrectly predicted at the surface and at the outer edge of the boundary layer and therefore power laws are best used for the determination of integral quantities.

It follows from equation (6.37) that

$$u_e/u_\tau = C_1(u_\tau\delta/\nu)^{1/n} = C_1^{n/(n+1)}\,(u_e\delta/\nu)^{1/(n+1)} \tag{6.38}$$

It will be recalled (Section 2.11) that the momentum integral equation in the form of equation (2.50) applies to both turbulent and laminar boundary layers. For zero pressure gradient this equation reduces to

$$\mathrm{d}\theta/\mathrm{d}x = \tau_w/\rho u_e^2 = (u_\tau/u_e)^2 \tag{6.39}$$

where $$\theta = \int_0^\delta \frac{u}{u_e}\left(1 - \frac{u}{u_e}\right)\mathrm{d}z = \delta\int_0^1 \eta^{1/n}\,(1 - \eta^{1/n})\mathrm{d}\eta$$
$$= \delta\{n/[(n+1)(n+2)]\} \tag{6.40}$$

Here $\eta = z/\delta$. From equations (6.38) to (6.40) it follows that

$$\left(\frac{u_\tau}{u_e}\right)^2 = \frac{\mathrm{d}\theta}{\mathrm{d}x} = \frac{\mathrm{d}\delta}{\mathrm{d}x}\,\frac{n}{(n+1)(n+2)} = C_1^{-2n/(n+1)}\left(\frac{u_e\delta}{\nu}\right)^{-2/(n+1)}$$

and on integrating for δ with respect to x we obtain

$$\delta/x = C_2 R_x^{-2/(n+3)} \tag{6.41}$$

where $$C_2 = \left[\frac{(n+2)(n+3)}{n}\,C_1^{-2n/(n+1)}\right]^{(n+1)/(n+3)}$$

and $R_x = u_e x/\nu$, and we have assumed $\delta = 0$ when $x = 0$.

If follows that the local skin friction coefficient is

$$c_f = 2\tau_w/\rho u_e^2 = 2(u_\tau/u_e)^2 = C_3\,R_x^{-2/(n+3)} \tag{6.42}$$

where $$C_3 = 2C_1^{-2n/(n+1)}\,C_2^{-2/(n+1)} = 2C_2 n/[(n+2)(n+3)]$$

We can deduce that $$\theta/x = C_4\,R_x^{-2/(n+1)} \tag{6.43}$$

where $$C_4 = \frac{C_2 n}{(n+1)(n+2)} = \frac{C_3(n+3)}{2(n+1)}$$

We can determine the overall skin friction coefficient for one side of a flat plate of chord c with fully turbulent boundary layer as

$$C_F = \frac{1}{c}\int_0^c c_f\,.\,\mathrm{d}x = \frac{n+3}{n+1}\,C_3\,R^{-2/(n+3)} = 2C_4\,R^{-2/(n+3)} \tag{6.44}$$

where $R = u_e c/\nu$ is the overall Reynolds number.

Finally, we note from equation (6.43) that

$$R_\theta = u_e\theta/\nu = C_4\ R_x^{(n+1)/(n+3)}$$

and therefore we can relate R_θ and c_f, using equation (6.42), thus

$$c_f = K'\ R_\theta^{-2/(n+1)} \tag{6.45}$$

where $K' = C_3\ C_4^{2/(n+1)} = [(n+3)/(2n+2)]^{2/(n+1)}\ C_3^{(n+3)/(n+1)}$

$$= \frac{2(n+1)}{n+3}\ C_4^{(n+3)/(n+1)}$$

If we take $n = 7$ we find good empirical fits in accordance with the form of these relations are

$$\left.\begin{aligned} &\delta/x = 0.37R_x^{-1/5}, \quad \theta/x = 0.037R_x^{-1/5}, \quad c_f = 0.0592R_x^{-1/5} \\ &C_F = 0.074R^{-1/5}, \quad c_f = 0.026R_\theta^{-1/4} \end{aligned}\right\} \tag{6.46}$$

Here the constants differ a little from those predicted using the pipe flow value of C_1 for $n = 7$ and give slightly better agreement with experimental results.

For $n = 9$ good fits are given by

$$\left.\begin{aligned} &\delta/x = 0.27R_x^{-1/6}, \quad \theta/x = 0.023R_x^{-1/6}, \quad c_f = 0.0375R_x^{-1/6} \\ &C_F = 0.045R^{-1/6}, \quad c_f = 0.0176R_\theta^{-1/5} \end{aligned}\right\} \tag{6.47}$$

It will be seen that even when R_x is a modest 10^6 (flight values can range up to 5×10^8) δ/x is about 0.03. This illustrates the fact that though the turbulent boundary layer is thicker than the corresponding laminar boundary layer it is still quite thin and this justifies *a posteriori* the applications of the associated assumptions of boundary layer theory. The above skin friction relations are illustrated in Fig. 6.8.

6.7 Alternative deductions from velocity defect relations

Von Karman's extension of the logarithmic law of the wall in the form of a velocity defect relation in the case of zero pressure gradient flow over a flat plate [equation (6.29)] was similarly combined by him with the momentum integral equation (6.39) and after some algebra and approximation it yielded relations of the form[6.16]

$$R_x = K_1\zeta^2 \exp\left[\frac{\zeta - \phi(1)}{A}\right] \tag{6.48}$$

$$R_\theta = \left(K_1 - \frac{K_2}{\zeta}\right) \exp\left[\frac{\zeta - \phi(1)}{A}\right] \tag{6.49}$$

where $\zeta = u_e/u_\tau = (2/c_f)^{1/2}, \quad K_1 = \int_0^1 f(\eta)\mathrm{d}\eta$

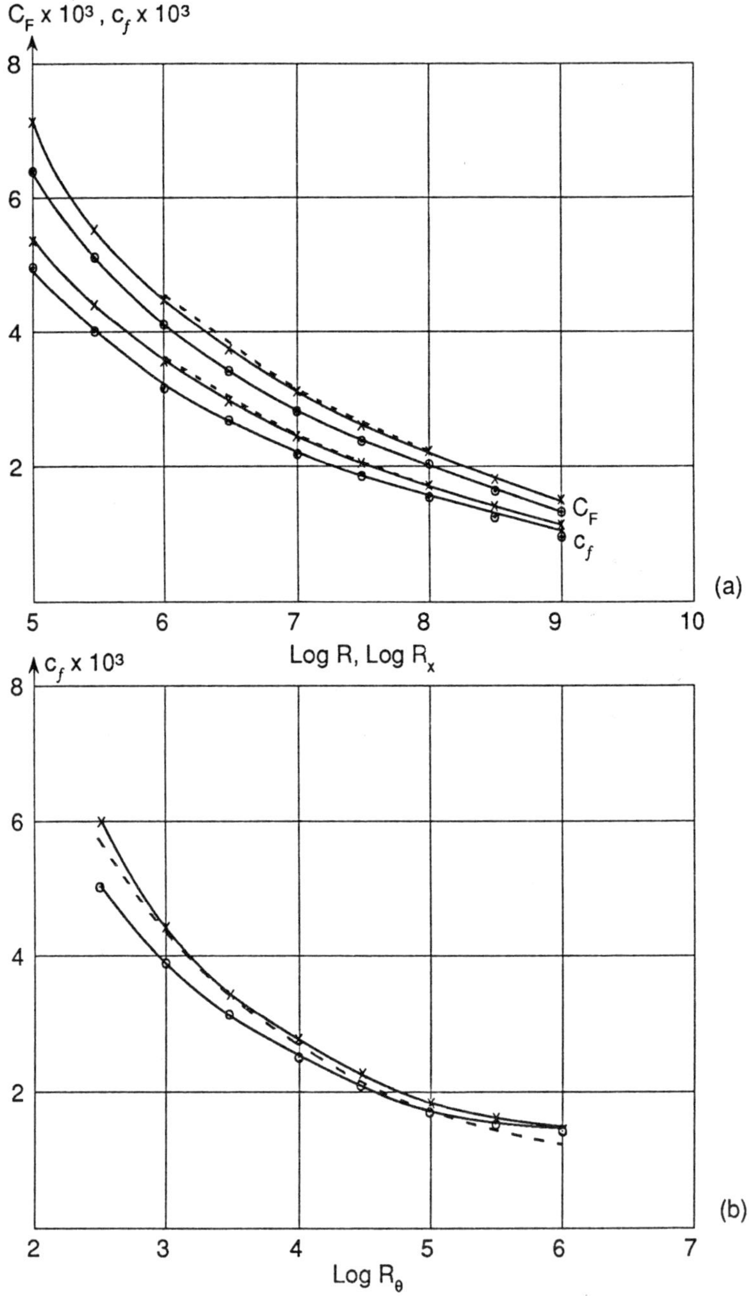

Fig. 6.8 Turbulent boundary layer on flat plate at zero incidence: (a) C_F against log R and c_f against log R_x; (b) c_f against log R_θ.

×——× Prandtl–Schlichting relations ⊙——⊙ Coles' relations

– – – – – 1/9th power law

$$K_2 = \int_0^1 f^2(\eta)\mathrm{d}\eta, \quad A = 1/K$$

Hence $\quad \zeta = \phi(1) + A \ln (R_x/K_1\zeta^2)$

or $\quad c_f^{-1/2} = A' + B' \log_{10}(R_x c_f) \qquad (6.50)$

where $\quad A' = [\phi(1) - A \ln (2K_1)]/\sqrt{2}, \quad B' = (A \ln 10)/\sqrt{2}$

It was found, however, that the best fits to the available experimental results were obtained by taking $A' = 1.7$ and $B' = 4.15$.

Working on similar lines for the overall skin friction coefficient Schoenherr[6.21] obtained the formula

$$C_F^{-1/2} = 4.13 \log_{10}(R\, C_F) \qquad (6.51)$$

Equations (6.50) and (6.51) are not very convenient for direct use so Prandtl[6.22] evaluated the values of c_f, C_F, and R_θ for values of R ranging from about 10^5 to 1.5×10^9 and a comprehensive range of values of x/c. Subsequently, Schlichting fitted these results with the following convenient interpolation formulae:

$$c_f = (2 \log_{10} R_x - 0.65)^{-2.3} \qquad (6.52)$$

$$C_F = 0.455/(\log_{10} R)^{2.58} \qquad (6.53)$$

Equations (6.52) and (6.53) are often referred to as the *Prandtl–Schlichting relations*. They are widely accepted because they have been found to be in good accord with experimental data acquired over many years since their formulation.

Likewise, relations for c_f and C_F as functions of the appropriate Reynolds number can be deduced from Coles' form of the velocity defect relation [equation (6.34)]. Comparisons with the relations derived from the 1/9 power law and the Prandtl–Schlichting relations are illustrated in Fig. 6.8.

Since ζ is of the order of 20–30 the expression for R_θ in equation (6.49) can be approximated by a relation of the form:

$$R_\theta = C' \exp(D'\zeta) \qquad (6.54)$$

Squire and Young[6.23] determined the constants C' and D' to give the best fit to the Prandtl–Schlichting relation for C_F over a range of Reynolds numbers from 10^6 to 10^8; these were

$$C' = 0.2454 \qquad D' = 0.3914$$

We note that $\quad \delta^*/\delta = \int_0^1 (1 - u/u_e)\mathrm{d}\eta$

$$= \frac{1}{\zeta}\int_0^1 f\,\mathrm{d}\eta = K_1/\zeta$$

where $\eta = z/\delta$

Also $\quad \theta/\delta = \int_0^1 \frac{u}{u_e}\left(1 - \frac{u}{u_e}\right)d\eta = K_1/\zeta - K_2/\zeta^2$

Hence the form parameter

$$H = \delta^*/\theta = 1/[1 - K_2/(K_1\zeta)] \tag{6.55}$$

Coles[6.24] has evaluated H for a range of values of R_θ and his results are shown in Table 6.1. We see that H is a slowly decreasing function of R and we note that the more approximate 1/7 and 1/9 power laws give constant values for H of 1.29 and 1.22, respectively. [The empirical analysis of Winter and Gaudet referred to in Section 6.10 shows the same trends but their results differ from those of Table 6.1 by a few percent in the value of H for a given R_θ (see Fig. 6.12).]

6.8 The drag of a smooth flat plate at zero incidence with a partly laminar and partly turbulent boundary layer

From the momentum integral equation (2.50) it is evident that θ must be a continuous function of x since any discontinuity in θ implies an infinite shear stress there to balance the infinite value of $d\theta/dx$. This requirement applies, in particular, to the transition region, and if we assume that transition from laminar to turbulent flow occurs at a point then the value of θ at the beginning of the turbulent boundary layer must be the same as that at the end of the laminar boundary layer.

Consider a boundary layer on a flat plate at zero incidence with the transition point T at $x = x_t$ aft of the leading edge, as illustrated in Fig. 6.9. We assume that the turbulent boundary layer develops aft of T as if it had started at some fictitious leading edge position O′ in the fully turbulent state such that $O'T = x_0$, say. We must now equate the value of θ for the fictitious turbulent boundary layer at T with the value there for

Table 6.1

R	R_θ (approx.)	H
10^5	3.3×10^2	1.56
10^6	2.2×10^3	1.41
10^7	1.45×10^4	1.32
10^8	1.12×10^5	1.27
10^9	7.1×10^5	1.23

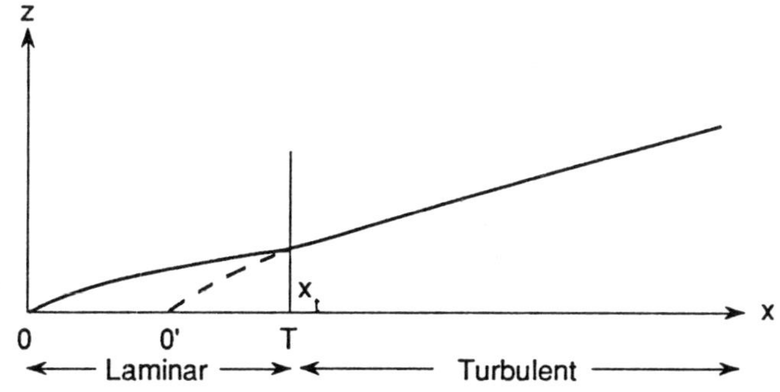

Fig. 6.9 Sketch of partly laminar and partly turbulent boundary layer.

the real laminar boundary layer starting at the leading edge O. From equation (3.14) the latter is given by

$$\theta_t = 0.664(u_e x_t/\nu)^{-1/2}\, x_t \qquad (6.56)$$

By way of illustration we shall assume for the turbulent boundary layer a 1/7 power law mean velocity distribution so that [see equation (6.46)]

$$\theta/x = 0.037 R_{\bar{x}}^{-1/5} \qquad (6.57)$$

where $\bar{x} = x - x_t + x_0$. Hence

$$\theta_t = 0.037 x_0 (u_e x_0/\nu)^{-1/5}$$

Therefore from equation (6.56)

$$\begin{aligned} x_0/c &= [0.664(x_t/c)^{1/2} R^{-1/2}/0.037]^{5/4}\, R^{1/4} \\ &= 36.9(x_t/c)^{5/8}\, R^{-3/8} \end{aligned} \qquad (6.58)$$

Also, since $\mathrm{d}\theta/\mathrm{d}x = c_f/2$, it follows that for a flat plate of chord c the overall skin friction coefficient

$$C_F = \frac{1}{c}\int_0^c c_f\, \mathrm{d}x = 2\theta_c/c \qquad (6.59)$$

where θ_c is the value of θ at the trailing edge ($x = c$). Hence from equation (6.57)

$$C_F = 0.074 R^{-1/5}[1 - (x_t - x_0)/c]^{4/5} \qquad (6.60)$$

where $R = u_e c/\nu$.

We can therefore determine C_F from equations (6.58) and (6.59), given the transition point position x_t and the Reynolds number R. For example, if $x_t/c = 0.4$ and $R = 10^7$ we see from equation (6.58) that

$$x_0/c = 36.9(0.4)^{5/8}\, 10^{-21/8} = 0.049$$

and from equation (6.60) $C_F = 0.074\ 10^{-2/5}\ (0.649)^{4/5} = 0.00208$.

The corresponding value of C_F for a fully turbulent boundary layer is 0.00295 so that the presence of the laminar part of the boundary layer from the leading edge to $0.4c$ has reduced the skin friction drag by about 30%.

The local skin friction coefficient is given by [see equations (3.10) and (6.46)]

$$c_f = 0.664 R_x^{-1/2}, \qquad \text{for } x < x_t$$

and $$c_f = 0.0592 R^{-1/5}\{[x - (x_t - x_0)]/c\}^{-1/5}, \quad \text{for } x > x_t$$

In principle, the same approach can be adopted using one of the velocity defect relations for the mean velocity distribution in the boundary layer. The algebra is somewhat more involved, and need not be reproduced here but the results differ very little from those arrived at above.

6.9 Two parameter relations

Relations between local quantities such as c_f and θ (or R_θ), as for example equations (6.45) and (6.54), can be of value for predicting the boundary layer development in cases of non-uniform pressure with moderate and smoothly changing pressure gradients, providing that u_e in these quantities is taken to be the local free stream velocity. This is because such relations are relatively insensitive to moderate streamwise pressure variations when the boundary layers are close to equilibrium.

More generally, we can expect even greater accuracy by making use of a two parameter relation that enables us to relate the local values of c_f, θ and H. For example, the Ludwieg–Tillmann relation[6.25] was based on the analysis of a series of experiments in the boundary layer on a flat plate in the presence of non-uniform pressure distributions in which the skin friction was inferred from heat transfer measurements using heated elements. This relation is

$$c_{fe} = 2\tau_w/\rho u_e^2 = 0.246 R_\theta^{-0.268}\ 10^{-0.678H} \tag{6.61}$$

It is found to be in reasonable agreement with the available results for boundary layers not close to separation, particularly for self-preserving boundary layers. However, we note that it predicts separation ($c_f = 0$) when $H = \infty$; this is unrealistic, it is found experimentally that the value of H at separation is not strictly constant but tends to lie in the range 2.5–3.0. It must also be noted that the above relation is based on data for which $R_\theta \leqslant 4 \times 10^4$, which corresponds to $R_x < 3 \times 10^7$ approx., and $H < 2$. Its validity for larger values of R_x and H has not been adequately explored.

Coles' relation (Section 6.5) is also a two parameter relation. Thus, from it integral quantities such as δ^*/δ, θ/δ and H can be determined in terms of $u_\tau\delta/\nu$ and Π, and hence c_f can be determined as a function of R_θ and H.

Another two parameter relation is that due to Sarnecki[6.26] later adapted by Thompson.[6.27] Sarnecki sugested that the mean velocity profile could be regarded as a combination of the external potential flow (u_p) and fully turbulent flow (u_t) weighted by the intermittency factor (γ). Thus, he postulated

$$u = \gamma u_t + (1 - \gamma)u_p$$

with $\quad u_t = u_\tau(5.4 + 5.7 \log_{10} z^+), \quad u_p = u_e$

However, the chosen distribution of γ as a function of distance from the wall was somewhat different from that of Klebanoff (see Section 6.3). Thompson did some extensive calculations of velocity profiles for wide ranges of the parameters u_τ and γ and presented charts of u/u_e as a function of z/θ as well as of c_f for useful ranges of H and R_θ.

More recently Gaudet[6.28] has adapted a variant of Coles' relation in which the wake term has been modified to satisfy the condition of zero velocity gradient at the edge of the boundary layer. Fig. 6.10 compares

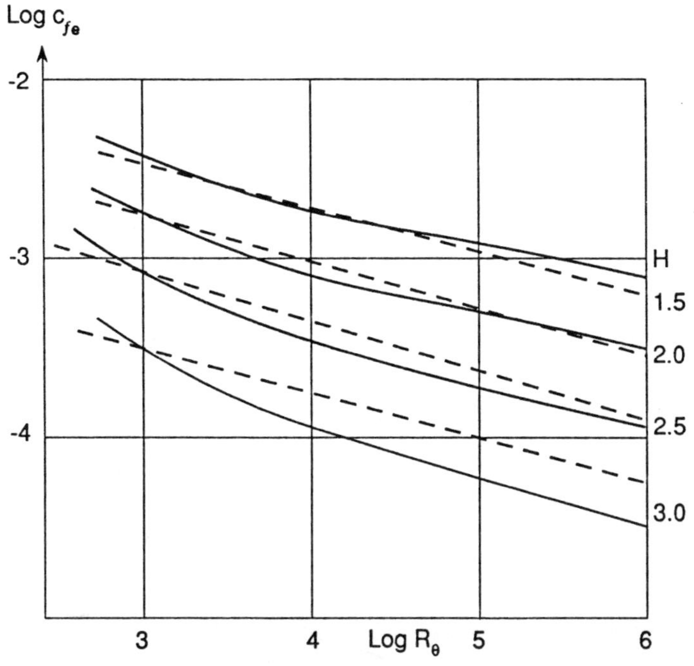

Fig. 6.10 c_{fe} against R_θ for various H.
——— Gaudet[6.23] ----- Ludwieg and Tillmann[6.25]

the values of c_{fe} as a function of R_θ for various values of H characteristic of an adverse pressure gradient ($H \geqslant 1.5$) as determined by the Ludwieg–Tillmann and Gaudet relations. Results for the Thompson relation tend to lie between the two shown. It will be seen that for $H < 2.5$ the three methods agree fairly well over the whole R_θ range considered, but for $H = 3.0$ they diverge markedly particularly for the higher values of R_θ. We have noted that the Ludwieg–Tillmann relation was based on data for which $H < 2$ and $R_\theta < 4 \times 10^4$ so results predicted by it outside these ranges must be viewed with some reserve. All three methods can be expected to agree fairly well for the smaller values of H characteristic of favourable and small adverse pressure gradients and for values of $R_\theta > 10^3$, as much of the data on which they were based are within those ranges.

However, the assumption in all such methods that the local flow is determined by the local values of two parameters implies that the boundary layer is never far from self-preserving (or equilibrium) and this is not the case when conditions are changing rapidly with downstream distance (x) and lag effects in the response of the turbulence to these conditions become significant.

6.10 Effect of compressibility on the boundary layer temperature distribution; the Crocco relation

Equation (2.61) relates the total temperature in a laminar boundary layer in zero pressure gradient and with the Prandtl number $\sigma = 1$ to the velocity distribution. It will be shown (Section 7.3) that this relation, namely

$$T_H = T[1 + \tfrac{1}{2}(\gamma - 1)M^2] = K_1 u + K_2, \quad \text{say}$$

applies to a turbulent boundary layer with the same provisos, where T and u are mean quantities. The constants K_1 and K_2 are determined by the boundary conditions. Thus with $T = T_w$ at $z = 0$, where $u = 0$ it follows that $K_2 = T_w$ and so

$$T/T_e = T_w/T_e + K_1(u/u_e)(u_e/T_e) - u^2/(2c_p T_e)$$

since $a^2 = (\gamma - 1)c_p T$. For zero heat transfer $(\partial T/\partial z)_w = 0$ and then $K_1 = 0$ and $T_w = T_r$, where T_r is the recovery temperature, so that

$$T_r/T_e = 1 + \tfrac{1}{2}(\gamma - 1)M_e^2$$

With heat transfer we see that at the outer edge of the boundary layer

$$1 = (T_w/T_e) + (K_1 u_e/T_e) - \tfrac{1}{2}M_e^2(\gamma - 1)$$

or $$K_1 u_e/T_e = 1 - (T_w/T_e) + \tfrac{1}{2}M_e^2(\gamma - 1) = (T_r - T_w)/T_e$$

Hence $$\frac{T}{T_e} = \frac{T_w}{T_e} + \frac{T_r - T_w}{T_\theta}\frac{u}{u_e} - \frac{T_r - T_e}{T_e}\left(\frac{u}{u_e}\right)^2 \quad (6.62)$$

This is often referred to as *Crocco's relation*. Now if $\sigma \neq 1$ then experiment and theory[6.29] lend some support to the relation

$$T_r/T_e = 1 + \tfrac{1}{2}(\gamma - 1)M_e^2\, r \quad (6.63)$$

where r, the recovery factory, is often taken as $= \sigma^{1/3} = 0.896$ for air. It has therefore been postulated that equation (6.62) holds for σ near unity so that

$$T/T_e = T_w(1 - u/u_e) + [1 + \tfrac{1}{2}r(\gamma - 1)M_e^2]u/u_e - \tfrac{1}{2}r(\gamma - 1)M_e^2u^2/u_e^2 \quad (6.64)$$

This is known as the *modified Crocco relation*.

The careful analysis of the available experimental results by Fernholz and Finley[6.29, 6.30] lends some support to this relation although it must be stressed that static temperature cannot be measured directly, it must be deduced from the measurement of other quantities, and the interpretation of the experimental data in a boundary layer is difficult and the accuracy can be low. We will have more to say about this later.

6.11 The effects of compressibility on skin friction, heat transfer and velocity profiles for zero pressure gradient

The work of Fernholz and Finley referred to above provides a comprehensive survey and analysis of the then available data on turbulent boundary layers in two dimensional compressible flow. In what follows we shall concentrate on the main integral boundary layer characteristics for the case of zero pressure gradient and highlight the major points of practical importance to be culled from their work and the other available literature.

We saw in Section 3.3.5 that for the laminar boundary layer there was an intermediate temperature T_m given by

$$T_m = T_e[0.45 + 0.55(T_w/T_e) + 0.09(\gamma - 1)M_e^2\, \sigma^{1/2}]$$

such that the relations between c_{fm} and R_{xm} and between C_{Fm} and R_m were formally the same as in incompressible flow. Here

$$c_{fm} = 2\tau_w/(\rho_m u_e^2), \quad R_{xm} = u_e x/\nu_m, \quad C_{Fm} = \frac{1}{c}\int_0^c c_{fm}\, dx, \quad R_m = u_e c/\nu_m$$

and suffix m denotes quantities evaluated at the temperature T_m. We find empirically that for the turbulent boundary layer on a flat plate with zero pressure gradient this concept of an intermediate temperature can also be applied with acceptable accuracy at least for main stream Mach numbers less than about 3. The formulae for T_m suggested by various authors differ a little in the constants involved, one that is widely used is that due to Sommer and Short[6.31], namely

$$T_m = T_e[0.55 + 0.45(T_w/T_e) + 0.2(T_r - T_e)/T_e] \tag{6.65}$$

where the recovery temperature T_r is given by equation (6.63). Hence

$$\left.\begin{aligned} T_m/T_e &= 0.55 + 0.45(T_w/T_e) + 0.1(\gamma - 1)M_e^2\,\sigma^{1/3} \\ &= 0.55 + 0.45(T_w/T_e) + 0.04M_e^2\,\sigma^{1/3}, \quad \text{with } \gamma = 1.4 \end{aligned}\right\} \tag{6.66}$$

This equation can be used in combination with any of the accepted incompressible skin friction/Reynolds number relations, such as those discussed in Sections 6.6 and 6.7, to determine corresponding compressible flow relations.

For example, if we accept the 1/7th power law relations of equation (6.46) then we infer

$$c_{fm} = 0.0592R_{xm}^{-1/5}$$

so that $\quad c_f\rho_e/\rho_m = 0.0592R_x^{-1/5}(\rho_m/\rho_e)^{-1/5}\,(\mu_e/\mu_m)^{-1/5}$

or $\quad c_f = 0.0592R_x^{-1/5}(T_e/T_m)^{(4-\omega)/5} \qquad (6.67)$

where we have made use of the relations $\rho \propto 1/T$ and $\mu \propto T^{\omega}$. If we take $\omega = 0.89$ (as for air at moderate temperatures) then we have

$$c_f = c_{fi}(T_e/T_m)^{0.622} \tag{6.68}$$

where suffix i denotes quantities at the same Reynolds number in incompressible flow. Similarly, we find

$$C_F = C_{Fi}(T_e/T_m)^{0.622} \tag{6.69}$$

The corresponding relations for $n = 9$ are

$$c_f/c_{fi} = C_F/C_{Fi} = (T_e/T_m)^{0.685} \tag{6.70}$$

For zero heat transfer $T_w = T_r$ and then

$$\begin{aligned} T_m = 0.35T_e + 0.65T_r &= T_e[1 + 0.325(\gamma - 1)M_e^2\,\sigma^{1/3}] \\ &= T_e[1 + 0.1165M_e^2] \end{aligned} \tag{6.71}$$

for $\gamma = 1.4$ and $\sigma = 0.72$. A plot of $c_f/c_{fi} = C_F/C_{Fi}$, for $n = 9$, $\omega = 0.89$ and $\sigma = 0.72$ for ranges of values of M_e and T_w/T_e, as well as for zero heat transfer, is presented in Fig. 6.11. It is as well to recall that the best choice of n depends on the Reynolds number range of interest, $n = 9$ applies to the range from about 10^6 to 10^8. However, we can infer as a general result that any increase in T_m, either due to an increase in wall temperature or an increase in free stream Mach number, results in a reduction of c_f. This is largely due to the associated reduction in density near the wall and a thickening of the viscous sub-layer.

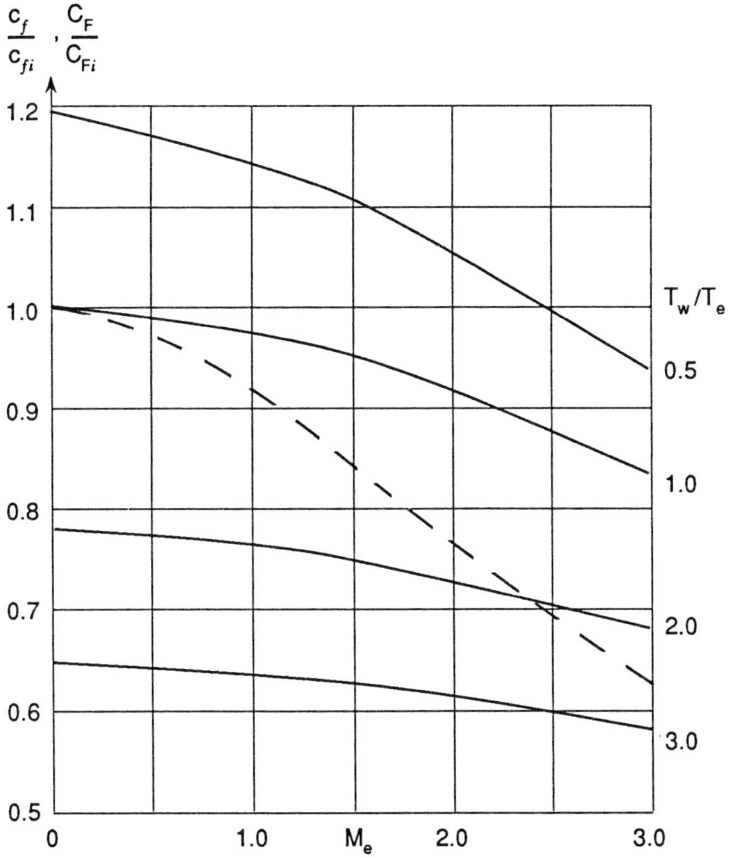

Fig. 6.11 Effects of Mach number and wall temperature on the skin friction of a flat plate at zero incidence based on the 1/9th power law and mean temperature concept; $\sigma = 0.72$.
——— T_w/T_e = const. - - - - - zero heat transfer

For the heat transfer we recall the definitions of the local Stanton number and of the Nusselt number of Section 3.3.5, viz.

$$\mathrm{St} = q/[\rho_e u_e c_p(T_w - T_r)], \qquad \mathrm{Nu} = Q/[k_e(T_w - T_r]$$

where q = local rate of heat transfer per unit area, and Q is the rate of heat transfer per unit span from one side of a plate of chord c. Then theory shows that for laminar flow [equations (3.65) and (3.67)]

$$\mathrm{St} = \tfrac{1}{2}c_f\,\sigma^{-2/3}, \qquad \mathrm{Nu} = \tfrac{1}{2}C_F\,\sigma^{1/3}\,\mathcal{R}$$

Here we have dropped l to denote the local Stanton number since we are not now considering the overall Stanton number of Section 3.2.3.

For the turbulent boundary layer a similar argument based on the analogy between heat and momentum transfer by the turbulent eddying motion – the so-called Reynolds analogy – results in relations of the form

$$\mathrm{St} = \tfrac{1}{2}c_f/s, \qquad \mathrm{Nu} = \tfrac{1}{2}C_F\,\mathcal{R}\sigma/s \tag{6.72}$$

where s is a factor – the Reynolds analogy factor – that depends largely on the turbulent Prandtl number, i.e. $\sigma_t = \mu_t c_p/k_t$, such that s will be very close to unity if σ_t and σ are close to unity. In the absence of a pressure gradient we may expect σ_t to be much closer to unity than σ, because of the dominance of the turbulent eddy transfer mechanism in imposing a strong similarity for the rates of exchanges of momentum and heat. The available experimental data are somewhat scattered reflecting the experimental difficulties, but it seems that for air no great error is incurred by assuming $s = 1$ in equation (6.72). We readily deduce from equation (6.72) that if the mean temperature concept applies in conjunction with the power law velocity distributions then:

$$\left.\begin{aligned}\mathrm{St}/\mathrm{Sti} = \mathrm{Nu}/\mathrm{Nui} = c_f/c_{fi} = C_F/C_{Fi} &= (T_e/T_m)^{0.622} \quad \text{for } n = 7\\ &= (T_e/T_m)^{0.685} \quad \text{for } n = 9\end{aligned}\right\} \tag{6.73}$$

It must be emphasised that in view of the simplified nature of the above analysis the probable error of the derived relations is unlikely to be better than $\pm$ 10%, and then only if n is matched to the Reynolds number.

A detailed statistical analysis of the results of a number of the then existing predictive methods as well as experimental data for zero pressure gradient turbulent boundary layers led Spalding and Chi[6.32] in 1964 to formulate a set of interesting and simple relations in the form:

$$c_f\,.\,F_c = f_1(R_x\,.\,F_{R_x}), \qquad C_F\,.\,F_c = f_2(R\,.\,F_{R_c}), \qquad c_f\,.\,F_c = f_3(R_\theta\,.\,F_{R_\theta}) \tag{6.74}$$

where F_c, F_{R_x} and F_{R_θ} are empirically derived tabulated functions of M_e and T_w/T_e, $F_{R_c} = F_{R_x}$ for $x = c$, and $F_{R_\theta} = F_{R_x}\,.\,F_c$. The basic functions f_1, f_2 and f_3 can be determined from Table 6.2 and some sample values of F_c and F_{R_θ} are presented in Table 6.3. In incompressible flow and with $T_w = T_e$ the basic skin friction Reynolds number relations that result lie between the Prandtl–Schlichting and the Coles relations shown in Fig. 6.8, but somewhat closer to the former. The standard deviation from the data on which the relations are based is quoted as about 10–11%; this, however, reflects the unknown experimental errors in the data and the overall accuracy is probably better than this figure suggests. Thus, if one is given M_e and T_w/T_e one can determine $F_{R_c} = F_{R_x}$ and F_c from Table 6.3, and given $R_c = R_x$, then from the product $F_{R_x}\,.\,R_x$ and Table 6.2 one can obtain $F_c\,.\,C_F$ and hence C_F.

An alternative but later approach on not dissimilar lines is that of Winter and Gaudet[6.33] subsequently improved by Gaudet.[6.34] Their work

Table 6.2

$F_c \cdot c_f$	$F_c \cdot C_F$	$F_{R_\theta} \cdot R_\theta$	$F_{R_x} \cdot R_x$
0.0090	0.01409	70.91	1.006×10^4
0.0055	0.007345	319.4	8.697×10^4
0.0050	0.006526	462.3	1.417×10^5
0.0045	0.005747	716.0	2.492×10^5
0.0040	0.005006	1208	4.828×10^5
0.0035	0.004299	2283	1.062×10^6
0.0030	0.003621	5030	2.778×10^6
0.0025	0.002967	13860	9.340×10^6
0.0020	0.002333	5.245×10^4	4.651×10^7
0.0015	0.001716	3.955×10^5	4.610×10^8
0.0010	0.001117	2.878×10^7	5.758×10^{10}

Table 6.3

T_w/T_e	M_e			
	0	1	2	3
			F_c	
0.5	0.7286	0.7580	0.8446	0.9839
1.0	1.000	1.0295	1.1167	1.2581
3.0	1.866	1.8956	1.9836	2.1278
			F_{R_θ}	
0.5	2.779	3.1524	4.2071	5.8121
1.0	1.000	1.1348	1.5141	2.0923
3.0	0.1980	0.2247	0.2999	0.4143
			F_{R_x}	
0.5	3.8142	4.1588	4.9811	5.9072
1.0	1.000	1.1058	1.3559	1.6631
3.0	0.1061	0.1185	0.1512	0.1947

was confined to the case of zero heat transfer and zero pressure gradient. A key concept is that of kinematic quantities denoted by the index i. Thus

$$\left.\begin{aligned} \delta^{*i} &= \int_0^\delta \left(1 - \frac{u}{u_e}\right)\mathrm{d}z, \quad \theta^i = \int_0^\delta \frac{u}{u_e}\left(1 - \frac{u}{u_e}\right)\mathrm{d}z \\ \delta_E^i &= \int_0^\delta \frac{u}{u_e}\left(1 - \frac{u^2}{u_e^2}\right)\mathrm{d}z, \quad H^i = \delta^{*i}/\theta^i \end{aligned}\right\} \tag{6.75}$$

It will be seen that the kinematic quantities differ from the corresponding physical ones of compressible flow (see Section 2.10) insofar as they do not involve density or temperature variations, but in incompressible flow (zero Mach number and zero heat transfer) kinematic and physical quantities coincide. Winter and Gaudet's analysis of their own fairly extensive and careful experimental results over a range of Mach numbers up to about 3, as well as the results of other workers, showed that H^i could be regarded as a unique function of R^i_θ $(= u_e\theta^i/\nu_e)$ independent of Mach number. This function is illustrated in Fig. 6.12. Their analysis of experimental velocity profiles led them to suggest that the Von Karman profile [equation (6.28)] applied for the kinematic quantities in the form:

$$u/u^i_\tau = \frac{1}{K}\ln\,(zu^i_\tau/\nu_e) + \phi(z/\delta) \qquad (6.76)$$

where $u^i_\tau = (c^i_f/2)^{1/2}\,u_e$

c^i_f is a kinematic quantity to be determined. Thus, with the aid of the momentum integral equation this leads, as in Section 6.7, to relations between c^i_f and R^i_x or R^i_θ and hence a relation can be determined between

$$C^i_F = (1/c)\int_0^c c^i_f\,\mathrm{d}x \quad \text{and} \quad R^i_c \qquad (6.77)$$

These relations are the same as in incompressible flow. It is then postulated and confirmed that the effects of compressibility can be accounted for by the introduction of simple factors f_c, f_x and f_θ which are functions of M_e only, and which relate actual and kinematic quantities as follows:

$$c^i_f = f_c\,.\,c_f, \quad C^i_F = f_c\,.\,C_F, \quad R^i_\theta = f_\theta R_\theta, \quad R^i_x = f_x\,.\,R_x \qquad (6.78)$$

Here $f_x = f_\theta/f_c$.

Winter and Gaudet's analysis, subsequently modified by Gaudet, resulted in the following empirical forms for the compressibility factors:

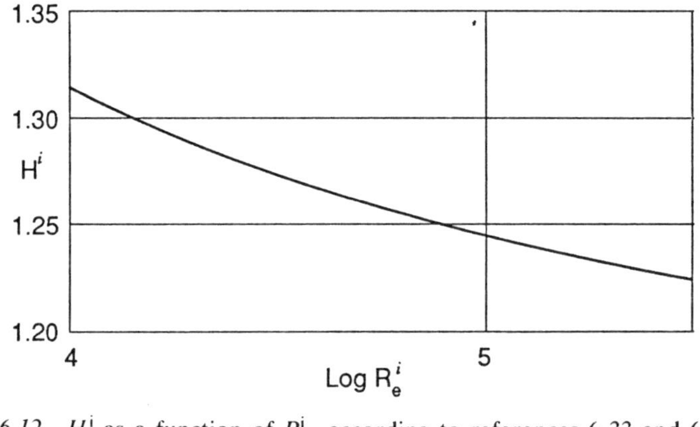

Fig. 6.12 H^i as a function of R^i_θ, according to references 6.33 and 6.34.

$$\left.\begin{aligned} f_c = 1 + 2M_e^2/(20 + M_e), \quad f_\theta = 1 + 0.056M_e^2 \\ f_H = H^i/H = (1 + M_e^2)^{-1} \end{aligned}\right\} \qquad (6.79)$$

Figure (6.13) shows the basic kinematic skin friction/Reynolds number relations derived by Gaudet in the form

$$c_f^i = f_t \,.\, c_f, \qquad C_F^i = f_c C_F$$

as functions of $R_x^i = f_x \,.\, R_x$. Comparison with the Prandtl–Schlichting relations demonstrates good agreement within the likely experimental error.

The use of these relations to determine the frictional coefficients for a given Reynolds number and Mach number is fairly straightforward. Thus, from the given M_e we can determine f_c, f_θ and f_x from the above relations. Knowing R_x we can determine R_x^i from equation (6.78), hence $c_f^i = f_c \,.\, c_f$ from Fig. 6.13, and so we can determine c_f.

The velocity profile in the logarithmic law region is then given by

$$u/u_\tau^i = \frac{1}{K} \ln (u_\tau^i z/\nu_e) + \phi(0)$$

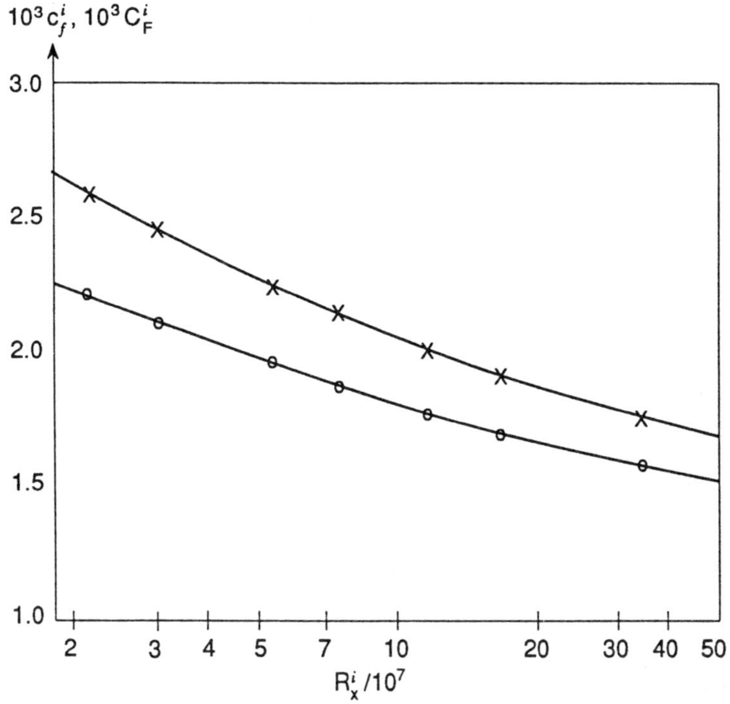

Fig. 6.13 $c_f^i = f_c \,.\, c_f$ and $C_F^i = f_c \,.\, C_F$ as functions of $R_x^i = f_x \,.\, R_x$, according to reference 6.34.

o———o c_f^i x———x C_F^i.

and Gaudet suggests that $K = 0.41$ and $\phi(0) = 5.2$.

For the remainder of the boundary layer velocity profile in the form of equation (6.76) Gaudet suggests [as an improvement on equation (6.30)] that

$$\phi(z/\delta) - \phi(0) = 1.124[1 - \cos(1.196\pi z/\delta)] \tag{6.80}$$

This expression for ϕ satisfies the requirement $\partial u/\partial z = 0$ at $z = \delta$. An expression for δ, the boundary layer thickness can be derived as follows. From equation (6.76) it follows that

$$(u_e - u)/u_\tau^i = -\frac{1}{K}\ln(z/\delta) + \phi(1) - \phi(z/\delta) = F(z/\delta), \quad \text{say} \tag{6.81}$$

and hence

$$\delta^{*i} = \int_0^\delta (1 - u/u_e)\,.\,dz = u_\tau^i C_1 \delta/u_e$$

where $\quad C_1 = \int_0^1 F d(z/\delta)$

Therefore $\quad \delta = \delta^{*i}(2/c_f^i)^{1/2}/C_1$

The value of C_1 depends on how the outer edge of the boundary layer is defined. If we specify it as where $u/u_e = 0.999$ then $C_1 = 3.3$. We can deduce that

$$\begin{aligned}\delta &= (H^i/H)(R_\theta^i/R_\theta)\,H\theta(2/f_c\,.\,c_f)^{1/2}/C_1 \\ &= (f_H\,.\,f_\theta/f_c^{1/2})(H/H_i)H_i\,.\,\theta/C_1\,.\,c_f^{1/2}\end{aligned}$$

where suffix i refers to the basic incompressible flow. Now, if we accept the inferences of the mean temperature concept, then $\theta/x \propto c_f$, and so

$$\delta/\delta_i = (f_H\,.\,f_\theta/f_c^{1/2})(H/H_i)(c_f/c_{fi})^{1/2} \tag{6.82}$$

where δ_i is the value of δ in incompressible flow and the same R_x. An estimate of H/H_i can be obtained by assuming an appropriate law for the velocity distribution and a temperature distribution [equation (6.62)]. For this purpose it was thought adequate to assume a power law for the velocity distribution and a constant total temperature distribution applicable to the case of zero heat transfer and $\sigma = 1$. The ratio (c_f/c_{fi}) can be most simply derived from the mean temperature concept plus power law relation, equation (6.70). Thus, with the aid of the above expressions for the compressibility factors f_H, f_θ and f_c the ratio δ/δ_i was estimated for a range of Mach numbers up to 3. The results showed little variation with Mach number, the small differences from unity of a few per cent were well within the probable error of the approximations involved. This result seems to be consistent with the few available experimental data, and is in marked contrast with the corresponding changes with Mach number for the laminar boundary layer. Again, the effects of heat transfer on δ for a turbulent boundary layer do not seem to be of evident significance to judge from the all too sparse experimental evidence available, but the matter is open for further investigation.

However, the laminar sub-layer thickness can be expected to show a marked variation with wall temperature (and therefore Mach number). It can be argued that the sub-layer will be dominated by conditions at the wall and hence one can expect

$$z^+ = u_{\tau w} z_l/\nu = \text{const.} = O(10)$$

where Z_l is the sub-layer thickness and

$$u_{\tau w} = (\tau_w/\rho_w)^{1/2}$$

Again, the available experimental evidence is sparse but is supportive of this conclusion. Therefore, we can infer [using equation (6.67)] that

$$z_l/z_{li} = (T_w/T_e)^{\omega+1/2}\,(T_m/T_e)^{(4-\omega)/10}$$

Hence, the sub-layer thickness will increase with T_w.

Yet another approach, which has received encouraging empirical support, although its theoretical basis is open to question, is a simple extension of Prandtl's mixing length theory to compressible flow. This was first independently mooted by Young and van Driest in 1951[6.35, 6.36] and later developed in some detail by Fernholz[6.37] in 1969.

One starts by accepting the relations [see equation (6.16) *et seq.*]

$$\tau = \tau_w = \rho l^2(\partial u/\partial z)^2 \qquad l = Kz$$

as valid for the law of the wall region outside the laminar sub-layer in compressible flow, so that

$$\partial u/\partial z = (u_{\tau w}/Kz)(T/T_w)^{1/2}$$

where $u_{\tau w} = (\tau/\rho_w)^{1/2}$. If we now make use of the modified Crocco relation between temperature and velocity (6.64) we obtain an integrable equation for u which yields after a little algebra:

$$u^*/u_{\tau w} = \frac{1}{K}\ln\,(zu_{\tau w}/\nu_w) + B^* \tag{6.83}$$

where $$u^* = (u_e/b)\sin^{-1}\{[2b^2(u/u_e) - a]/(a^2 + 4b^2)^{1/2}\} \tag{6.84}$$

$$B^* = -\frac{1}{K}ln\,(z_1 u_{\tau w}/\nu_w) + (u_e/u_{\tau w}b)\sin^{-1}\{[(2b^2 u_1/u_e) - a]/(a^2 + 4b^2)^{1/2}\}$$

and $$a = (T_r - T_w)/T_w, \qquad b^2 = (T_r - T_e)/T_w$$

The suffix l as in Section 6.4 denotes a position near the edge of the laminar sub-layer below which equation (6.83) will not apply, and as before it must be determined empirically, but we can expect that $z_l^+ = z_l u_{\tau w}/\nu_w$ will be of the order of 10.

The analysis by Fernholz and Finley[6.29] of the available experimental results shows much support for equation (6.83) as a description of the logarithmic law region with $K = 0.40$ and $B^* = 5.10$, well in the range of the accepted values for incompressible flow, so that equation (6.83) can be regarded as the compressible generalisation of equation (6.19).

An attempt to develop a companion outer region velocity defect law in terms of $(u_e^* - u^*)/u_{\tau w}$ on similar lines to those described for incompressible flow in Section 6.5 was less successful, but for further details the reader is referred to reference 6.29.

6.12 The effects of distributed surface roughness

For a detailed review of the topic of excrescences and their aerodynamic effects the reader is referred to reference 6.38.

6.12.1 Frictional drag

Our basic understanding of the effects of distributed roughness in a turbulent boundary layer in incompressible flow derives from a classical and comprehensive range of experiments made by Nikuradse[6.39] on pipes of circular section with sand roughnesses of various sizes fixed to their inner surfaces. The sand grains in each case investigated were closely uniform and fixed in a densely packed arrangement. Wide ranges of grain size and pipe radius were investigated, and the flow was 'fully developed pipe flow' i.e. the boundary layers extended to the pipe axis so the boundary layer thickness was equal to the pipe radius and therefore did not change with axial distance x. In such a flow there is a downstream pressure gradient to balance the wall friction, and the latter can be deduced from the former.

It was found that for each grain size there was a critical pipe Reynolds number (defined by $u_m d/\nu$, where u_m is the average velocity across a pipe section and d is the pipe diameter) below which no effect was evident on the wall friction. The wall surface was then referred to as *hydraulically smooth*. It is inferred that such cases arise when the roughnesses are fully immersed in the viscous sub-layer and they do not therefore shed eddies so that the axial momentum flux is unchanged by the roughnesses. Consistent with this is the fact that the maximum roughness height (k) for hydraulic smoothness is such that $k^+ = u_\tau k/\nu$ is about 5.

With increase of the Reynolds number beyond the critical the wall friction increased above that of the smooth pipe. This was because the roughnesses increasingly protruded above the decreasing viscous sub-layer and so shed eddies which augmented the axial momentum flux defect. The drag increment was therefore the sum of the pressure (or form) drag of the excrescences and the accompanying changes of the local wall shear stresses. Eventually a second critical Reynolds number was reached above which the frictional drag coefficient remained constant, showing that the form drag of the excrescences, which is characteristically insensitive to changes of Reynolds number, was then dominant. For the sand roughnesses tested this second critical Reynolds number was such that

$$k_s^+ = u_\tau k_s/\nu \simeq 70$$

Here the suffix s is to show that we are referring specifically to the sand roughness, for other forms of roughness we can expect a different value for this critical condition. When this condition is reached we refer to the flow as *fully developed roughness flow*. Between the hydraulically smooth flow regime and the fully developed roughness regime we speak of the flow as being *intermediate rough* (see Fig. 6.14).

These pipe flow results can be readily adapted to enable us to infer the effects of roughness on the frictional characteristics of the turbulent boundary layer on a flat plate at zero incidence, if we assume that the effects of the small pressure gradient in pipe flow can be neglected. The boundary layer thickness, δ, unlike the pipe radius, is a growing function of x, the distance from the leading edge. Hence, with a given uniformly distributed roughness over the plate we can expect an initial region of

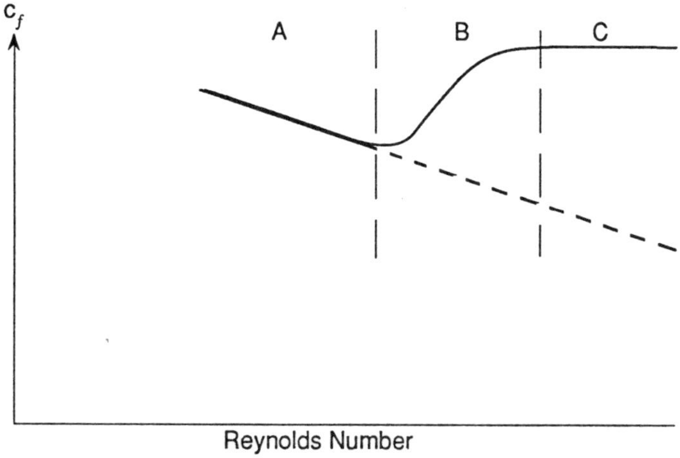

Fig. 6.14 Sketch of c_f variation of a rough surface with Reynolds number.
——— rough - - - - - smooth.
A – hydraulically smooth flow; B – intermediate rough flow;
C – fully developed roughness flow for sand roughness.

fully developed roughness flow followed by a region of intermediate rough flow, and if the plate chord is sufficiently long there will finally be a region of hydraulically smooth flow. Figure 6.15 shows the critical conditions in the form of plots of log (x/k_s) against log R_x derived for sand roughness from the above relations.

Prandtl and Schlichting[6.40] adapted Nikuradse's results to determine the local and overall skin friction coefficients for sand roughened plates for wide ranges of the roughness height/plate chord (k_s/c) and of the plate Reynolds number (R_c). Figure 6.16 shows the overall results in a convenient form with $\Delta D/D$ as a function of R_c for various values of c/k_s, where D is the smooth surface drag and ΔD is the drag increase due to the roughness. An advantage of this presentation is that we may expect that it can be used for determining the effects of roughness on aerofoils and wings at small angles of incidence, since the effects of the moderate pressure gradients on the ratio $\Delta D/D$ are likely to be much smaller than on ΔD and D separately.

The above discussion refers to incompressible flow. The effects of compressibility and heat transfer have not been thoroughly explored but

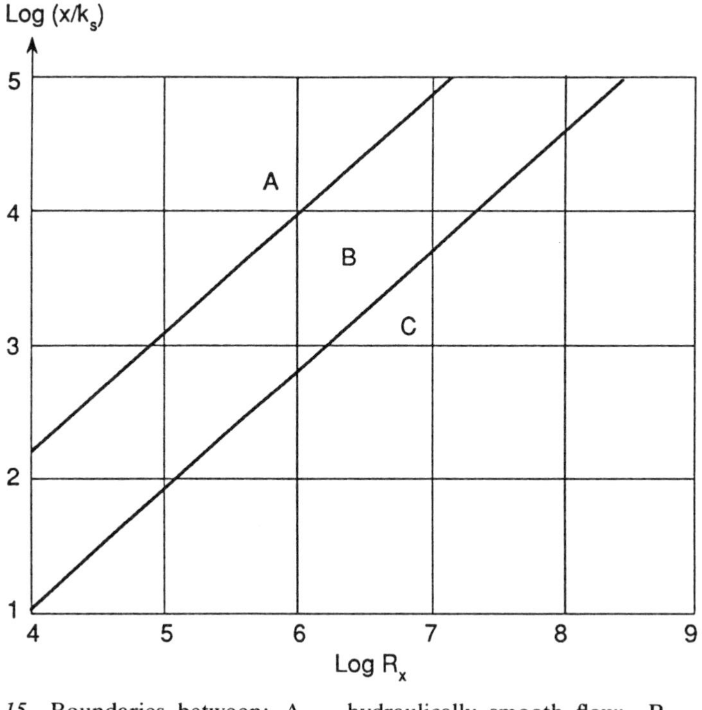

Fig. 6.15 Boundaries between: A – hydraulically smooth flow; B – intermediate rough flow; C – fully developed roughness flow for sand roughness.

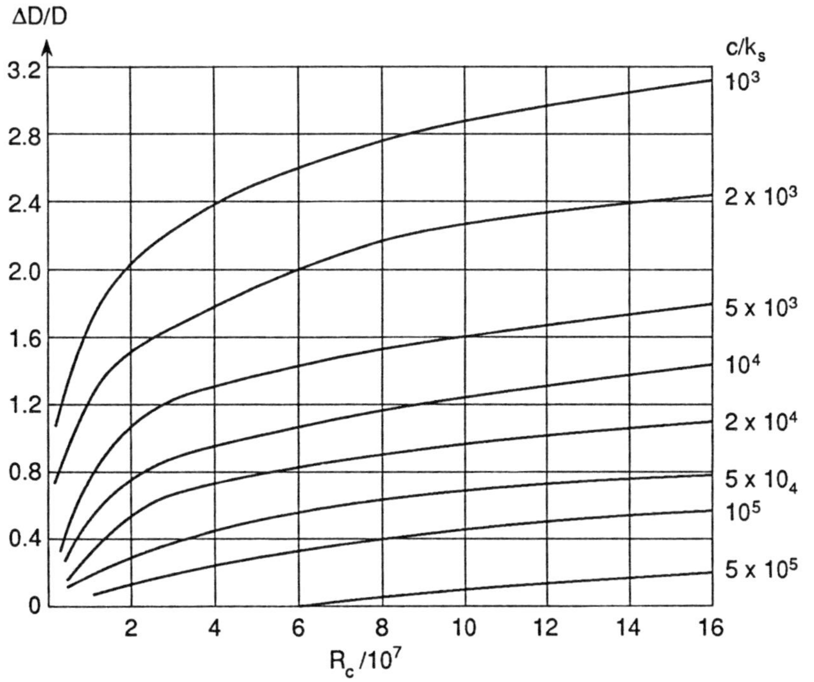

Fig. 6.16 Fractional drag increase due to sand roughness.

it can be plausibly argued that the above relations will still apply if the relevant density and viscosity values are the values at the wall, since they apply to the flow region in which the roughnesses operate. Some experimental results in support of this argument have been obtained by Berg.[6.41] Thus, we replace ρ and ν by ρ_w and ν_w, respectively, and u_τ by $u_{\tau w} = (\tau_w/\rho_w)^{1/2}$, so that the critical sand roughness heights are given by $u_{\tau w} \,.\, k_s/\nu_w = \text{const.}$, independent of Mach number. However, we recall that u_τ will itself vary with wall temperature. If, to assess this, we accept the Sommer and Short mean temperature concept and the power law velocity distribution then we have seen that [equation (6.68) and (6.70)]

$$\begin{aligned} \tau_w &\propto (T_e/T_m)^{0.622}, \quad \text{for } n = 7 \\ &\propto (T_e/T_m)^{0.685}, \quad \text{for } n = 9 \end{aligned}$$

where [equation (6.66)]

$$T_m/T_e = 0.55 + 0.45T_w/T_e + 0.036M_e^2$$

For the purpose in hand it is probably sufficient to take a mean value of the exponent and write

$$\tau_w \propto (T_e/T_m)^{0.65}$$

Hence the variation with Mach number of a critical value of k_s will be given by

$$\begin{aligned} k_s/k_{si} &= (\nu_w/\nu_{wi})(u_{\tau wi}/u_{\tau w}) = (\mu_w/\mu_{wi})(\rho_{wi}/\rho_w)^{1/2}(\tau_{wi}/\tau_w)^{1/2} \\ &\fallingdotseq (T_w/T_e)^{\omega+1/2}(T_m/T_e)^{0.325} \\ &= (T_w/T_e)^{1.39}(0.55 + 0.45T_w/T_e + 0.036M_e^2)^{0.325} \end{aligned} \tag{6.85}$$

Here we have taken air as the fluid with $\omega = 0.89$ and $\sigma = 0.72$. For zero heat transfer we have

$$T_w/T_e = 1 + \tfrac{1}{2}(\gamma - 1)M_e^2\, r = 1 + 0.179M_e^2, \quad \text{for air}$$

Hence $$k_s/k_{si} = [1 + 0.179M_e^2]^{1.39}[1 + 0.215M_e^2]^{0.325} \tag{6.86}$$

Thus, for $M_e = 1$, $k_s/k_{si} = 1.34$; and for $M_e = 2$, $k_s/k_{si} = 2.58$. These results reflect the increase of the viscous sub-layer thickness with Mach number and/or wall temperature, referred to in Section 6.11.

In reference 6.41 it was deduced from experimental data that $\Delta D/D$, regarded as a function of R_c, as in Fig. 6.16, was dependent on the parameter $u_{\tau w} \,.\, k_s/\nu_w$, independently of the Mach number. Hence, one can infer that Fig. 6.16 applies to compressible flow if for k_s one reads

$$k_{sc} = k_s(\nu_e/\nu_w)(\rho_e/\rho_w)^{1/2} = k_s(T_e/T_w)^{1.39} \tag{6.87}$$

For zero heat transfer

$$k_{sc} = k_s/[1 + \tfrac{1}{2}(\gamma - 1)M_e^2\, r]^{1.39}$$

which for $M_e = 1$ yields $k_{sc} = 0.8k_s$, and for $M_e = 2$ gives $k_{sc} = 0.47k_s$.

These results show that the proportional effect on drag for a given roughness decreases with increase of Mach number and/or increase of wall temperature; this is a consequence of the associated reduction of density near the wall. However, in supersonic flow the above analysis only applies to roughnesses small enough to be immersed in the subsonic region of the boundary layer, otherwise they may generate shock waves which would add to the drag increment.

6.12.2 *Velocity distribution*

If the roughnesses are small enough to be immersed in the boundary layer then the eddies they generate rapidly get absorbed into the local turbulence without significantly altering the general structure and pattern of the turbulence in the boundary layer. Their main effect, therefore, is an increase in the mean shear stress at and near the wall, but the dimensional reasoning underlying the derivation of the law of the wall and the velocity defect relation (see Section 6.3) remains valid for the roughened surface boundary layer.

Hence, we infer for the law of the wall region that

$$u^+ = \frac{1}{K} \ln z^+ + B \tag{6.88}$$

where K is the Von Karman constant as before (0.4–0.41), but the value of the constant B now depends on the roughness size and geometry. To go further, we note that there must be a lower limit of z, z_0, say, defining the region in which equation (6.88) applies, and z_0 can be expected to be a function of the roughness height, u_τ, and ν as well as of the roughness shape. Therefore, for roughnesses of similar shape we can write

$$z_0/k = f(u_\tau k/\nu), \quad \text{say}$$

where f is a function of the roughness shape. Hence

$$[u - u(z_0)]/u_\tau = \frac{1}{K} \ln (z/z_0) = \frac{1}{K} \ln (z/kf)$$

We can similarly expect that $u(z_0)/u_\tau$ is a function of $(u_\tau k/\nu)$ and so

$$u/u_\tau = \frac{1}{K} \ln (z/k) + h(u_\tau k/\nu) \tag{6.89}$$

where h is a function of $u_\tau k/\nu$ and of the roughness shape and distribution. Nikuradse's results provide support for this relation and the resulting function h for his closely packed sand grains is illustrated in Fig. 6.17. We may note that since for $u_\tau k/\nu < 5$ we get the smooth surface relation with $K = 0.4$, $B = 5.5$ (Nikuradse's values), it follows that then

$$h(u_\tau k_s/\nu) = \frac{1}{0.4} \ln (u_\tau k_s/\nu) + 5.5$$

In the fully developed roughness regime

$$h(u\tau k_s/\nu) = \text{const.} = 8.5$$

so that then

$$u/u_\tau = \frac{1}{0.4} \ln (z/k_s) + 8.5 \tag{6.90}$$

This is in accord with the argument that for the fully developed roughness flow the velocity profile and skin friction are independent of the Reynolds number. We can write equation (6.89) in the form

$$u/u_t = \frac{1}{K} \ln (zu_\tau/\nu) + B - \Delta u/u_\tau \tag{6.91}$$

where $\Delta u/u_\tau = B + \frac{1}{K} \ln (u_\tau k/\nu) - h(u_\tau k/\nu) = \text{function of } (u_\tau k/\nu)$

for roughnesses of similar shape, and it represents the departure of the

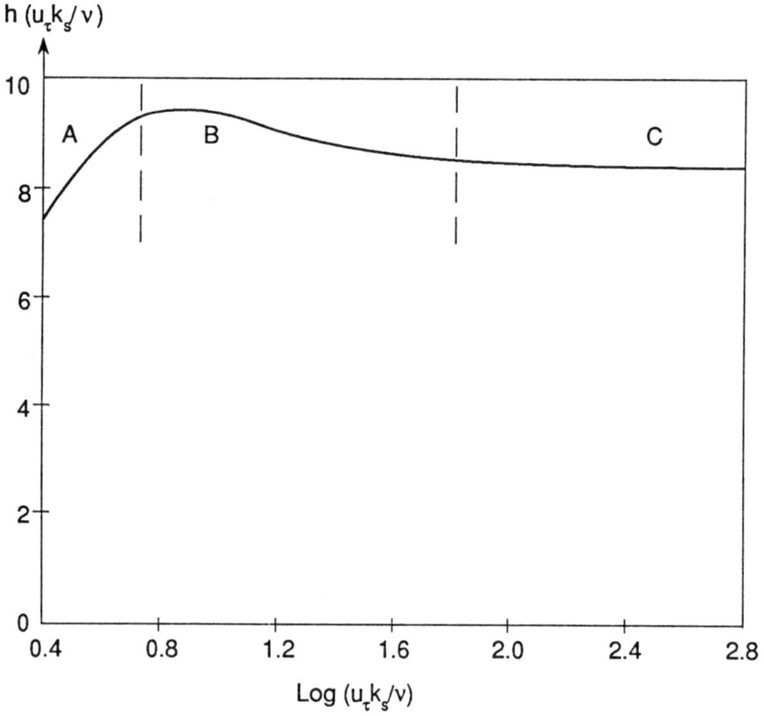

Fig. 6.17 The function $h(u_\tau k_s/\nu)$ for sand roughness.

velocity profile in the logarithmic law of the wall region from the corresponding smooth surface profile.

Hence for that region the plot of u/u_τ as a function of zu_τ/ν is linear with slope $1/K$ and displaced parallel to the corresponding smooth surface line with a displacement $\Delta u/u_\tau$ which is a function of $u_\tau k/\nu$. This result is amply confirmed by the experimental results obtained by a number of different workers.[6.38]

For fully developed roughness flow we see from the above expression for $\Delta u/u_\tau$ that

$$\Delta u/u_\tau = \frac{1}{K} \ln (u_\tau k/\nu) + D \tag{6.92}$$

where D is a constant dependent only on the roughness form. For example, Blanchard[6.42] found that a good fit to his experimental results for abrasive paper surfaces was

$$\Delta u/u_\tau = \frac{1}{K} \ln (u_\tau k/\nu) - 1.04$$

The arguments that led to the velocity defect relation (Section 6.5) for the smooth wall apply unchanged to rough walls, as the roughnesses do not affect the flow outside the law of the wall region, apart from the change of the scaling velocity u_τ. A proviso is that the roughnesses are small enough to remain well immersed in the boundary layer. Hence, we deduce as before that

$$(u_e - u)/u_\tau = f(z/\delta), \quad \text{say} \tag{6.93}$$

where $$f(z/\delta) = -\frac{1}{K} \ln (z/\delta) + \phi(1) - \phi(z/\delta)$$

and $\phi(z/\delta)$ can be obtained from the various empirical fits such as equation (6.30), Coles' velocity distribution, or equation (6.80).

6.12.3 Equivalent sand roughness concept

The value of Nikuradse's sand roughness results can be readily enhanced to provide basic material for other forms of distributed roughness provided that for each such form one can define an equivalent sand roughness. A number of workers have demonstrated the validity of the concept of an equivalent sand roughness at least for the fully developed roughness regime for a wide range of different kinds of roughness.[6.43–6.45]

To determine the equivalent sand roughness from experimental data for a particular form of roughness we note that in equation (6.89) the function h is a constant in fully developed roughness flow which depends only on the type of roughness. We will denote this constant h_{fr}. If k_s denotes the equivalent sand roughness height having the same velocity profile and u_τ for the same external conditions then from equation (6.89) and (6.90) it follows that

$$\frac{1}{K} \ln (k/k_s) = 8.5 - h_{fr} \tag{6.94}$$

Hence, since h_{fr} can be determined from the measured velocity distribution for the particular roughness we can then determine from this equation the equivalent sand roughness k_s. It must be noted that the equivalence need not necessarily hold for the intermediate roughness regime.

Based on the available experimental data various attempts have been made to establish a method for predicting the equivalent sand roughness for a given roughness in terms of its geometry and distribution.

The most effective method to date seems to be that of Grabow and White.[6.45] They have derived an empirical correlation between

$$\alpha = k/k_s \quad \text{and} \quad \Lambda = (l_r/k)(A_s/A_p)^{4/3}$$

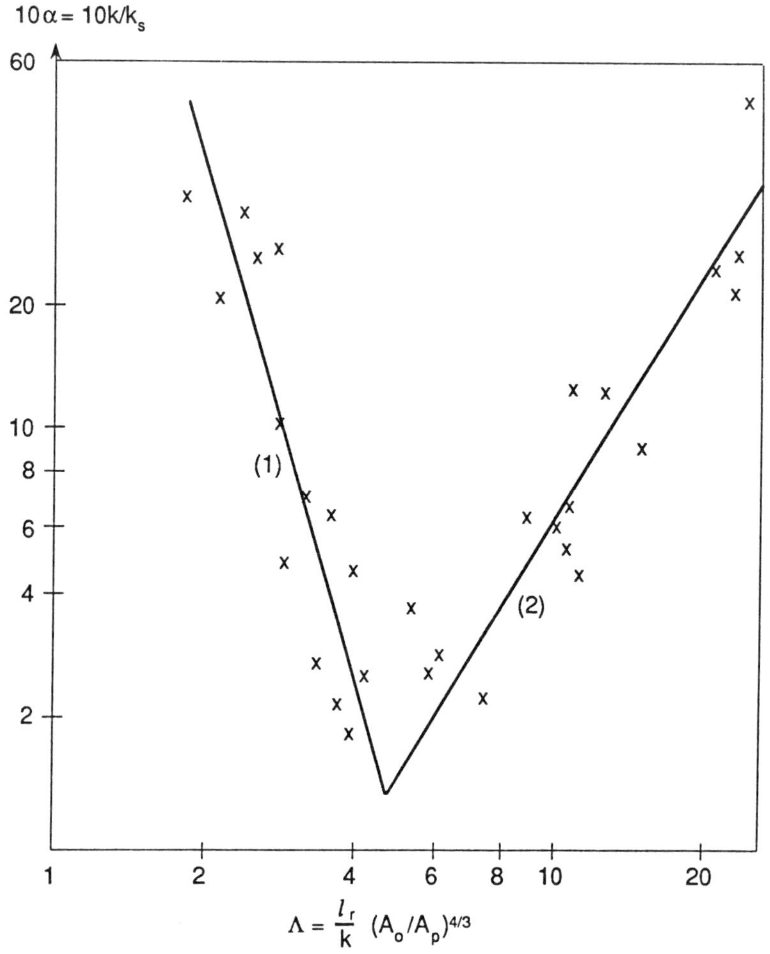

Fig. 6.18 Equivalent sand roughness.[6.46]
(1) $\alpha = 61\Lambda^{-3.78}$ (2) $\alpha = 0.0072\Lambda^{1.90}$
× experimental results for the various forms of roughness tested.

where

l_r = mean distance between the roughness elements,
k = mean roughness height,
A_p = maximum cross-sectional area of a typical roughness element normal to the flow,
A_s = the surface area of a typical roughness element forward of the maximum cross-section.

The correlation is illustrated in Fig. 6.18.

It is of interest to note that there is a minimum value of α ($\simeq 0.15$) at a value of $\Lambda \simeq 5$. This is because for a given form of roughness there is a density of packing, which is not the closest possible, for which the drag increment is a maximum. With a closer packing roughnesses become more immersed in the wakes of roughnesses upstream of them and the shielding effect tends to reduce the overall drag increment; but with a reduced density and hence a reduced number of roughnesses per unit area the overall drag increment is again reduced.

With the aid of this correlation and the standard sand roughness results, as in Fig. 6.16, it becomes possible to make an estimate of the drag effects of any specified form of roughness in a turbulent boundary layer in zero pressure gradient, and the method can be extended to roughnesses on wings and bodies at small incidences, because as remarked above the ratio $\Delta D/D$ can be expected to be relatively insensitive to small pressure gradients. The overall accuracy of the method is not high as is indicated by the scatter of the experimental data in Fig. 6.18 (note the logarithmic scales), but it offers a useful guide for initial predictive purposes.

References

6.1 Klebanoff, P.L. (1955) 'Characteristics of turbulence in a boundary layer with zero pressure gradient'. NACA Rep. 1247.
6.2 Motzfield, H. (1938) 'Frequenzanalyse turbulenter Schwankungen'. *ZAMM*, **18**, 362.
6.3 Townsend, A.A. (1976) *The Structure of the Turbulent Boundary Layer*. (2nd. Ed.) CUP.
6.4 Boussinesq, J. (1877) 'Essai sur la théorie des eaux courantes'. *Mem. pres. Acad. Sci.*, Paris, XXIII, **46**.
6.5 Boussinesq, J. (1896) 'Théorie de l'écoulement tourbillonant et tumulteux des liquides dans les lits rectilignes'. *Comptes Rendus de l'Acad. des Sciences*, **CXXII**, p. 1293.
6.6 Squire, H.B. (1948) 'Reconsideration of the theory of free turbulence'. *Phil. Mag.*, **39**, p. 1.
6.7 V. Karman, Th. (1930) 'Mechanische Ähnlichkeit und Turbulenz'. Nachr. Ges. Wiss., Göttingen, *Math. Phys. Kl.*, **58**.
6.8 Reichardt, H. (1933) 'Messungen turbulenter Schwankungen'. *ZAMM*, **13**, p. 177.
6.9 Prandtl, L. (1939) 'Über die ausgebildete Turbulenz'. *ZAMM*, **18**, p. 358.
6.10 Taylor, G.I. (1932) 'The transport of vorticity and heat through fluids in turbulent motion'. *Proc. Roy. Soc. A*, **135**, p. 685.
6.11 Van Driest, E.L. (1956) 'On turbulent flow near a wall'. *JAS*, **23**, p. 1007.
6.12 Clauser, F.H. (1956) 'The turbulent boundary layer'. *Adv. App. Mech.*, **4**, p. 1.
6.13 Preston, J.H. (1954) 'The determination of turbulent skin friction by means of pitot tubes'. *J. Aero. Soc.*, **58**, p. 109.

6.14 Smith, D.W. and Walker, J.H. (1959) 'Skin friction measurements in incompressible flow'. NACA TR R-26.
6.15 Cebeci, T. and Smith, A.M.O. (1974) *Analysis of Turbulent Boundary Layers*. Acad. Press, NY.
6.16 V. Karman, Th. (1934) 'Turbulence and skin friction'. *JAS*, **1**, p. 1.
6.17 Clauser, F.H. (1954) 'Turbulent boundary layers in adverse pressure gradient'. *JAS*, **21**, p. 91.
6.18 Coles, D. (1956) 'The law of the wake in the turbulent boundary layer'. *JFM*, **1**, p. 191.
6.19 Spence, A. (1956) 'The development of turbulent boundary layers'. *JAS*, **23**, 1, p. 3.
6.20 Nikuradse, J. (1932) 'Gesetzmässigkeit der turbulenten Strömung in glatten Rohren'. *Forschungs-Arb. Ing.-Wesen*, **356**.
6.21 Schoenherr, R.E. (1932) 'Resistance of flat surfaces moving through a fluid'. *Trans. Soc. Nav. Arch. & Mar. Eng.*, **40**, p. 279.
6.22 Prandtl, L. (1927) 'Über den Reibungswiderstand strömender Luft'. *Erg. AVA*, Göttingen, III series; also IV series (1932).
6.23 Squire, H.B. and Young, A.D. (1938) 'The calculation of the profile drag of aerofoils'. ARC R & M 1838.
6.24 Coles, D. (1954) 'The problem of the turbulent boundary layer'. *ZAMP*, **5**, p. 181.
6.25 Ludwieg, H. and Tillmann, W. (1949) 'Untersuchungen über die Wandschubsspannung in turbulenten Reibungsschichten'. *Ing. Arch.*, **17**, p. 288; also NACA TM 1285.
6.26 Sarnecki, A.J. (1959) 'The turbulent boundary layer on a permeable surface'. Ph.D Dissertation, Camb. Univ.
6.27 Thompson, B.G.J. (1967) 'A new two-parameter family of mean velocity profiles for incompressible boundary layers on smooth walls'. ARC R & M 3463.
6.28 Gaudet, L. (1986) 'Integral boundary layer parameter relationships and a skin friction law derived from a velocity profile family for two dimensional incompressible flow'. RAE Tech. Memo. Aero 2080.
6.29 Fernholz, H.H. and Finley, P.J. (1980) 'A critical commentary on mean flow data for two dimensional compressible turbulent boundary layers'. AGARDograph 253.
6.30 Fernholz, H.H. and Finley, P.L. (1981) 'A further compilation of compressible boundary layer data with a survey of turbulence data'. AGARDograph 263.
6.31 Sommer, S.C. and Short, B.J. (1956) 'Free flight measurements of skin friction of turbulent boundary layers with high rates of heat transfer at high supersonic speeds'. *JAS*, **23**, 6, p. 536.
6.32 Spalding, B. and Chi, W. (1964) 'The drag of a compressible turbulent boundary layer on a smooth plate with and without heat transfer'. *JFM*, **18**, 1, p. 117.
6.33 Winter, K.G. and Gaudet, L. (1973) 'Turbulent boundary layer studies at high Reynolds numbers and Mach numbers between 0.2 and 0.8'. ARC R & M 3712.

6.34 Gaudet, L. (1986) 'Experimental investigation of the turbulent boundary layer at high Reynolds numbers and a Mach number of 0.8'. *Aero. J.*, **90**, p. 83.
6.35 Young, A.D. (1951) 'The equations of motion and energy and the velocity profile in a turbulent boundary layer in compressible flow'. Coll. of Aeronautics, Cranfield, Rep. 42.
6.36 Van Driest, E.R. (1951) 'Turbulent boundary layers in compressible fluids'. *JAS*, **18**, p. 145.
6.37 Fernholz, H.H. (1969) 'Geschwindigskeitprofile, Temperaturprofile und halb-empirische Gesetze in kompressiblen turbulenten Grenzschichten bei konstantem Druck'. *Ing. Arch.*, **38**, p. 311.
6.38 Young, A.D. and Paterson, J.H (1981) 'Aircraft excrescence drag'. AGARDograph 264.
6.39 Nikuradse, J. (1933) 'Strömungsgesetzen rauhen Rohren'. *Forsch. Arb. Ing. Wes.*, 161.
6.40 Prandtl, L. and Schlichting, H. (1934) *Das Widerstand Gesetz rauher Platten*. Werft-Reederei-Hafen, 1–4.
6.41 Berg, D.E. (1979) 'Surface roughness effects on Mach 6 turbulent boundary layers'. *AIAA J*, **17**, 9, p. 929.
6.42 Blanchard, A. (1977) 'Analyse expérimentale et théorique de la structure de la turbulence d'une couche limite sur paroi rugueuse'. Univ. of Poitiers Thesis 97.
6.43 Schlichting, H. (1936) 'Experimentelle Untersuchungen zum Rauhigkeitsproblem'. *Ing. Arch.*, **7**, p. 1.
6.44 Young, A.D. (1950) 'The drag effect of roughness at high sub-critical speeds'. *J. RAeS*, **18**, p. 534.
6.45 Streeter, V.L. (1935) 'Frictional resistance in artificially roughened pipes'. *Proc. Amer. Soc. Civil Eng.*, **61**, p. 163.
6.46 Grabow, R.M. and White, C.O. (1975) 'Surface roughness effects on nosetip ablation characteristics'. *AIAA J*, p. 605.

Chapter 7

The Equations of Motion and Energy for a Turbulent Boundary Layer

7.1 Introductory remarks

In Chapter 2 we derived in general terms the main conservation equations of mass (continuity), momentum (The Navier–Stokes equations) and energy for a viscous fluid in motion. Thc assumption is now made that these equations apply to a turbulent flow for which we can express the instantaneous value of a quantity as the sum of its mean or time average, as defined in equation (5.1), and a turbulent part, as in equation (5.2). The mean value of the latter is by definition zero, as in equation (5.4). The conservation equations can then in turn be time-averaged, again over a period that is assumed long compared with a typical turbulent fluctuation but short compared with a typical time scale of the mean motion. As can be expected from the discussion of Section 5.2, we find that the eddy or Reynolds stresses play a key role in the resulting mean momentum equations, and likewise in the mean energy equation the work done by these stresses as well as eddy heat transport contributions are of major importance.

Again, as in the case of the laminar boundary layer, the resulting equations can then be simplified by means of the boundary layer assumptions which derive from the relative thinness of the boundary layer at the Reynolds numbers of aeronautical interest. In this chapter we shall follow this procedure to derive the mean turbulent boundary layer equations in two dimensional flow and will offer a few deductions that readily follow.

Here it may be remarked that we find a welcome degree of similarity between the corresponding equations for laminar and turbulent boundary layers. The dominance over almost all the turbulent boundary layer of the eddy transfer terms reflects the powerful mixing mechanism due to turbulence for both momentum and heat energy which ensures a close correlation between the velocity and temperature profiles in the presence of little or no pressure gradient. We have seen in Section 2.11 that there is likewise a close relationship between these profiles for the laminar boundary layer if the Prandtl number is near unity, in that case the molecular transport effects for momentum and heat are also similar if much less intense. However, in the presence of a pressure gradient, which acts differently on the vector momentum from its effect on the scalar heat, we must expect the similarity between the profiles to decrease.

As in Chapter 5, we shall denote the turbulent parts of quantities by a prime, mean terms will be left unmarked or denoted by a bar where the distinction is needed. Again, as in Chapter 2, tensor notation will be used where convenient.

7.2 The equation of continuity and the Navier–Stokes equations

In terms of the mean and turbulent components of density, velocity, etc. equation (2.1) is

$$\frac{\partial}{\partial t}(\rho + \rho') + \frac{\partial}{\partial x_\alpha}[(\rho + \rho')(u_\alpha + u'_\alpha)] = 0$$

On taking the mean of this equation we get

$$\frac{\partial \rho}{\partial t} + \frac{\partial}{\partial x_\alpha}(\rho u_\alpha) + \frac{\partial}{\partial x_\alpha} + \overline{(\rho' u'_\alpha)} = 0 \qquad (7.1)$$

For steady flow this becomes

$$\frac{\partial}{\partial x_\alpha}(\rho u_\alpha) + \frac{\partial}{\partial x_\alpha}\overline{(\rho' u'_\alpha)} = 0 \qquad (7.2)$$

and for incompressible flow

$$\partial u_\alpha / \partial x_\alpha = 0 \qquad (7.3)$$

Similarly, the mean equations of momentum (2.5) are, in the absence of body forces

$$\left.\begin{aligned} \overline{(\rho + \rho')\frac{\partial}{\partial t}(u_\alpha + u'_\alpha)} &= \rho\frac{\partial u_\alpha}{\partial t} + \overline{\rho'\frac{\partial u'_\alpha}{\partial t}} \\ &= \frac{\partial}{\partial x_\beta}(\tau_{\alpha\beta} + \overline{\tau'_{\alpha\beta}} - \rho u_\alpha u_\beta - \rho\overline{u'_\alpha u'_\beta} \\ &\quad - \overline{\rho' u'_\alpha} \,.\, u_\beta - \overline{\rho' u'_\beta} \,.\, u_\alpha \\ &\quad - \overline{\rho' u'_\alpha u'_\beta}) - u_\alpha(\partial\rho/\partial t) \\ &\quad - \overline{u'_\alpha \,.\, \partial\rho'/\partial t} \end{aligned}\right\} \qquad (7.4)$$

where

$$\tau_{\alpha\beta} = -(p + \tfrac{2}{3}\mu\Delta)\delta_{\alpha\beta} + \mu e_{\alpha\beta}, \qquad e_{\alpha\beta} = (\partial u_\alpha/\partial x_\beta + \partial u_\beta/\partial x_\alpha)$$

$$\delta_{\alpha\beta} = 0, \quad \text{if } \alpha \neq \beta; \qquad \delta_{\alpha\beta} = 1, \quad \text{if } \alpha = \beta$$

and $$\overline{\tau'_{\alpha\beta}} = -\delta_{\alpha\beta}\,\overline{\mu'\Delta'} + \overline{\mu' e'_{\alpha\beta}}, \qquad \Delta' = \partial u'_\alpha/\partial x_\alpha$$
$$e'_{\alpha\beta} = \frac{\partial u'_\alpha}{\partial x_\beta} + \frac{\partial u'_\beta}{\partial x_\alpha}$$

We see that the turbulence terms involve the Reynolds stresses of incompressible flow ($\rho\overline{u'_\alpha u'_\beta}$), although now ρ is a variable, together with terms involving the fluctuations in ρ, associated with fluctuations in pressure and temperature through the equation of state (see Section 2.5), and in μ, associated with fluctuations in temperature through the viscosity–temperature relation (see Section 1.3).

For two dimensional flow, equation (7.4) becomes

$$\begin{aligned}\rho\frac{\partial u}{\partial t} + \overline{\rho'\frac{\partial u'}{\partial t}} = &-\frac{\partial p}{\partial x} - \frac{\partial}{\partial x}\left(\tfrac{2}{3}\mu\Delta - 2\mu\frac{\partial u}{\partial x}\right)\\ &- \frac{\partial}{\partial x}(\rho u^2 + \rho\overline{u'^2} + 2\overline{\rho' u'}u) - \frac{\partial}{\partial x}\left(\tfrac{2}{3}\overline{\mu'\Delta'} - 2\overline{\mu'\frac{\partial u'}{\partial x}}\right)\\ &+ \frac{\partial}{\partial z}\left[\mu\left(\frac{\partial u}{\partial z} + \frac{\partial w}{\partial x}\right) + \overline{\mu'\left(\frac{\partial u'}{\partial z} + \frac{\partial w'}{\partial x}\right)}\right]\\ &- \frac{\partial}{\partial z}(\rho u w + \rho\overline{u'w'} + \overline{\rho'u'}\,w + \overline{\rho'w'}\,u\\ &+ \overline{\rho'u'w'}) - u\frac{\partial\rho}{\partial t} - \overline{u'\frac{\partial\rho'}{\partial t}}\end{aligned}$$

plus a similar equation for the z direction obtained by interchanging u and w, and x and z.

We now introduce the usual boundary layer assumptions, viz. $\delta/x \ll 1$, w/u is of order δ/x, and $\partial/\partial x$ of a mean quantity is small compared with $\partial/\partial z$ of that quantity. With the aid of equation (7.2) and after some algebra, the above equation then becomes, with $\mathrm{D}/\mathrm{D}t \equiv \partial/\partial t + u_\alpha\,\partial/\partial x_\alpha$,

$$\left.\begin{aligned}\rho\frac{\mathrm{D}u}{\mathrm{D}t} + \frac{\partial}{\partial t}\overline{\rho'u'} = &-\frac{\partial p}{\partial x} + \frac{\partial}{\partial z}\left(\mu\frac{\partial u}{\partial z}\right) - \frac{\partial}{\partial x}(\rho\overline{u'^2} + \overline{\rho'u'}\,u)\\ &- \frac{\partial}{\partial z}(\rho\overline{u'w'} + \overline{\rho'u'w'}) - \overline{\rho'u'}\frac{\partial u}{\partial x}\\ &- \overline{\rho'w'}\frac{\partial u}{\partial z} - \frac{\partial}{\partial x}\left(\frac{2}{3}\overline{\mu'\Delta'} - 2\overline{\mu'\frac{\partial u'}{\partial x}}\right)\\ &+ \frac{\partial}{\partial z}\left[\overline{\mu'\left(\frac{\partial u'}{\partial z} + \frac{\partial w'}{\partial x}\right)}\right]\end{aligned}\right\} \quad (7.5)$$

We now further assume that:

(1) u' and w' are of the same order of magnitude as are $\overline{\rho'u'}$ and $\overline{\rho'w'}$ so that $\overline{\rho'u'}\,\partial u/\partial x$ can be neglected compared with $\overline{\rho'w'}\,\partial u/\partial z$.
(2) $\overline{\rho'u'w'}$ is small compared with $\rho\overline{u'w'}$.

(3) The terms in the last two groups of equation (7.5) involving μ' are small compared with the corresponding quantities involving μ, and these in turn are small compared with the eddy fluctuation terms except in the viscous sub-layer.

(4) $\partial(\rho\overline{u'^2})/\partial x$ is small compared with $\partial(\rho\overline{u'w'})/\partial z$ and $\overline{\rho'u'}\ \partial u/\partial x$ is small compared with $\overline{\rho'w'}\ \partial u/\partial z$.

Then equation (7.5) reduces to

$$\rho\frac{\mathrm{D}u}{\mathrm{D}t} + \frac{\partial}{\partial t}(\overline{\rho'u'}) = -\frac{\partial p}{\partial x} + \frac{\partial}{\partial z}\left(\mu\frac{\partial u}{\partial z}\right) - \frac{\partial}{\partial z}(\rho\overline{u'w'}) - \overline{\rho'w'}\frac{\partial u}{\partial z} \tag{7.6}$$

For steady flow this becomes

$$\rho\frac{\mathrm{D}u}{\mathrm{D}t} = -\frac{\mathrm{d}p}{\mathrm{d}x} + \frac{\partial}{\partial z}\left(\mu\frac{\partial u}{\partial z} - \rho\overline{u'w'}\right) - \overline{\rho'w'}\frac{\partial u}{\partial z} \tag{7.7}$$

Following the same procedure for the equation of motion in the z direction we find that it reduces to

$$\partial p/\partial z + \partial(\rho\overline{w'^2})/\partial z = 0 \tag{7.8}$$

so that the pressure change across the boundary layer is

$$\Delta p = p_\mathrm{e} - p_\mathrm{w} = \mathrm{O}(\rho_\mathrm{e}\overline{w'^2})_\mathrm{r} = p_\mathrm{e}\gamma M_\mathrm{e}^2\overline{w_\mathrm{r}'^2}/u_\mathrm{e}^2 = p_\mathrm{e}\mathrm{O}[M_\mathrm{e}^2\delta^2/x^2] \tag{7.9}$$

where suffix r denotes a representative value in the boundary layer, and we assume that $w' = \mathrm{O}/w)$. Hence, for moderate Mach numbers the pressure change is generally small enough to be neglected, as in laminar flow. Again, as for laminar flow, the effect of surface curvature κ is to introduce another term in equation (7.8) thus

$$\partial p/\partial z + \frac{\partial}{\partial z}(\rho\overline{w'^2}) = \kappa\rho u^2$$

so that $\quad \Delta p/p_\mathrm{e} = \mathrm{O}[\gamma\, M_\mathrm{e}^2\, \overline{w_\mathrm{r}'^2}/u_\mathrm{e}^2] + \mathrm{O}[\kappa\rho_\mathrm{e}u_\mathrm{e}^2\delta/p_\mathrm{e}]$

The additional term on the RHS can be neglected only if $\kappa\delta M_\mathrm{e}^2$ is small compared with unity. This implies that the limit on surface curvature for the neglect of the normal pressure gradient becomes less as the Mach number increases. This reflects the fact that the isobars in the local external flow can then become increasingly inclined to the stream direction with a corresponding increasing pressure gradient component normal to the wall; and it has been shown by Myring and Young,[7.1] that the isobars inside the boundary layer are very nearly linear extensions of the external isobars. For example, if the external flow is a supersonic simple wave flow the external isobars coincide with the local Mach wave fronts and lie at an angle $\sin^{-1}(1/M_\mathrm{e})$ to the flow direction. In that case,

$\partial p/\partial z = (\partial p/\partial x)(M_e^2 - 1)^{1/2}$, which increases rapidly with Mach number for a given wall curvature. In consequence, some reconsideration has to be given to the way quantities such as the displacement and momentum thicknesses should be defined, for details see references 7.1 and 6.29. Further, as for laminar flow (see Section 2.8), another requirement for the flat plate boundary layer equations to apply when the surface is curved is that $(\delta c/R)\mathrm{d}\kappa/\mathrm{d}x \ll 1$.

In certain circumstances, notably regions of rapid changes with x (as when approaching separation) usually associated with high flow or surface curvature, assumption (4) above becomes dubious and it is desirable to retain the normal stress term $\partial(\rho\overline{u'^2})/\partial x$ as well as the curvature terms and to take account of the rate of change with z of the external inviscid flow near the surface. Such terms are referred to as second order terms, for a detailed discussion of them see Lock and Firmin.[7.2] However, for attached boundary layers not close to separation their effects are negligible.

For incompressible flow, equation (7.7) becomes

$$\rho\frac{\mathrm{D}u}{\mathrm{D}t} = -\frac{\mathrm{d}p}{\mathrm{d}x} + \frac{\partial}{\partial z}\left(\mu\frac{\partial u}{\partial z} - \rho\overline{u'w'}\right) = -\frac{\mathrm{d}p}{\mathrm{d}x} + \frac{\partial}{\partial z}(\tau_l + \tau_t) \qquad (7.10)$$

Reverting to compressible flow, we can denote the mass-weighted velocity normal to the wall as w_m so that

$$\rho w_m = \overline{(\rho + \rho')(w + w')} = \rho w + \overline{\rho' w'}$$

Hence, the equation of continuity (7.1) for two dimensional flow can be written

$$\frac{\partial\rho}{\partial t} + \frac{\partial(\rho u)}{\partial x} + \frac{\partial(\rho w_m)}{\partial z} = 0 \qquad (7.11)$$

which is the same as for laminar flow [see equation (2.22)] if we replace w_m by w. In incompressible flow the two equations are identical.

Likewise, the equation of motion (7.6) can be written

$$\rho\frac{\partial u}{\partial t} + \rho u\frac{\partial u}{\partial x} + \rho w_m\frac{\partial u}{\partial z} = -\frac{\partial p}{\partial x} + \frac{\partial\tau}{\partial z} \qquad (7.12)$$

where $\tau = \mu\dfrac{\partial u}{\partial z} - \rho\overline{u'w'} = \tau_l + \tau_t.$

We see that this equation is also formally the same as for the laminar boundary layer, equation (2.32), except that there $\tau = \tau_l$ and $w_m = w$. This is not surprising since the equation of motion expresses the physical equivalence of the mass times acceleration of an element of fluid to the sum of the forces acting on it resulting from the normal stresses (or pressure) and the shear stress components. We note that outside the viscous sub-layer $\tau_t \gg \tau_l$ and so $\tau \simeq \tau_t = -\rho\overline{u'w'}$.

7.3 The energy equation

Following the same procedure, we start with the energy equation in the form of equation (2.18), express the instantaneous quantities as the sum of mean and fluctuating components, then take the mean of the equation to derive the time averaged form and finally we apply the boundary layer approximations and neglect terms that are manifestly of a smaller order than the retained terms in the light of the assumptions described in Section 7.1.

We finally arrive at the following equation for two dimensional flow:

$$\rho c_p\left(\frac{\partial T}{\partial t} + u\frac{\partial T}{\partial x} + w_m\frac{\partial T}{\partial z}\right) = \frac{\partial p}{\partial t} + u\frac{\partial p}{\partial x} + (\tau_l + \tau_t)\frac{\partial u}{\partial z} + \frac{\partial}{\partial z}\left(k\frac{\partial T}{\partial z} - c_p\rho\overline{w'T'}\right) \tag{7.13}$$

It is of interest to identify the terms involving the turbulence contributions (including w_m), namely,

$$\tau_t\frac{\partial u}{\partial z} = -\rho\overline{u'w'}\frac{\partial u}{\partial z}, \quad \text{and} \quad \frac{\partial}{\partial z}(-c_p\rho\overline{w'T'}) - c_p\overline{\rho'w'}\frac{\partial T}{\partial z}$$

The first is the contribution to the overall rate of energy change of a small fluid element due to the rate of work per unit volume done on the element by the dominant Reynolds shear stress $-\rho\overline{u'w'}$ in the presence of the velocity gradient $\partial u/\partial z$. The second group are eddy heat transfer contributions analogous to the eddy shear stresses and they can be similarly arrived at from a basic consideration of the net heat transport due to the turbulent motion into a small element of fluid of sides parallel to the axes.

7.4 The total energy equation

The boundary layer equations of motion and energy (7.5) and (7.13) are

$$\rho\frac{Du}{Dt} = -\frac{\partial p}{\partial x} + \frac{\partial \tau}{\partial z}$$

and $$\rho\frac{Di}{Dt} = \frac{\partial p}{\partial t} + u\frac{\partial p}{\partial x} + \tau\frac{\partial u}{\partial z} + \frac{\partial}{\partial z}\left(k\frac{\partial T}{\partial z} - c_p\rho\overline{w'T'}\right)$$

where $$\frac{D}{Dt} = \frac{\partial}{\partial t} + u\frac{\partial}{\partial x} + \frac{\rho w + \overline{\rho' w'}}{\rho}\frac{\partial}{\partial z}$$

$i = c_p T$ for a perfect gas, $\tau = \tau_l + \tau_t$.

We follow the procedure of Section 2.11 for laminar flow and multiply the equation of motion by u and add it to the equation of energy to obtain

$$\rho\frac{D}{Dt}\{i + \tfrac{1}{2}u^2\} = \frac{\partial p}{\partial t} + \frac{\partial}{\partial z}\left(u\tau + k\frac{\partial T}{\partial z} - c_p\rho\overline{w' T'}\right)$$

We write $i + \frac{1}{2}u^2 = c_p T + \frac{1}{2}u^2 = c_p T_H$, where T_H is often called the total temperature, and $c_p T_H$ is referred to as the total energy, although strictly both should include the contribution of the kinetic energy due to the transverse velocity component, $\frac{1}{2}w^2$, which can however be neglected for the flow in an attached boundary layer.

The above equation can be written

$$\rho c_p\frac{DT_H}{Dt} = \frac{\partial p}{\partial t} + \frac{\partial}{\partial z}\left\{k\frac{\partial}{\partial z}[T_H + (\sigma - 1)\tfrac{1}{2}u^2] - u\rho\overline{u' w'} - c_p\rho\overline{w' T'}\right\} \tag{7.14}$$

Here σ (the Prandtl number $= \mu c_p/k$) is assumed constant. From the definition of T_H it follows that

$$c_p T'_H = c_p T' + uu' + \text{2nd order terms}$$

and so $$c_p\rho\overline{w' T'_H} = c_p\rho\overline{w' T'} + \rho u\overline{u' w'}$$

Hence, equation (7.14) can be written

$$\rho c_p\frac{DT_H}{Dt} = \frac{\partial p}{\partial t} + \frac{\partial}{\partial z}\left\{k\frac{\partial}{\partial z}[T_H + (\sigma - 1)\tfrac{1}{2}u^2] - c_p\rho\overline{w' T'_H}\right\} \tag{7.15}$$

Here we may note that the experimental evidence shows T'_H to be small for adiabatic walls.

For steady flow and $\sigma = 1$, equation (7.15) reduces to

$$\rho c_p\frac{DT_H}{Dt} = \frac{\partial}{\partial z}\left(k\frac{\partial T_H}{\partial z} - c_p\rho\overline{w' T'_H}\right) \tag{7.16}$$

and the equation of motion for zero pressure gradient is

$$\rho\frac{Du}{Dt} = \frac{\partial}{\partial z}\left(\mu\frac{\partial u}{\partial z} - \rho\overline{u' w'}\right) \tag{7.17}$$

Comparing equations (7.16) and (7.17) we see that they are similar and since $(T_H - T_w)$ and u have analogous boundary conditions they admit of a solution

$$T_H - T_w = K_1 u, \qquad \overline{T'_H w'} = K_1 \overline{u' w'} \qquad (7.18)$$

where K_1 is a constant. At the outer edge of the boundary layer we see that

$$K_1 u_e / T_e = 1 + \tfrac{1}{2} M_e^2 (\gamma - 1) - T_w / T_e = (T_r - T_w)/T_e \qquad (7.19)$$

where T_r is the wall temperature for zero heat transfer $= T_e[1 + \frac{1}{2}(\gamma - 1)M_e^2]$ for $\sigma = 1$. Hence

$$T_H / T_e = T_w / T_e + (T_r - T_w) u / u_e T_e$$

or
$$\frac{T}{T_e} = \frac{T_w}{T_e} + \frac{T_r - T_w}{T_e}\frac{u}{u_e} - \frac{T_r - T_e}{T_e}\frac{u^2}{u_e^2} \qquad (7.20)$$

As anticipated in Section 6.10, we have recovered the Crocco relation. We note that we have done so strictly for the case $\sigma = 1$ and zero pressure gradient. However, as remarked in Section 6.10, the analysis of experimental results by Fernholz and Finley[5.29] demonstrates that the relation (7.20) applies with acceptable scatter when σ is near unity if for T_r we understand the appropriate recovery temperature, namely

$$T_r = T_e[1 + \tfrac{1}{2}(\gamma - 1) M_e^2\, r]$$

where r is the recovery factor, often taken as $\sigma^{1/3} = 0.896$ for air. With this understanding equation (7.20) is the modified Crocco relation.

Fernholz and Finley also showed that this relation gave acceptable agreement with experiment in the presence of moderate pressure gradients but the available evidence cannot be regarded as conclusive and there is no theoretical justification for it.

However, whatever the pressure gradient a possible solution of equation (7.16) is

$$T_H = \text{const.} = T_e[1 + \tfrac{1}{2}(\gamma - 1)M_e^2] \quad \text{and} \quad T'_H = 0 \qquad (7.21)$$

This must correspond to the case of an insulated wall since $(\partial T/\partial z)_w = 0$. Experiments lend good support to the near constancy of T_H for insulated walls, although when $\sigma \neq 1$, $T_w = T_r \neq T_{He}$ and there must be compensating small departures from constancy elsewhere in the boundary layer for reasons given below. The condition $T'_H = 0$ is a sufficient condition but not a necessary one and there is evidence to show that although T'_H is small for insulated walls it cannot be regarded as always negligible. For a discussion of alternative conditions see Gaviglio.[7.3]

If, as in Section 2.11, we integrate equation (7.15) for steady flow between $z = 0$ and $z = h$, where $h > \delta$, such that $T'_H = 0$, $\partial T_H/\partial z = \partial u/\partial z = 0$ at $z = h$, we get

$$\int_0^h \rho \frac{D}{Dt}(c_p T_H)\mathrm{d}z = -k_w \frac{\partial (T)_w}{\partial z} \qquad (7.22)$$

If we now integrate this relation with respect to x between x_1 and x_2, we deduce, as in Section 2.11

$$\left[\int_0^h \rho u c_p (T_H - T_{He}) dz\right]_{x_1}^{x_2} = -\int_{x_1}^{x_2} k_w (\partial T/\partial z)_w \, dx \tag{7.23}$$

Thus, as in the case of laminar flow, we see that the difference between the integrated fluxes of the total energy increment relative to the free stream value across sections of the boundary layer at stations x_1 and x_2 is equal to the rate at which heat is transferred to the fluid from the wall between those stations. Hence, if the surface is insulated

$$\int_0^h \rho u c_p (T_H - T_{He}) dz = 0$$

or

$$\int_0^h \rho u c_p T_H dz \Big/ \int_0^h \rho u \, dz = c_p T_{He} \tag{7.24}$$

Thus, for the turbulent boundary layer, as for the laminar boundary layer, the mean rate of flux of total energy across a normal section divided by the mean rate of mass flow $= c_p T_{He}$. We have seen that with $\sigma \neq 1$ and zero heat transfer the recovery temperature at the wall T_r is a little less than T_{He}. Hence, it follows that somewhere in the boundary layer $T_H > T_{He}$ by a small compensating amount.

7.5 The momentum and energy integral equations for steady flow

It will be readily evident that the analysis leading to the momentum integral equation for a laminar boundary layer [equation (2.48)] applies essentially unchanged for a turbulent boundary layer. Therefore, we again obtain for steady mean flow, accepting the assumptions listed in Section 7.4,

$$\frac{d\theta}{dx} + \frac{1}{u_e}\frac{du_e}{dx}\theta(H + 2) + \frac{\theta}{\rho_e}\frac{d\rho_e}{dx} = \frac{\tau_w}{\rho_e u_e^2} \tag{7.25}$$

Here the quantities are mean or time averaged. Since $p/\rho = \mathcal{R}T$ (the equation of state for a perfect gas) and since outside the boundary layer

$$c_p T_e + \tfrac{1}{2}u_e^2 = c_p T_e[1 + \tfrac{1}{2}(\gamma - 1)M_e^2] = \text{const.}$$

it follows that

$$\frac{1}{\rho_e}\frac{d\rho_e}{dx} = \frac{1}{p}\frac{dp}{dx} - \frac{1}{T_e}\frac{dT_e}{dx} = -\frac{1}{u_e}\frac{du_e}{dx}M_e^2$$

and so equation (7.25) can be written

$$\frac{d\theta}{dx} + \frac{\theta}{u_e}\frac{du_e}{dx}(H + 2 - M_e^2) = \frac{\tau_w}{\rho_e u_e^2} \tag{7.26}$$

In incompressible flow this becomes

$$\frac{\mathrm{d}\theta}{\mathrm{d}x} + \frac{\theta}{u_\mathrm{e}}\frac{\mathrm{d}u_\mathrm{e}}{\mathrm{d}x}(H + 2) = \frac{\tau_\mathrm{w}}{\rho_\mathrm{e} u_\mathrm{e}^2} \tag{7.27}$$

If it is necessary to include second order terms in the equations of motion then additional terms, involving curvature, the normal eddy stresses and the pressure change across the boundary layer, correspondingly enter the momentum integral equation. For details see Lock and Firmin.[7.2] These terms can become significant when separation is imminent.

The kinetic energy integral equation (2.52) likewise applies to the turbulent boundary layer in the form

$$\tfrac{1}{2}\frac{\mathrm{d}}{\mathrm{d}x}(\rho_\mathrm{e} u_\mathrm{e}^3 \delta_\mathrm{E}) + \rho_\mathrm{e} u_\mathrm{e}^2 \frac{\mathrm{d}u_\mathrm{e}}{\mathrm{d}x}(\delta^* - \delta^{*\mathrm{i}}) = \frac{2}{\rho_\mathrm{e} u_\mathrm{e}^3}\int_0^h \tau \frac{\partial u}{\partial z}\,\mathrm{d}z$$

where $\tau = \tau_l + \tau_\mathrm{t} = \mu \dfrac{\partial u}{\partial z} - \rho\overline{u'w'}$

Hence
$$\frac{\mathrm{d}\delta_\mathrm{E}}{\mathrm{d}x} + \frac{\delta_\mathrm{E}}{u_\mathrm{e}}\frac{\mathrm{d}u_\mathrm{e}}{\mathrm{d}x}\left(3 + \frac{2\delta_\mathrm{T}}{\delta_\mathrm{E}} - M_\mathrm{e}^2\right) = \frac{2}{\rho_\mathrm{e} u_\mathrm{e}^3}\int_0^h \tau \frac{\partial u}{\partial z}\,\mathrm{d}z \tag{7.28}$$

Here δ_E is as defined in equation (2.43)

$$\delta^{*\mathrm{i}} = \int_0^h [1 - (u/u_\mathrm{e})]\mathrm{d}z$$

and δ_T is a temperature (or enthalpy) thickness defined by

$$\delta_\mathrm{T} = \int_0^h \frac{\rho u}{\rho_\mathrm{e} u_\mathrm{e}}\left(\frac{T}{T_\mathrm{e}} - 1\right)\mathrm{d}z$$

In incompressible flow over an adiabatic wall this becomes

$$\frac{\mathrm{d}\delta_\mathrm{E}}{\mathrm{d}x} + 3\frac{\delta_\mathrm{E}}{u_\mathrm{e}}\frac{\mathrm{d}u_\mathrm{e}}{\mathrm{d}x} = \frac{2}{\rho_\mathrm{e} u_\mathrm{e}^3}\int_0^h \tau \frac{\partial u}{\partial z}\,\mathrm{d}z \tag{7.29}$$

References

7.1 Myring, D.F. and Young, A.D. (1968) 'The isobars in boundary layers at supersonic speeds'. *Aero. Qu.*, **XIX**, 2, p. 105.
7.2 Lock, R.C. and Firmin, M.C.P. (1981) 'Survey of techniques for estimating viscous effects in external aerodynamics'. *Numerical Methods in Aeronautical Fluid Dynamics* (ed. Roe P.L.) Acad. Press, NY.
7.3 Gaviglio, J. (1987) 'Reynolds analogies and experimental study of heat transfer in supersonic flow'. *Int. J. Heat Mass Transfer*, **30**, 5, p. 911.

Chapter 8

Boundary Layer Drag: Prediction by Integral Methods

8.1 Introductory remarks

In Section 1.8 we described the various contributions to the drag of a body in motion in a fluid. It can be inferred that when the drag due to lift (vortex or induced drag) plus that due to any shock waves generated by the body (wave drag) are deducted from the pressure drag there remains a residual drag component due to the effective change of shape of the body associated with the presence of the boundary layer on the body. That residual component, which we call boundary layer pressure drag, plus the drag due to the frictional shear stresses at the surface (skin friction drag) comprise the boundary layer drag or profile drag.

We have also noted in Section 2.10 that the effective change of shape of the body due to the boundary layer is a consequence of the effect of the boundary layer in displacing fluid away from the surface, insofar as the local velocity component parallel to the surface is reduced, and the velocity component normal to the surface is increased as compared with inviscid flow. This displacement is equivalent to displacing the surface outwards a distance δ^* – the displacement thickness. We shall now consider this displacement effect in a little more detail and see how it results in a contribution to the pressure drag. We shall go on to relate the overall boundary layer drag to the wake momentum thickness far downstream, and we shall then introduce boundary layer prediction methods mainly based on solutions of the momentum integral equation of Section 2.11.

8.2 Boundary layer pressure drag

For simplicity we shall begin by considering the steady flow of an incompressible fluid past a two dimensional streamline shape such as an aerofoil at a small angle of incidence. Then from the equation of continuity

$$\partial u/\partial x + \partial w/\partial z = 0$$

we see that $w_\delta = -\int_0^\delta (\partial u/\partial x)\mathrm{d}z$

where w_δ is the value of w at the outer edge of the boundary layer ($z = \delta$). Therefore

$$w_\delta = -\frac{\mathrm{d}}{\mathrm{d}x}\left[\int_0^\delta u\ \mathrm{d}z\right] + u_e\frac{\mathrm{d}\delta}{\mathrm{d}x} = \frac{\mathrm{d}}{\mathrm{d}x}[u_e(\delta^* - \delta)] + u_e\frac{\mathrm{d}\delta}{\mathrm{d}x}$$

where $\delta^* = \int_0^\delta [1 - (u/u_e)]\mathrm{d}z$

Hence $w_\delta = \mathrm{d}(u_e\delta^*)/\mathrm{d}x - \delta\ \mathrm{d}u_e/\mathrm{d}x$ (8.1)

Now suppose the surface is displaced outwards normal to itself by a small distance δ_E^*, say, and consider the flow of an inviscid fluid past the surface with the same undisturbed main stream velocity u_∞ and with the same coordinate system as in the original viscous flow. We use suffix E to denote this equivalent inviscid flow. Then after again integrating the equation of continuity from $z = 0$ to $z = \delta$, we obtain

$$w_{\delta E} = \mathrm{d}(u_{eE}\ \delta_E^*)/\mathrm{d}x - \delta\ \mathrm{d}(u_{eE})/\mathrm{d}x \tag{8.2}$$

Here u_{eE} is the velocity close to the surface in the inviscid flow and we have neglected the change in u_{eE} with z from $z = \delta_E^*$ to $z = \delta$. If we now postulate that the displacement δ_E^* is equivalent in its effect on the external flow to the boundary layer in the real flow, so that in both flows the velocity magnitude and direction are the same at $z = \delta$, it follows that

$$u_{eE} = u_e \quad \text{and} \quad w_{\delta E} = w_\delta$$

and hence, from equations (8.1) and (8.2)

$$\delta_E^* = \delta^*$$

Therefore the equivalent surface displacement is the displacement thickness.

This argument is readily generalised to compressible flow in the absence of shock waves, with the definition of δ^* given in Section 2.10. It applies strictly to cases where the boundary layer is thin in relation to a linear dimension of the body and the change in $\rho_E u_E$ over a distance normal to the wall equal to δ can be neglected. This is equivalent to regarding the pressure change over that distance as negligible. It cannot be applied unmodified to cases of large surface curvature or large streamline curvature such as in a region of flow separation where $\mathrm{d}\delta^*/\mathrm{d}x$ can be large, or in a region of shock-wave boundary layer interaction or of a trailing edge. For such cases the concept of a displacement thickness and the associated equivalent inviscid flow still applies but the definition of δ^* has to be changed to

$$\delta^* = \frac{1}{\rho_{Ew}u_{Ew}}\int_0^\delta (\rho_E u_E - \rho u)\mathrm{d}z \tag{8.3}$$

to allow for the possible variation of $\rho_E u_E$ normal to the wall over the boundary layer height. The questions that need to be considered when such secondary effects have to be taken into account are complex and cannot be developed here, for further details the reader is referred to Lock and Williams.[8.1]

We note that the wake behind the aerofoil is a prolongation of the boundary layers on its surfaces, and the displacement thickness distribution along the wake downstream of the trailing edge defines an infinite extension of the equivalent shape.

Consider the simplest case of a symmetrical wing section at zero incidence with a finite trailing edge angle and its equivalent shape, including the wake extension, as illustrated in Fig. 8.1. We know that a body in steady sub-critical inviscid flow has no drag, since there is no mechanism for the dissipation of mechanical energy. We also know that for the aerofoil in such a flow the velocity at the trailing edge is zero and the pressure attains the full stagnation value with a pressure coefficient $c_p = 1$. However, if we now consider the inviscid flow past the displaced surface and its wake extension, equivalent to the real flow past the aerofoil as far as the flow outside the boundary layers and wake is concerned, then we see that the sharp trailing edge is in the equivalent flow replaced by smooth streamlines that merge with the wake equivalent. Hence the pressure as the trailing edge is approached does not rise to the full stagnation value, but reaches a value much closer to the free stream value. In fact, with an attached boundary layer on the aerofoil the trailing edge pressure coefficient is about 0.1–0.2. With some flow separation ahead of the trailing edge the pressure coefficient can become negative.

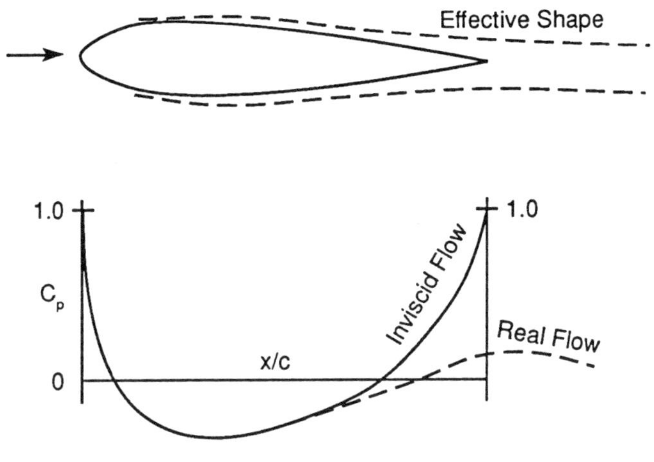

Fig. 8.1 Effective shape (not to scale) of an aerofoil due to the boundary layer and wake displacement effect, and the corresponding pressure distribution change.

The overall effect of the boundary layer and wake is therefore seen to be a reduction of pressure over the rear of the wing as compared with that of inviscid flow over the undisplaced surface, whilst the pressure over the forward part of the wing is practically unaffected. Consequently, instead of the zero drag of inviscid flow the wing experiences a pressure drag directly attributable to the presence of the boundary layers, hence the name *boundary layer pressure drag*.

This concept of boundary layer pressure drag is quite general and applies to all bodies, three dimensional as well as two dimensional, in a real fluid.

8.3 Relation between boundary layer drag and momentum thickness far downstream

Consider an aerofoil in steady compressible but shock-free flow as illustrated in Fig. 8.2. ABCD is a large rectangle and AB and CD are far upstream and downstream of the aerofoil, respectively, so that there the value of the static pressure is that of the undisturbed flow, p_∞. At AB the velocity and density are the same as in the undisturbed flow, u_∞ and ρ_∞, respectively. AB and CD are normal to the undisturbed stream velocity direction so that if D is the drag per unit span of the aerofoil it must equal the net reduction in the rate of flux of momentum in that direction (the direction of the x axis) across the faces of ABCD. Since the pressures at the faces AB and CD are equal they do not contribute to this momentum flux. Hence

$$D = \int_{\mathrm{A}}^{\mathrm{B}} \rho_\infty u_\infty^2 \, \mathrm{d}z - \int_{\mathrm{C}}^{\mathrm{D}} \rho u^2 \, \mathrm{d}z - \int_{\mathrm{B}}^{\mathrm{C}} \rho w u_\infty \, \mathrm{d}x - \int_{\mathrm{A}}^{\mathrm{B}} \rho w u_\infty \, \mathrm{d}x$$

Here we assume that the faces BC and AD are sufficiently far from the aerofoil for $u = u_\infty$ there.

Now continuity requires zero net flux of mass across the faces of ABCD. Hence

$$\int_{\mathrm{A}}^{\mathrm{B}} \rho_\infty u_\infty \, \mathrm{d}z - \int_{\mathrm{C}}^{\mathrm{D}} \rho u \, \mathrm{d}z - \int_{\mathrm{B}}^{\mathrm{C}} \rho w \, \mathrm{d}x - \int_{\mathrm{A}}^{\mathrm{D}} \rho w \, \mathrm{d}x$$

therefore $$D = \int_{\mathrm{C}}^{\mathrm{D}} \rho u (u_\infty - u) \, \mathrm{d}z \qquad (8.4)$$

But outside the wake the total pressure must be constant and since the static pressure $= p_\infty$ on CD it follows that $u = u_\infty$ outside the wake. Therefore, we can write

$$D = \int_{\mathrm{w(CD)}} \rho u (u_\infty - u) \, \mathrm{d}z \qquad (8.5)$$

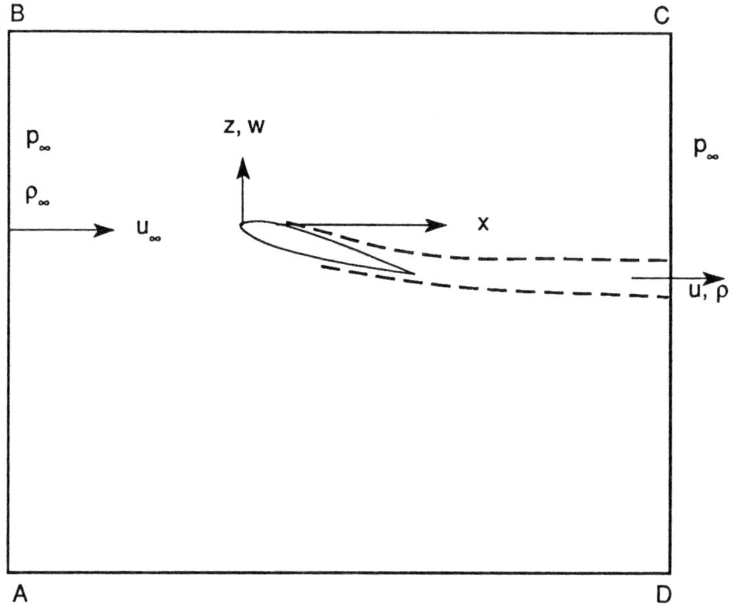

Fig. 8.2 Control volume to illustrate the relation between aerofoil drag and the momentum flux defect far downstream.

where the integral is indicated as confined to the section of the wake cut by CD. Thus, the drag is equal to the integral of the mass flow rate across an element dz of CD times the velocity decrement suffered by the flow across the element relative to the velocity far upstream.

Therefore the drag coefficient is

$$C_D = \frac{D}{\frac{1}{2}\rho_\infty u_\infty^2 c} = \frac{2}{c}\int_{W(CD)} \frac{\rho u}{\rho_\infty u_\infty}\left(1 - \frac{u}{u_\infty}\right)dz = 2\theta_\infty/c \qquad (8.6)$$

where θ_∞ is the value of the momentum thickness of the wake far downstream.

If there are shock waves present on the wing then equation (8.6) applies unchanged if θ is determined by integration to the outer edge of the wakes of the shocks so as to encompass the rate of momentum loss due to the shocks. In that case D is the sum of the boundary layer drag and the wave drag. [In viscous–inviscid interaction methods it is usual to define θ in terms of the differences between the actual flow and an equivalent inviscid flow, as in the discussion in Section 8.2, and the latter flow includes any shock waves and their effects down to the displacement surface. In that case θ continues to be regarded as the momentum thickness of the boundary layer, or its wake, and D in equation (8.6) is the boundary layer drag.[8.1]]

8.4 Some initial comments on differential and integral methods

As we shall see we can link the value of θ at the trailing edge of a wing (θ_c) to its value far downstream (θ_∞) by solving the flow equations of the wake, hence we can then use equation (8.6) to determine the boundary layer drag coefficient. To determine θ at any station on the aerofoil surface given the distributions of u_e and ρ_e we can either solve the time averaged boundary layer equations of motion and energy (*differential methods*) or we can solve one or more of the integral equations of Section 7.4 (*integral methods*).

We have seen in Chapter 7 that in the case of turbulent flow the time averaged equations involve the unknown turbulent eddy transport terms and these must in some way be related to the mean flow to solve those equations. In the absence of a satisfactory physical understanding and theory of turbulence such relations must be empirically based. The particular empirical forms chosen are referred to as 'closure' relations, and the process of deriving them, based on simplifying assumptions regarding the turbulence structure, is referred to as 'turbulence modelling'. In principle, differential methods provide detailed information of the boundary layer flow, e.g. velocity and turbulent eddy stress distributions but require a correspondingly detailed input of the initial conditions.

Integral methods more simply aim to determine overall quantities such as θ, c_f and H by satisfying the integral equations, but additional relations are needed between these quantities for this purpose and these relations again involve empiricisms and more or less plausible assumptions. However, the latter need only involve overall quantities and do not require any detailed turbulence modelling. Integral methods do not lead to flow details of the kind provided by differential methods, but by the same token they do not require very detailed initial conditions. They are much simpler to use and they make significantly smaller demands on computing resources. Within their more limited aims their accuracy appears to be no poorer than that for differential methods.

In the remainder of this chapter we shall discuss some representative integral methods for attached boundary layers, and in the subsequent chapter we shall similarly discuss some differential methods.

8.5 Some integral methods for incompressible flow

8.5.1 Simple methods based on power law relations and constant H

Consider steady flow over a thin aerofoil at small angles of incidence for which the pressure gradients are weak to moderate. Then the integral momentum equation is [equation (7.27)]

$$\frac{\mathrm{d}\theta}{\mathrm{d}x} + \frac{\theta}{u_e}\frac{du_e}{\mathrm{d}x}(H + 2) = \frac{\tau_w}{\rho u_e^2} = \frac{c_{fe}}{2} = \frac{1}{\zeta^2}$$

where $\zeta = (2/c_{fe})^{1/2}$

$$c_{fe} = \frac{2\tau_w}{\rho u_e^2} = c_f(u_e/u_\infty)^2$$

This equation can be solved for θ, given the external velocity u_e as a function of x, if we have two more relations between θ, H and τ_w (or ζ). The form parameter H is relatively insensitive to moderate pressure gradients and the simplest assumption is to take H = const. The value 1.4 is often used as a reasonable mean value over a wide range of R_x and typical wing pressure distributions at small angles of incidence. As we shall see, the resulting value of θ is fairly insensitive to the value of H chosen. For the other required relation the simplest procedure is to assume that a zero pressure gradient relation between c_f and R_θ, as obtained for the flow over a flat plate at zero incidence, holds locally. For example, we can use the relation based on the power law velocity profile [equation (6.45)]

$$c_{fe} = 2\tau_w/\rho u_e^2 = K' \, R_\theta^{-2/(n+1)} \tag{8.7}$$

where K' is a constant dependent on n, and $R_\theta = u_e\theta/\nu$. Then the momentum integral equation can be written

$$\frac{\mathrm{d}\theta}{\mathrm{d}x} + \frac{\theta}{u_e}\frac{du_e}{\mathrm{d}x}(H + 2) = \frac{K'}{2}(u_e\theta/\nu)^{-2/(n+1)}$$

or $$\theta^{2/(n+1)}\frac{\mathrm{d}\theta}{\mathrm{d}x} + \theta^{(n+3)/(n+1)}\frac{\mathrm{d}u_e}{\mathrm{d}x}(H + 2) = \frac{K'}{2}(u_e/\nu)^{-2/(n+1)}$$

Hence

$$\frac{\mathrm{d}}{\mathrm{d}x}\left[\frac{n+1}{n+3}\theta^{(n+3)/(n+1)}\, u_e^g\right] = \frac{K'}{2}(u_e/\nu)^{-2/(n+1)}\, u_e^g$$

where $g = (H + 2)(n + 3)/(n + 1)$. Therefore on integration with respect to x we get

$$(\theta^{(n+3)/(n+1)}\, u_e^g)_x - (\theta^{(n+3)/(n+1)}\, u_e^g)_t = \frac{n+3}{n+1}\frac{K'}{2}\int_{x_t}^{x} u_e^g(u_e/\nu)^{-2/(n+1)}\,\mathrm{d}x \tag{8.8}$$

where the suffix t refers to the transition point from which the integration starts. The fact that u_e^g appears on both sides of equation (8.8) implies that the resulting value of θ is fairly insensitive to small changes in g (or in H).

If we now understand the velocity u_e to be made non-dimensional by dividing it by u_∞ and all lengths made non-dimensional in terms of the chord c, then this equation takes the non-dimensional form

$$[\theta^{(n+3)/(n+1)}\,u_e^g]_{x_t}^{x} = \frac{n+3}{n+1}\frac{K'}{2}R^{-2/(n+1)}\int_{x_t}^{x} u_e^{g-2/(n+1)}\,dx \tag{8.9}$$

where $R = u_\infty c/\nu$. We see therefore that we can determine θ as a function of x by a simple integration of u_e^f with respect to x, where $f = g - 2/(n+1)$. Having θ as a function of x we can determine c_{fe}, using equation (8.7), so that

$$c_f = c_{fe}u_e^2 = K'\,R^{-2/(n+1)}\,u_e^{2n/(n+1)}\,\theta^{-2/(n+1)} \tag{8.10}$$

For example, with $n = 7$, $K' = 0.0260$, and with $H = 1.4$ we get $g = 4.25$. Then

$$\left.\begin{aligned}[\theta^{5/4}\,u_e^{4.25}]_{x_t}^{x} &= 0.0163R^{-1/4}\int_{x_t}^{x} u_e^4\,dx\\ c_f &= 0.0260R^{-1/4}\,u_e^{7/4}\,\theta^{-1/4}\end{aligned}\right\} \tag{8.11}$$

With $n = 9$, $K' = 0.0176$, and with $H = 1.4$, $g = 4.08$ and so

$$\left.\begin{aligned}[\theta^{6/5}\,u_e^{4.08}]_{x_t}^{x} &= 0.0106R^{-1/5}\int_{x_t}^{x} u_e^{3.88}\,dx\\ c_f &= 0.0176R^{-1/5}\,u_e^{9/5}\,\theta^{-1/5}\end{aligned}\right\} \tag{8.12}$$

It is assumed that θ is continuous at the transition point, to avoid unrealistic infinite shear stresses there, and its value there can be determined from a calculation of the initial laminar boundary layer as described in Section 4.2. As with the basic flat plate relations the $n = 7$ relations are applicable for R in the range 5×10^5 to 10^7, whilst the $n = 9$ relations are applicable for R in the range 10^6 to 10^8.

8.5.2 Simple method based on log law velocity profile

An earlier method of almost equal simplicity was developed by Squire and Young[6.23] making use of the log law deduction [equation (6.54)]

$$u_e\theta/\nu = R_\theta = 0.2454\,\exp\,(0.3914\zeta) \tag{8.13}$$

Combined with the momentum integral equation and the assumption $H = 1.4$ this leads to the following equation expressed in terms of non-dimensional quantities

$$\left.\begin{aligned}\frac{d\zeta}{dx} + \frac{6.13}{u_e}\frac{du_e}{dx} &= R\,u_e\,F(\zeta)\\ \text{where}\quad F(\zeta) &= 10.411\zeta^{-2}\exp\,[-0.3914\zeta]\end{aligned}\right\} \tag{8.14}$$

This equation can be readily solved numerically to yield ζ as a function of

x starting from the transition point, and hence the distributions of θ and c_f can be deduced.

8.5.3 Approximate solution of wake momentum integral equation

Having determined θ as a function of x over the aerofoil surfaces, we have seen in Section 8.3 that we need to relate θ at the trailing edge, θ_c, to θ far downstream to deduce the boundary layer drag coefficient, C_D. For this we must solve either the momentum equation of the wake or its momentum integral equation. Here we pursue the latter course. The momentum integral equation is (since τ_w is zero)

$$\frac{d\theta}{dx} + \frac{\theta}{u_e}\frac{du_e}{dx}(H + 2) = 0 \tag{8.15}$$

We assume that as the aerofoil incidence is small the curvature of the wake is everywhere small enough for the pressures just above and below it to be regarded as equal. In that case we can add the values of θ for the upper and lower surfaces of the aerofoil at the trailing edge to provide the starting value of θ for the wake, and we can assume that u_e is the same at both the outer edges of the wake for any given station x, where x is measured along the centre line of the wake. However, we can no longer assume that H is constant, the wake velocity profile spreads and becomes less peaky with increase of x, and consequently H decreases from its value at the trailing edge, 1.4, say, to the value unity far downstream. At the same time u_e/u_∞ increases from a value a little less than unity at the trailing edge (in the absence of separation there) to unity far downstream.

We can write equation (8.15) as

$$\frac{\theta_\infty}{\theta}\frac{d}{dx}(\theta/\theta_\infty) + \frac{H+2}{u_e/u_\infty}\frac{d}{dx}(u_e/u_\infty) = 0$$

or
$$\frac{d}{dx}[\ln(\theta/\theta_\infty)] + \frac{d}{dx}[(H + 2)\ln(u_e/u_\infty)] = \ln(u_e/u_\infty)\frac{dH}{dx}$$

which when integrated from the trailing edge (suffix c) to $x = \infty$ leads to

$$\ln(\theta_c/\theta_\infty) + (H_c + 2)\ln(u_{ec}/u_\infty) = \int_1^{H_c}\ln(u_e/u_\infty)dH$$

Hence
$$\theta_\infty = \theta_c(u_{ec}/u_\infty)^{H_c+2}\exp\left[\int_1^{H_c}\ln(u_\infty/u_e)dH\right] \tag{8.16}$$

To proceed further we need to relate u_e and H. Experiments have suggested that for the aerofoil at small incidences considered here

$$\ln(u_\infty/u_e) = \text{const.}\,(H - 1)$$

so that
$$\frac{\ln(u_\infty/u_e)}{\ln(u_\infty/u_{e0})} = \frac{H - 1}{H_c - 1}$$

and hence $$\exp\left[\int_1^{H_c} \ln(u_\infty/u_e)\mathrm{d}H\right] = \exp\left[\int_1^{H_c} \ln(u_\infty/u_{ec})\frac{(H-1)}{(H_c-1)}\mathrm{d}H\right]$$

$$= \exp\,[\ln\,(u_\infty/u_{ec})(H_c - 1)/2] = (u_\infty/u_{ec})^{(H_c-1)/2}$$

Therefore equation (8.16) becomes

$$\left.\begin{aligned}\theta_\infty &= \theta_c(u_{ec}/u_\infty)^{(H_c+5)/2} \\ &= \theta_c(u_{ec}/u_\infty)^{3.2}, \quad \text{for } H_c = 1.4\end{aligned}\right\} \tag{8.17}$$

From equations (8.6) and (8.17) we have finally

$$C_D = 2\theta_c(u_{ec}/u_\infty)^{3.2} \tag{8.18}$$

Therefore from the calculated value of θ at the trailing edge, summed for the upper and lower surfaces, and the specified value of u_{ec}/u_∞ we can determine the boundary layer drag coefficient C_D.

The above analysis was used by Squire and Young in 1938[6.23] to calculate the values of C_D for a range of aerofoils of the type suitable for the relatively slow propeller driven aircraft of that time, with the maximum thickness about 0.3c back from the leading edge. Some representative results of the calculations for ranges of Reynolds number, transition position (taken at a mean value for upper and lower surfaces) and section thickness are illustrated in Fig. 8.3. Comparison of the results with available experimental data showed satisfactory agreement for the types of aerofoils and small incidences assumed for the method.

An extension of the method to axi-symmetric flow and hence to the drag of streamline bodies of revolution at zero incidence was subsequently made by Young[8.2] who again covered practical ranges of fineness ratio, Reynolds number and transition position.

It will be evident that the calculations provide not only estimates of the total boundary layer drag but can also yield estimates of the skin friction drag, since the calculated distribution of c_f can be integrated with respect to the undisturbed stream direction over the surface. The difference then yields the boundary layer pressure drag. The latter is difficult to determine experimentally for attached flows since integration of the pressure coefficient, c_p, with respect to z over the surface leads to the boundary layer pressure drag as the small difference between two relatively large quantities.

However, the calculated results for the aerofoils showed that

$$\frac{\text{boundary layer pressure drag}}{\text{boundary layer drag}} = \left(\frac{t}{c}\right), \text{ approx.}$$

where t is the maximum thickness of the aerofoil. For streamline bodies of revolution it was likewise found that

$$\frac{\text{boundary layer pressure drag}}{\text{boundary layer drag}} = \frac{0.4d}{l}, \text{ approx.}$$

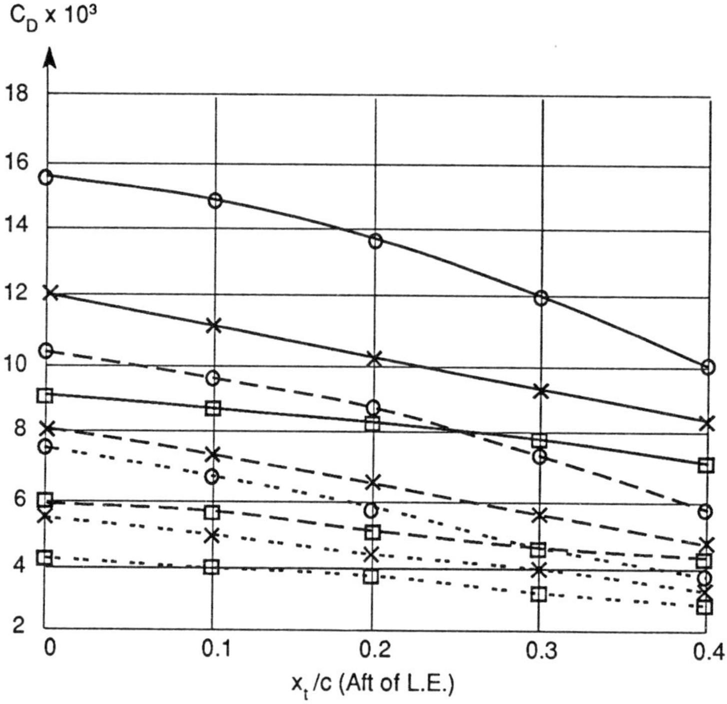

Fig. 8.3 Calculated C_D as a function of transition position, t/c (max) and Reynolds number for conventional aerofoil sections.[6.23] The key is as follows:

line type	R	symbol	t/c (max)
————	10^6	⊙	0.20
– – – – –	10^7	x	0.10
-- -- -- -- --	10^8	⊡	0

where d is the maximum diameter and l is the length of the body.

These relations illustrate the fact that the boundary layer pressure drag is a very small part of the boundary layer drag for attached flow over a streamline shape.

8.5.4 Methods with H as a variable

8.5.4.1 Early methods

The methods described in the preceding section are remarkably simple in concept and application, but it has been noted that they are limited in the range of cases to which they can be applied with acceptable accuracy. The assumption of a constant H in the boundary layer and the use of zero pressure gradient relations between τ_w and θ on the RHS of the momentum integral equation effectively eliminate possible pressure distribution effects on the velocity profiles and on the turbulent eddy stress profiles. This limits the methods to thin aerofoils at small angles of incidence involving only small pressure gradients. To improve on these methods we must therefore as a first step seek auxiliary relations between θ, H and τ_w that more closely reflect such pressure distribution effects.

A move in this direction, first due to Ludwieg and Tillmann[6.25] and later developed by Maskell[8.3], is to replace equation (8.7) by

$$2\tau_w/\rho_e u_e^2 = G(H)R_\theta^{-2/(n+1)} \tag{8.19}$$

where $G(H)$ is for the present undetermined, but it can be determined empirically as, for example, from the Ludwieg–Tillmann relation [equation (6.61)]. Therefore, the momentum integral equation then becomes

$$\frac{d\theta}{dx} + \frac{\theta}{u_e}\frac{du_e}{dx}(H+2) = G(H)R_\theta^{-2/(n+1)}$$

and hence

$$\begin{aligned}\frac{d}{dx}[\theta\, R_\theta^{2/(n+1)}] &= \frac{n+3}{n+1}\left\{G(H) - \frac{\theta}{u_e}\frac{du_e}{dx}R_\theta^{2/(n+1)}\left[(H+2) - \frac{2}{(n+3)}\right]\right\}\\ &= \frac{n+3}{n+1}\{G(H) + \Gamma[H + (2n+4)/(n+3)]\}\end{aligned} \tag{8.20}$$

Here $$\Gamma = -\frac{\theta}{u_e}\frac{du_e}{dx}R_\theta^{2/(n+1)} \tag{8.21}$$

is Buri's turbulent boundary layer parameter;[8.4] it is closely analogous to the Pohlhausen laminar boundary layer parameter [see equation (4.4)]

$$\lambda = \frac{\delta^2}{\nu}\frac{du_e}{dx} = \frac{\theta}{u_e}\frac{du_e}{dx}R_\theta(\delta/\theta)^2$$

An analysis of experimental data by Spence[8.5] showed that the RHS of equation (8.20) is very nearly a linear function of Γ so that

$$\frac{d}{dx}[\theta\, R_\theta^{2/(n+1)}] = a + b\Gamma, \quad \text{say} \tag{8.22}$$

where a and b are constants. This can be written

$$\frac{d}{dx}[\theta\, R_\theta^{2/(n+1)}\, u_e^b] = a\, u_e^b$$

and hence $$[\theta\, R_\theta^{2/(n+1)}\, u_e^b]_{x_t}^{x} = a\int_{x_t}^{x} u_e^b\, dx$$

If we now write this equation in terms of non-dimensional quantities using u_∞ and c as standard velocity and length, respectively, as before, then we get

$$[\theta^{(n+3)/(n+1)}\, u_{\mathrm{e}}^{\mathrm{f}}]_{x_{\mathrm{t}}}^{x} = a\, R^{-2/(n+1)} \int_{x_{\mathrm{t}}}^{x} u_{\mathrm{e}}^{b}\, \mathrm{d}x \tag{8.23}$$

where $f = b + 2/(n + 1)$.

Equation (8.23) is identical in form to equation (8.9) and since it must reduce to (8.9) for small pressure gradients it follows that $f = g$, and $a = (n + 3)K'/2(n + 1)$.

thus, for $n = 7$, $f = g = 4.25$, $b = 4.0$, $a = 0.0163$
whilst for $n = 9$, $f = g = 4.08$, $b = 3.88$, $a = 0.0106$

As a small but acceptable empirical modification of the set of values with $n = 9$, Spence suggested

$$f = g = 4.2, \quad b = 4.0, \quad a = 0.0106$$

and these lead to

$$[\theta^{6/5}\, u_{\mathrm{e}}^{4.2}]_{x_{\mathrm{t}}}^{x} = 0.0106 R^{-1/5} \int_{x_{\mathrm{t}}}^{x} u_{\mathrm{e}}^{4}\, \mathrm{d}x \tag{8.24}$$

The accuracy of such relations as (8.23) for deriving θ appears to be good except for conditions close to separation, and this is presumably because the relation is relatively insensitive to small variations of b, since u_{e}^{b} appears on both sides of the equation, and variations in H would mainly influence the result by their effect on b. However, to obtain c_{f} with adequate accuracy equation (8.19) is too approximate and we need to use an established two-parameter relation between c_{fe}, H and θ such as the Ludwieg–Tillmann, Thompson or Gaudet relations (see Section 6.9) plus a further auxiliary relation between these quantities. In the 1950s a number of authors sought the latter relation in the form of an empirically derived differential equation for $\mathrm{d}H/\mathrm{d}x$ in terms of H and Γ. An analysis by Thompson[8.6] of their resulting predictions when compared with experimental data demonstrated that none of these methods was consistently reliable, significant errors were found when the pressure gradients were very large or changing rapidly. However, the most reliable was the entrainment relation of Head[8.7]. The basic approach of this relation has been widely accepted and generalised to compressible flow and it will therefore now be described.

8.5.4.2 The entrainment equation

We denote the rate of volume flow (per unit span) across a section of the boundary layer as Q so that

$$Q = \int_0^{\delta} u\, \mathrm{d}z = \int_0^{\delta} u_{\mathrm{e}}\, \mathrm{d}z - \int_0^{\delta} u_{\mathrm{e}}(1 - u/u_{\mathrm{e}})\mathrm{d}z = u_{\mathrm{e}}(\delta - \delta^*)$$

and hence
$$\frac{\mathrm{d}Q}{\mathrm{d}x} = \frac{\mathrm{d}}{\mathrm{d}x}[u_{\mathrm{e}}(\delta - \delta^*)] \tag{8.25}$$

We write $H_1 = (\delta - \delta^*)/\theta$ (8.26)

Head argued that the entrainment into the boundary layer arose from the interaction at the convoluted outer edge of the boundary layer of the large eddies and the adjacent potential flow and hence was determined by the parameters of the velocity defect region. He therefore postulated that the entrainment rate dQ/dx was a function of H_1, u_e and the length $(\delta - \delta^*)$ so that

$$\frac{d}{dx}[u_e(\delta - \delta^*)] = f(H_1, u_e, \delta - \delta^*), \quad \text{say}$$

Dimensional reasoning then leads to the non-dimensional relation

$$C_E = \frac{1}{u_e}\frac{dQ}{dx} = \frac{1}{u_e}\frac{d}{dx}[u_e(\delta - \delta^*)] = \frac{1}{u_e}\frac{d}{dx}(u_e\theta H_1) = F_1(H_1), \quad \text{say} \tag{8.27}$$

Head then derived a mean curve for $F_1(H_1)$ from some experimental data due to Newman[8.8] and Schubauer and Klebanoff[8.9]. Consistent with the assumption that the velocity profiles are uni-parametric he also derived empirically a relation between H_1 and H

$$H_1 = F_2(H), \quad \text{say} \tag{8.28}$$

However, from any of the two-parameter relations referred to in Section 6.9 the relation between H_1 and H can be derived with R_θ as an additional parameter, and the results obtained by Gaudet[6.28] are shown in Fig. 8.4. It is of interest to note that the simple form of power law due to Spence, referred to in Section 6.5, leads to the relation

$$H_1 = 2H/(H - 1)$$

For H less than about 3, i.e. for attached flows, this is close to Gaudet's results for the higher Reynolds numbers (R_θ in the range 10^5-10^6).

Therefore, given u_e as a function of x and the starting value of θ at transition, we can determine θ for a given station x from equation (8.23) or its variant (8.24). Then, given starting values of δ and δ^*, we can solve for H by a step by step process using equations (8.27) and (8.28). Alternatively, the momentum integral equation can be solved simultaneously with equations (8.27) and (8.28) by a step by step process using an additional relation between c_f, θ and H such as is given by any of the two-parameter analyses referred to above.

The use of the entrainment relation as an auxiliary relation has undoubtedly improved the range of applicability of integral methods and as already noted it has been widely adopted. However, the first of two important conferences held at Stanford,[8.10, 8.11] at which comparisons were made of the predictive abilities of a number of different methods

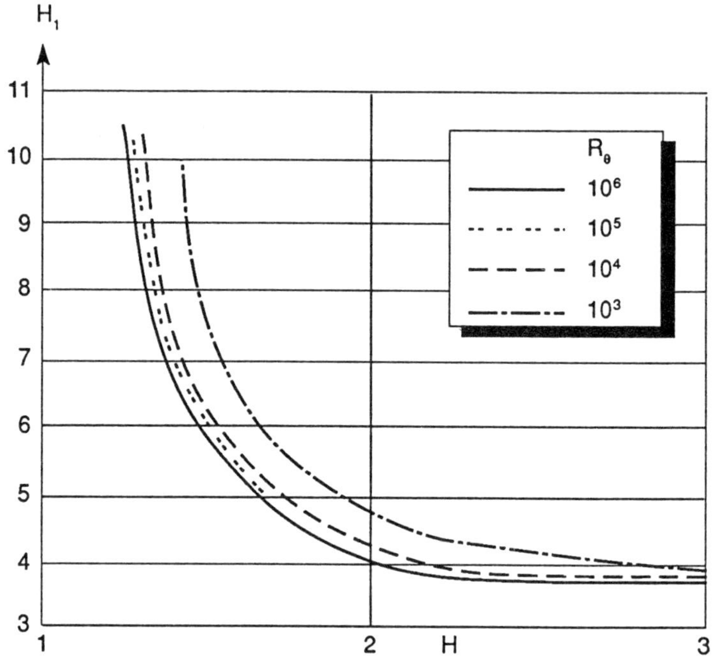

Fig. 8.4 H_1 against H relationship for various R_θ, according to reference 6.28.

tested against accepted experimental data, demonstrated that there was a need for further improvements. In particular, implicit in the methods so far described is the assumption that the local turbulence structure is determined solely by the mean local conditions, and the time lag of the turbulence structure in responding to changing external flow conditions is not taken into account. One can expect this lag to be especially important in conditions of a rapidly changing pressure distribution with streamwise distance.

As we shall see in the next chapter, differential methods involving the direct numerical solution of the equations of motion are less subject to criticism on these grounds and to that extent an important advantage can be claimed for them over integral methods.

However, the challenge of this situation led Head and Patel[8.12] to develop a modification of Head's entrainment method in which allowance for lag in the response of the turbulence structure was made by modulating the entrainment coefficient C_E with a function determined by a measure of the local departure from equilibrium conditions and tending to reduce that departure. The method is referred to as the *lag entrainment method* and in principle it is now widely adopted. We shall now briefly describe it.

8.5.4.3 Lag entrainment concept

Head and Patel chose as a parameter of departure from equilibrium

$$r_1 = \frac{1}{u_e}\frac{d}{dx}(u_e\theta)\Big/\left[\frac{1}{u_e}\frac{d}{dx}(u_e\theta)\right]_{eq} \tag{8.29}$$

where the suffix eq denotes equilibrium conditions for the same values of H and R_θ. They then postulated

$$C_E = (C_E)_{eq}\,F(r_1), \quad \text{say} \tag{8.30}$$

subject to

$$\begin{aligned} F(r_1) &= 1.0 \quad \text{when } r_1 = 1.0 \\ 0 < F(r_1) &< 1.0 \quad \text{when } r_1 > 1.0 \\ 1.0 < F(r_1) &< 2.0 \quad \text{when } r_1 < 1.0 \end{aligned}$$

To fit these conditions they chose

$$\left.\begin{aligned} F(r_1) &= (5 - 4r_1)/(3 - 2r_1) \quad \text{for } r_1 \leqslant 1.0 \\ &= 1/(2r_1 - 1) \quad \text{for } r_1 \geqslant 1.0 \end{aligned}\right\} \tag{8.31}$$

The momentum integral equation can be rearranged in the form

$$\frac{1}{u_e}\frac{d}{dx}(u_e\theta) = c_{fe}/2 - (H + 1)\frac{\theta}{u_e}\frac{du_e}{dx} \tag{8.32}$$

so that given c_{fe}, H and θ at a station x we can determine the change in $u_e\theta$ over a small step Δx.

Now

$$\begin{aligned} (C_E)_{eq} &= \frac{1}{u_e}\frac{d}{dx}[H_1(u_e\theta)]_{eq} \\ &= H_1\left[\frac{1}{u_e}\frac{d}{dx}(u_e\theta)\right]_{eq} \end{aligned} \tag{8.33}$$

since H_1 can be regarded as constant for an equilibrium flow. To determine

$$\left[\frac{1}{u_e}\frac{d}{dx}(u_e\theta)\right]_{eq}$$

for given values of H and R_θ, we can make use of relations derived from known solutions of equilibrium flow. Head and Patel made use of an empirical relation due to Nash

$$G = 6.1(\beta_p + 1.81)^{1/2} - 1.7 \tag{8.34}$$

where (see Section 6.5)

$$G = \frac{H - 1}{H}(2/c_{fe})^{1/2}, \qquad \beta_p = \frac{\delta^*}{\tau_w}\frac{dp}{dx} = -\frac{\delta^*}{\tau_w}\rho u_\tau\frac{du_e}{dx}$$

and for equilibrium flows G and β_p are constant. The momentum integral equation can be written

$$\frac{1}{u_e}\frac{d}{dx}(u_e\theta) = \frac{c_{fe}}{2}[1 + (H + 1)\beta_p/H]$$

so that
$$\left[\frac{1}{u_e}\frac{d}{dx}(u_e\theta)\right]_{eq} = \frac{c_{fe}}{2}[1 + (H + 1)\beta_p/H]_{eq} \tag{8.35}$$

For given H and R_θ we can determine c_{fe} from the equilibrium solutions obtained by Thompson and hence G can be determined and β_p then follows from equation (8.34). $(C_E)_{eq}$ can then be deduced from equations (8.35) and (8.33). If we know the value of $(1/u_e)d(u_e\theta)/dx$ at some point x, say, we can then determine r_1 and hence $F(r_1)$ using equations (8.29) and (8.31), and C_E follows from equation (8.30). We can then proceed by a step by step process as before. Alternatively, the relation between C_E and H_1, equation (8.27), can conveniently be rearranged as

$$C_E = \frac{1}{u_e}\frac{d}{dx}(u_e\theta H_1) = \theta\frac{dH_1}{dx} + \frac{H_1}{u_e}\frac{d}{dx}(u_e\theta)$$

or
$$\theta\frac{dH_1}{dx} = (C_E)_{eq}F(r_1) - \frac{H_1}{u_e}\frac{d}{dx}(u_e\theta)$$
$$= H_1\left[\frac{1}{u_e}\frac{d}{dx}(u_e\theta)\right]_{eq}[F(r_1) - r_1] \tag{8.36}$$

This equation can then be used with the momentum integral equation, in the form of equation (8.32), together with r_1, determined as above, to solve for H_1 and hence H and c_f by a numerical step by step process.

Head and Patel went on to apply the same approach to allow for small departures of the flow from strict two-dimensionality arising from some slight convergence or divergence of the external streamlines such as frequently occur in practice. Such departures may be expected to alter the closeness of the large eddies to each other and hence the entrainment rate. Thus, convergence would bring them closer together and hence would reduce the entrainment rate whilst divergence would be expected to have the opposite effect. This led Head and Patel to define a parameter r_2, the ratio of the experimental value of $(1/u_e)d(u_e\theta)/dx$ to its value in two dimensional flow, and to postulate a correction factor $F(r_2)$ relating the entrainment coefficient C_E to the corresponding value of C_E in two dimensions. For further details see reference 8.12.

The introduction of the lag entrainment concept undoubtedly improved the accuracy and extended the range of applicability of Head's integral method to put it on a par with the better differential methods as far as the prediction of overall quantities such as θ, c_f and H is concerned. Its relative simplicity and modest demands on computing resources make it an attractive method for most engineering purposes.

A subsequent variant of the lag entrainment method that has been much in use and later modified and extended to compressible flow (see

Section 8.6.2) is that of Green.[8.13, 8.14] The method makes use of the momentum integral equation (8.32) and the entrainment equation (8.27) with the latter rearranged to read

$$\theta \frac{dH}{dx} = \frac{dH}{dH_1}\left[C_E - \frac{H_1}{u_e}\frac{d}{dx}(u_e\theta)\right] \tag{8.37}$$

However, for the lag equation use is made of the turbulent kinetic energy transport equation. We shall refer to the latter equation in more detail when we come to consider differential methods (Chapter 9). Here we note that it can be derived from the Navier–Stokes equations for a mean steady flow in the form

$$\rho\left(u\frac{\partial\overline{e'}}{\partial x} + w\frac{\partial\overline{e'}}{\partial z}\right) - \tau\frac{\partial u}{\partial z} + \frac{\partial}{\partial z}(\overline{p'w'} + \overline{\rho e'w'}) + \rho\epsilon = 0 \tag{8.38}$$

advection production diffusion dissipation

where e' is the turbulence kinetic energy per unit mass and the terms in the equation are labelled according to the physical nature of their contribution to the overall rate of change of $\overline{e'}$ for a fluid particle.

The dissipation rate term

$$\epsilon = \frac{\nu}{2}\overline{\left(\frac{\partial u_i'}{\partial x_i} + \frac{\partial u_j'}{\partial x_i}\right)^2} \tag{8.39}$$

If the dissipating eddies are assumed to be isotropic then

$$\epsilon = \nu\overline{\left(\frac{\partial u_i'}{\partial x_j}\right)^2} = \nu\overline{\left[\left(\frac{\partial u'}{\partial x}\right)^2 + \left(\frac{\partial w'}{\partial z}\right)^2 + \left(\frac{\partial u'}{\partial z}\right)^2 + \left(\frac{\partial w'}{\partial x}\right)^2\right]} \tag{8.40}$$

With the aid of certain approximations Bradshaw *et al.*[8.15] transformed equation (8.38) into an equation for the shear stress τ in the development of his differential method which will be discussed in Section 9.5. Green further transformed this equation into an ordinary differential equation for $c_{\tau m} = \tau_{max}/\rho u_e^2$, at the point where τ_{max} occurs, in the form

$$\frac{\delta}{c_{\tau m}}\frac{dc_{\tau m}}{dx} = 2a_1\frac{u_e}{u}\left[\frac{\delta}{u_e}\frac{\partial u}{\partial z} - \frac{\delta}{L}c_{\tau m}^{1/2} - c_{\tau m}\frac{\partial\zeta}{\partial(z/\delta)} - \frac{2\delta}{u_e}\frac{du_e}{dx}\right] \tag{8.41}$$

where a_1 $(= \tau/\rho\overline{e'})$ is usually assumed constant $= 0.15$ with some experimental justification, ζ is a function of z/δ determined empirically to provide an approximate description of the diffusion term in equation (8.38) and L is a length scale given by

$$L = (\overline{e'})^{3/2}/\epsilon$$

so that L is related to the size of the dissipating eddies. For equilibrium flows $c_{\tau m}$ and the shapes of the non-dimensional shear stress and velocity profiles in terms of z/δ do not vary with x. These considerations led to the relation

$$\frac{\delta}{c_{\tau m}}\frac{dc_{\tau m}}{dx} = 2a_1\frac{u_e}{u}\frac{\delta}{L}(c_{\tau m\,eq}^{1/2} - c_{\tau m}^{1/2}) + \left(\frac{2\delta}{u_e}\frac{du_e}{dx}\right)_{eq} - \frac{2\delta}{u_e}\frac{du_e}{dx} \qquad (8.42)$$

With the use of an empirical relation between $c_{\tau m}$ and C_E this was then converted into an equation for C_E. This equation was of the form

$$\theta(H_1 + H)\frac{dC_E}{dx} = \frac{C_E(C_E + 0.2) + 0.8c_{f0}/3}{C_E + 0.01}[f(C_E, c_{f0})_{eq} - \lambda f(C_E, c_{f0})] + \left(\frac{\theta}{u_e}\frac{du_e}{dx}\right)_{eq} - \frac{\theta}{u_e}\frac{du_e}{dx} \qquad (8.43)$$

where $$f(C_E, c_{f0}) = 2.8\{0.32c_{f0} + 0.024C_E + 1.2C_E^2\}^{1/2} \qquad (8.44)$$

Here λ is a factor which is normally 1.0 but which can be modified to take account of secondary effects such as flow curvature or other factors affecting the dissipation process. In wakes where L/δ is approximately twice the value in a boundary layer λ is consequently taken as 0.5. Also c_{f0} is the value of c_{fe} for zero pressure gradient flow; the relation used was an empirical fit of the existing relations, such as Winter and Gaudet's[6.33]

$$c_{f0} = [0.01013/(\log_{10} R_\theta - 1.02)] - 0.00075 \qquad (8.45)$$

The relation between H_1 and H postulated was not that of Head but was based on some later experimental data and took the form

$$H_1 = 3.15 + 1.72/(H - 1) - 0.01(H - 1)^2 \qquad (8.46A)$$

Subsequent empirical formulae based on wider ranges of H have been devised by Lock[8.16] to include separated flows, viz.

$$H_1 = 2 + 1.5[1.12/(H - 1)]^{1.093} + 0.5[(H - 1)/1.12]^{1.093} \quad \text{for } 4 > H > 1.3 \qquad (8.46B)$$

and $$H_1 = 4 + (H - 4)/3 \quad \text{for } H > 4.$$

These relations show a minimum in H_1 at about $H = 2.8$, i.e. a value for which separation is likely. However, it must be emphasised that a universal and unique relation between H_1 and H is not to be expected as it implies that all turbulent boundary layer velocity profiles are uni-parametric.

For equilibrium flows the empirical relation between the form parameters G and β_p used was

$$G = \frac{(H - 1)}{H}(2/c_{fe})^{1/2} = 6.432(1 + 0.8\beta_p)^{1/2} \qquad (8.47)$$

instead of equation (8.34).

These relations combined with the momentum integral equation and the entrainment relation for equilibrium boundary layers led to

$$\left(\frac{\theta}{u_e}\frac{du_e}{dx}\right)_{eq} = \frac{1.25}{H}\left[\frac{c_{fe}}{2} - 0.0242\left(\frac{H-1}{H}\right)^2\right] \tag{8.48}$$

and

$$(C_E)_{eq} = H_1\left[0.0302\,\frac{(H+1)(H-1)^2}{H^3} - c_{fe}\left(0.25 + \frac{1.25}{H}\right)\right] \tag{8.49}$$

Finally, an empirical relation between c_{fe}, c_{f0} and H was used in the form

$$[(c_{fe}/c_{f0}) + 0.5][(H/H_0) - 0.4] = 0.9 \tag{8.50}$$

with H_0, the value of H for zero pressure gradient flow, given by

$$1 - (1/H_0) = 6.55(c_{f0}/2)^{1/2} \tag{8.51}$$

Equations (8.27), (8.37) and (8.43)–(8.51) can be solved numerically and simultaneously to give the distributions of θ, c_{fe} and H over the surface of a wing given the initial values at the transition location. For the wake the same equations were solved but with $c_{fe} = 0$. Also, as noted above, experimental results have indicated that the entrainment rate doubles in the wake as compared with conditions over the wing surface; this was met by assuming that there the factor λ in equation (8.43) = 0.5.

8.5.5 The use of the kinetic energy integral equation

The methods discussed above are representative of methods essentially devoted to the solution of the momentum integral equation. It was noted that they required two auxiliary equations and the methods differed in the nature and treatment of the auxiliary equations used.

A possibility not yet discussed which has been particularly favoured by German workers (see Truckenbrodt[8.17]) is the use of the kinetic energy integral equation (2.53) in the form

$$\frac{d}{dx}(u_e^3\,\delta_E/2) = \frac{1}{\rho}\int_0^\delta \tau\,\frac{\partial u}{\partial z}\,dz \tag{8.52}$$

where $\tau = \mu\,\partial u/\partial z - \rho\overline{u'w'} \simeq -\rho\overline{u'w'}$. This equation can be written

$$\frac{d\delta_E}{dx} + \frac{3\delta_E}{u_e}\frac{du_e}{dx} = \frac{2}{\rho u_e^2}\int_0^\delta \tau\,\frac{\partial u}{\partial z}\,dz = 2C_{DI} \tag{8.53}$$

where C_{DI} can be regarded as a dissipation coefficient. Examination of experimental data suggests that a power law relation, analogous to equation (8.19), can be used in the form

$$C_{\mathrm{DI}} = \beta\, R_{\delta\mathrm{E}}^{-b} \tag{8.54}$$

where β and b are functions of $\bar{H}$ where

$$\bar{H} = \exp \int_{H_{\mathrm{E0}}}^{H_{\mathrm{E}}} \frac{\mathrm{d}H_{\mathrm{E}}}{(H-1)H_{\mathrm{E}}} \tag{8.55}$$

and H_{E0} = value of H_{E} for zero pressure gradient flow, $R_{\delta_{\mathrm{E}}} = u_{\mathrm{e}}\delta_{\mathrm{E}}/\nu$. For equilibrium boundary layers β and b are assumed constant so that from equations (8.53) and (8.54) we deduce in terms of non-dimensional quantities

$$\frac{\mathrm{d}}{\mathrm{d}x}\left(\delta_{\mathrm{E}}^{b+1}\, u_{\mathrm{e}}^{3(1+b)}\right) = u_{\mathrm{e}}^{3+2b}\,(1+b)2\beta/R^{b}, \quad \text{where } R = u_{\infty}c/\nu$$

Hence

$$\left[\delta_{\mathrm{E}}^{b+1}\, u_{\mathrm{e}}^{3(1+b)}\right]_{x_{\mathrm{t}}}^{x} = \frac{2\beta(1+b)}{R^{b}} \int_{x_{\mathrm{t}}}^{x} u_{\mathrm{e}}^{3+2b}\,\mathrm{d}x \tag{8.56}$$

Therefore δ_{E} as a function of x can be determined by a simple integration of u_{e}^{3+2b} with respect to x. We note the similarity of this result with that of equation (8.9). For small pressure gradients $\bar{H} \simeq 1$ and experimental data suggest $b = 0.152$ and $\beta = 0.0053$. For the relations between δ_{E}, θ and c_{f} one can appeal to experimental data for equilibrium flows relating these quantities to the parameter $\bar{H}$. For further details see references 8.17 and 2.2.

8.6 Some integral methods for compressible flow

8.6.1 Simple methods using zero pressure gradient relations with H as a function of M_{e} and T_{w}

The two dimensional momentum integral equation (2.48) for compressible flow is

$$\frac{\mathrm{d}\theta}{\mathrm{d}x} + \frac{\theta}{u_{\mathrm{e}}}(H+2)\frac{\mathrm{d}u_{\mathrm{e}}}{\mathrm{d}x} + \frac{\theta}{\rho_{\mathrm{e}}}\frac{\mathrm{d}\rho_{\mathrm{e}}}{\mathrm{d}x} = \frac{\tau_{\mathrm{w}}}{\rho_{\mathrm{e}}u_{\mathrm{e}}^{2}}$$

which can be written [see equation 7.26)]

$$\frac{\mathrm{d}\theta}{\mathrm{d}x} + \frac{\theta}{u_{\mathrm{e}}}\frac{\mathrm{d}u_{\mathrm{e}}}{\mathrm{d}x}\left[(H+2) - M_{\mathrm{e}}^{2}\right] = \frac{\tau_{\mathrm{w}}}{\rho_{\mathrm{e}}u_{\mathrm{e}}^{2}} = \tfrac{1}{2}c_{\mathrm{fe}} = \tfrac{1}{2}c_{\mathrm{f}}\frac{\rho_{\infty}u_{\infty}^{2}}{\rho_{\mathrm{e}}u_{\mathrm{e}}^{2}} \tag{8.57}$$

where $c_{\mathrm{f}} = 2\tau_{\mathrm{w}}/\rho_{\infty}u_{\infty}^{2}$, and suffix ∞ refers to reference conditions (usually the undisturbed flow at ∞). Here $H = \delta^{*}/\theta$ as defined in equations (2.39) and (2.41).

Given u_{e} (or M_{e}) as a function of x, together with the reference conditions, our problem is as before to establish two auxiliary relations between θ, H and c_{fe} in order to be able to solve equation (8.57). We saw that in incompressible flow with zero pressure gradient the power law assumption for the velocity profile led to the relation [equation (6.45)]

$$c_{fe} = K' \, R_\theta^{-2/(n+1)}$$

where K' and n were related constants, but whose values were chosen to fit the Reynolds number range of interest. A suggested and simple generalisation of this relation for compressible flow is

$$c_{fe} = K' \, R_\theta^{-2/(n+1)} \, h(M_e) \tag{8.58}$$

where $h(M_e)$ is an as yet unspecified function of M_e, such that $h(0) = 1$. If we accept this relation then equation (8.57) becomes

$$\frac{d\theta}{dx} + \frac{\theta}{u_e}\frac{du_e}{dx}[(H+2) - M_e^2] = \tfrac{1}{2}K' \, R_\theta^{-2/(n+1)} \, h(M_e)$$

If we make all quantities non-dimensional by expressing lengths, velocities and the kinematic viscosity in terms of c (the chord), u_∞ and ν_∞, respectively, then

$$\frac{d\theta}{dx} + \frac{\theta}{u_e}\frac{du_e}{dx}(H + 2 - M_e^2) = \tfrac{1}{2}K' \, R^{-2/(n+1)} \, (u_e\theta/\nu_e)^{-2/(n+1)} \, h(M_e) \tag{8.59}$$

where $R = u_\infty c/\nu_\infty$. Hence

$$\theta^{2/(n+1)}\frac{d\theta}{dx} + \frac{\theta^{(n+3)/(n+1)}}{u_e}\frac{du_e}{dx}(H + 2 - M_e^2) = \tfrac{1}{2}K' \, R^{-2/(n+1)} \, (u_e/\nu_e)^{-2/(n+1)} \, h(M_e)$$

Let $$F(x) = \frac{1}{u_e}\frac{du_e}{dx}(H + 2 - M_e^2)\frac{n+3}{n+1}$$
$$G(x) = (u_e/\nu_e)^{-2/(n+1)} \, h(M_e) \tag{8.60}$$
then, when multiplied by

$$\exp \int_{x_t}^{x} F(x)dx$$

the above equation can be integrated to yield

$$\theta^{(n+3)/(n+1)} \exp \int_{x_t}^{x} F(x)dx = \tfrac{1}{2}\frac{(n+3)}{(n+1)} K' \, R^{-2/(n+1)} \int_{x_t}^{x} G(x)\left[\exp \int_{x_t}^{x} F(x)dx\right]dx \tag{8.61}$$

To determine $h(M_e)$ we note for the flow over a flat plate with zero pressure gradient and fully turbulent boundary layer that $F(x) = 0$, $u_e = 1$, and $\nu_e = 1$, so that equation (8.61) yields

$$\theta^{(n+3)/(n+1)} = \frac{n+3}{n+1}\frac{K'}{2} R^{-2/(n+1)} \, h(M_e)x$$

and so

$$
\begin{aligned}
c_{fe}/2 &= \frac{d\theta}{dx} \\
&= \left(\frac{n+3}{n+1}\right)^{-2/(n+1)} (K'/2)^{(n+1)/(n+3)} R_x^{-2/(n+3)} h(M_e)^{(n+1)/(n+3)} \\
&= \tfrac{1}{2} c_{fi}\, h(M_e)^{(n+1)/(n+3)}
\end{aligned}
$$

where c_{fi} is the value of c_f for the flat plate at zero Mach number and the same value of R_x.

Hence $$h(M_e) = (c_f/c_{fi})^{(n+3)/(n+1)} \tag{8.62}$$

where the coefficients c_f, c_{fi} are to be determined from the basic zero pressure gradient results such as are presented in Chapter 6.

For simplicity, we could adopt the relations derived from the intermediate temperature concept of Sommer and Short[6.31] (see Section 6.11) leading to

$$h(M_e) = (T_e/T_m)^{0.778} \tag{8.63}$$

for air with $n = 7$ (appropriate to moderate Reynolds numbers $< 10^7$), the intermediate temperature T_m is defined by equation (6.65). For air with $n = 9$ (appropriate to Reynolds numbers in the range $10^6 - 10^8$

$$h(M_e) = (T_e/T_m)^{0.822} \tag{8.64}$$

Alternatively, $h(M_e)$ can be obtained as a function of x, using equation (8.62) and the Spalding–Chi correlation [see equation (6.74) and Tables 6.2 and 6.3].

To complete the solution of the momentum integral equation we require to determine H as a function of x. Corresponding to the simple assumption that H = const. that we made for incompressible flow (Section 6.5.1) we can assume H is a function of M_e and T_w, only. Such an assumption can be expected to be valid only for cases where the pressure gradients and the consequent variations of Mach number M_e over the wing surface are small. For this purpose it is probably adequate to use the formula derived by Spence[8.18] after applying the modified Crocco temperature–velocity relation (7.20), namely

$$H = H_{i0}\, T_w/T_e + T_r/T_e - 1 \tag{8.65}$$

where H_{i0} is the value for H in incompressible flow and zero heat transfer at the same Reynolds number in terms of x as in the compressible flow.

This result was based on Spence's observation that for a flat plate at zero incidence the velocity profile was closely represented for wide ranges of M_e and T_w by

$$u/u_e = \phi(\eta/\Delta), \quad \text{say}$$

where $\eta = \int_0^z \rho/\rho_e \, dz, \qquad \Delta = \int_0^\delta \rho/\rho_e \, dz$

This observation of the existence of a transformation that resulted in a velocity distribution almost independent of Mach number and wall temperature in the absence of a pressure gradient encouraged Spence to develop another method of solution of the momentum integral equation by seeking for a more general transformation similar to the Stewartson–Illingworth transformation for laminar boundary layers (see Section 3.5). With such transformations the aim is to reduce the main equations to forms close to those of incompressible flow, so that the same modes of solution may be applied.

Thus he postulated new variables $\tilde{\theta}$ and $\tilde{u}_e$ where

$$\left.\begin{aligned} \tilde{\theta} &= (T_e/T_0)^k \, \theta = \xi \, . \, \theta, \quad \text{say} \\ \tilde{u}_e &= (a_0/a_e)^l \, u_e = (T_0/T_e)^{l/2} \, u_e \end{aligned}\right\} \tag{8.66}$$

Here suffix 0 refers to reservoir conditions in the external flow (i.e. corresponding to the flow brought to rest isentropically) k and l are constants open to choice.

If we now differentiate equation (8.66) with respect to x, noting

$$T_0 = T_e[1 + \tfrac{1}{2}(\gamma - 1)M_e^2], \qquad \frac{1}{M_e}\frac{dM_e}{dx} = \frac{(T_0/T_e)}{u_e}\frac{du_e}{dx}$$

we get $$\xi\frac{d\theta}{dx} = \frac{d\tilde{\theta}}{dx} - \theta\frac{d\xi}{dx} = \frac{d\tilde{\theta}}{dx} + (\gamma - 1)kM_e^2\frac{\tilde{\theta}}{u_e}\frac{du_e}{dx}$$

and

$$\frac{1}{\tilde{u}_e}\frac{d\tilde{u}_e}{dx} = \left[1 + \frac{l}{2}(\gamma - 1)M_e^2\right]\frac{1}{u_e}\frac{du_e}{dx}$$

Hence equation (8.57) becomes

$$\frac{d\tilde{\theta}}{dx} + \left[\frac{H + 2 - M_e^2 + (\gamma - 1)kM_e^2}{1 + \tfrac{1}{2}l(\gamma - 1)M_e^2}\right]\frac{\tilde{\theta}}{\tilde{u}_e}\frac{d\tilde{u}_e}{dx} = \xi\tau_w/\rho_e u_e^2 \tag{8.67}$$

For the case of zero heat transfer $T_w = T_r$ and (8.65) becomes

$$H + 1 = T_r(H_{i0} + 1)/T_e$$

so that

$$\begin{aligned} H + 2 - M_e^2 + (\gamma - 1)kM_e^2 &= \frac{T_r}{T_e}(H_{i0} + 1) + 1 + \left[2k - \frac{2}{(\gamma - 1)}\right]\frac{\gamma - 1}{2}M_e^2 \\ &= \frac{T_r}{T_e}(H_{i0} + 2), \text{ if } 2k - 2/(\gamma - 1) = r \end{aligned} \tag{8.68}$$

Here r is the recovery factor. If we also put $l = r$ then equation (8.67) becomes

$$\frac{d\tilde{\theta}}{dx} + \frac{\tilde{\theta}}{\tilde{u}_e}(H + 2)\frac{d\tilde{u}}{dx} = \frac{\xi\tau_w}{\rho_e u_e^2} \tag{8.69}$$

which as far as the LHS is concerned is formally the same as in incompressible flow.

For the case of an isothermal wall equation (8.67) can be reduced to

$$\frac{d\tilde{\theta}}{dx} + \left(\frac{T_w}{T_0}H_{i0} + 2\right)\frac{\tilde{\theta}}{\tilde{u}_e}\frac{d\tilde{u}_e}{dx} = \frac{\xi\tau_w}{\rho_e u_e^2} \tag{8.70}$$

if $\quad 2k + r - 2/(\gamma - 1) = 2, \quad \text{and} \quad l = 1 \qquad (8.71)$

We note that if $r = 1$, k and l take the Stewartson–Illingworth values, viz.

$$k = \tfrac{1}{2} + 1/(\gamma - 1), \qquad l = 1$$

To evaluate the RHS of equation (8.67) Spence assumed, as for the previous method discussed, that zero pressure gradient relations could be taken to hold locally and for this he made use of the mean temperature concept. However, he used Eckert's formula for the mean temperature, viz. $T_m = (T_w + T_r)/2$; here it is suggested in the light of more recent experimental data that the formula of Sommer and Short (6.65) is likely to prove more accurate. According to the mean temperature concept and using the power law relations for the basic incompressible flow we have [see equation (6.42)]

$$\tau_w/\rho_m u_e^2 = \tfrac{1}{2}C_3(u_e x/\nu_m)^{-2/(n+1)}$$

and after some algebra this yields [cf. equation (6.45)]

$$\xi\tau_w/\rho_e u_e^2 = \tfrac{1}{2}K'(T_e/T_0)^{\alpha}\,(T_m/T_e)^{-\beta}\,\phi/\tilde{\theta}$$

where $\quad \alpha = k + \dfrac{2}{n+1}\left(k - \dfrac{l}{2} - \dfrac{1}{\gamma - 1} + \omega\right), \qquad \beta = 1 - 2\omega/(n + 1)$

$$\phi = \tilde{\theta}(\tilde{u}_e\tilde{\theta}/\nu_e)^{-2/(n+1)}$$

After some further algebra equation (8.67) becomes

$$\frac{d\phi}{dx} + \frac{B\phi}{\tilde{u}_e}\frac{d\tilde{u}_e}{dx} = \tfrac{1}{2}K'\frac{(n+3)}{(n+1)}(T_e/T_0)^{\alpha}\,(T_m/T_e)^{-\beta} \tag{8.72}$$

where

$$\left.\begin{aligned} &\text{for zero heat transfer} \quad B = \frac{n+3}{n+1}(H_{i0} + 2) - \frac{2}{n+1} \\ &\text{for an isothermal wall} \quad B = \frac{n+3}{n+1}\left(\frac{T_w}{T_0}H_{i0} + 2\right) - \frac{2}{n+1} \end{aligned}\right\} \tag{8.73}$$

If as before we take H_{i0} as constant then B is constant and equation (8.72) can be integrated directly to give

$$[\phi\, u_e^B]_{x_t}^x = \tfrac{1}{2}K' \frac{n+3}{n+1} \int_{x_t}^x (T_e/T_0)^\alpha (T_m/T_e)^{-\beta} u_e^B \,dx \qquad (8.74)$$

This can be rewritten in terms of θ and M_e to yield

$$\left(\frac{\theta}{c}\right)^{(n+3)/(n+1)} M_e^{B+2/(n+1)} G(M_e) = \tfrac{1}{2}K' \frac{n+3}{n+1} R_a^{-2/(n+1)} \int_{x_t}^x M_e^B F(M_e) d(x/c) \qquad (8.75)$$

Here $R_a = a_0 c/\nu_0 = R[1 + 0.2M_\infty^2]^{3-\omega}/M_\infty$, for $\gamma = 1.4$

$$F(M_e) = (T_e/T_0)^\alpha (T_m/T_e)^{-\beta}, \qquad G(M_e) = (T_e/T_0)^s \qquad (8.76)$$

where $s = k(n+3)/(n+1) + 2[B(n+1) + 2](1-l)/(n+1)$.

By way of illustration, if we choose for air $n = 9$, $K' = 0.0176$ and $H_{i0} = 1.5$, the quantities appearing in equation (8.75) are as shown in Table 8.1.

In non-dimensional terms equation (8.74) then becomes

$$[\theta^{1.2} M_e^{B+0.2} G(M_e)]_{x_t}^x = 0.0106 R_a^{-0.2} \int_{x_t}^x M_e^B F(M_e) \,dx \qquad (8.77)$$

For the local skin friction coefficient c_{fe} we have

$$c_{fe} = 2\tau_w/\rho_e u_e^2 = K'(\rho_m/\rho_e)(\rho_e u_e \theta/\mu_m)^{-2/(n+1)}$$

which eventually gives

$$c_{fe} = K'(T_m/T_e)^{-1+2\omega/(n+1)} (T_e/T_0)^{-2(1/(\gamma-1)-\omega)/(n+1)} (\theta R_a)^{-2/(n+1)} \times [(1/M_\infty^2) + 0.2]^{1/(n+1)} u_e^{-2/(n+1)} \qquad (8.78)$$

For $n = 9$ and $K' = 0.0176$ we get

$$c_{fe} = 0.0176(T_m/T_e)^{-1+\omega/5} (T_e/T_0)^{(\omega/5)-(1/2)} (\theta R_a)^{-1/5} \times [(1/M_\infty^2) + 0.2]^{1/10} u_e^{-1/5} \qquad (8.79)$$

Table 8.1

	zero heat transfer	isothermal wall
B	4	$1.8(T_w/T_0) + 2.2$
$F(M_e)$	$(T_e/T_0)^{3.331} (T_m/T_e)^{-0.822}$	$(T_e/T_0)^{3.239} (T_m/T_e)^{-0.822}$
$G(M_e)$	$(T_e/T_0)^{3.753}$	$(T_e/T_0)^{3.661}$
T_m/T_e	$1 + 0.1158M_e^2$	$0.55 + 0.035M_e^2 + 0.45T_w/T_e$
T_e/T_0	$(1 + 0.2M_e^2)^{-1}$	$(1 + 0.2M_e^2)^{-1}$

Some calculated results[8.19] for the skin friction distributions on a biconvez wing 5% thick, at $M_\infty = 2.5$ and zero incidence, for various transition positions and wall temperatures are shown in Fig. 8.5, where $S_w = T_w/T_r - 1$. The laminar part of the boundary layer was calculated by the method described in Section 4.3.3. The strong effect of wall cooling in increasing the skin friction in the turbulent boundary layer is evident, it arises essentially from the associated increase of density close to the wall. It is to be noted that over such a wing at supersonic speeds the pressure gradient is favourable aft of the leading edge shock. For the laminar boundary layer the streamwise acceleration in this favourable pressure gradient is more sensitive than in the turbulent boundary layer to the changes of density associated with changes of wall temperature and this tends to reduce the skin friction when the wall is cooled so the overall effect is much smaller than with the boundary layer turbulent. It is evident that with far back transition there can be a slight increase of skin friction over the rear of the wing with wall cooling.

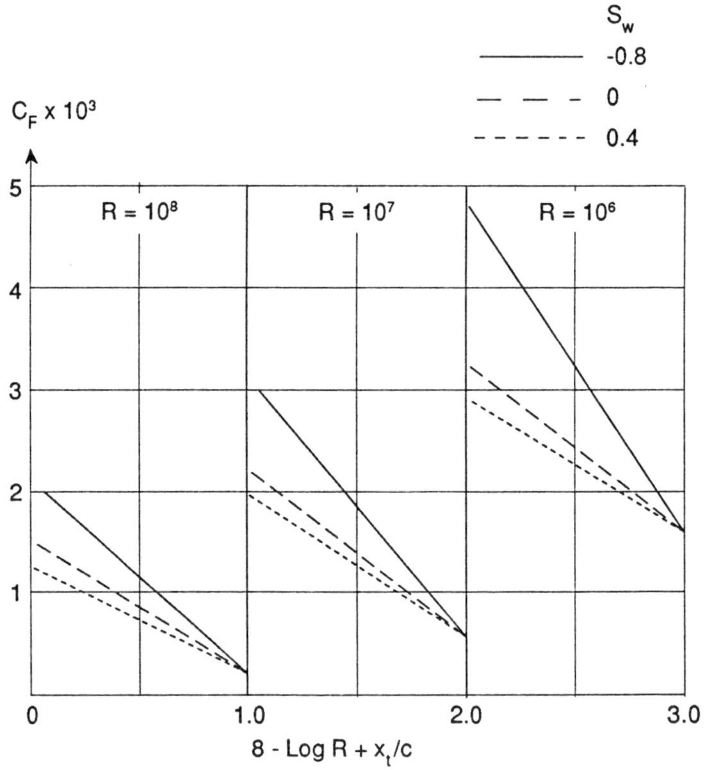

Fig. 8.5 c_f for one surface of a biconvex wing, $t/c = 0.05$, $M_\infty = 2.5$, with heat transfer.[8.19]

8.6.2 Extension of the Green lag-entrainment method to compressible flow

The extension of the Green lag entrainment method to compressible flow was made by Green, Weeks and Brooman[8.14] and it is one of the more widely adopted of the integral methods. The basic approach is essentially the same as that for incompressible flow already described in Section 8.5.4 and so we shall confine ourselves here to a brief summary of the distinguishing features for compressible flow.

Green[8.13] defined a modified form factor $\bar{H}$ where

$$\bar{H} = \int_0^\delta \frac{\rho}{\rho_e}\left(1 - \frac{u}{u_e}\right)dz \Big/ \int_0^\delta \frac{\rho u}{\rho_e u_e}\left(1 - \frac{u}{u_e}\right)dz = \frac{1}{\theta}\int_0^\delta \frac{\rho}{\rho_e}\left(1 - \frac{u}{u_e}\right)dz \tag{8.80}$$

Note that this is not the same as $H = \delta^*/\theta$ for compressible flow. However, Green argued that $\bar{H}$ played the same role in the entrainment relation as H does in incompressible flow. For zero heat transfer he showed that

$$H + 1 = (\bar{H} + 1)[1 + \tfrac{1}{2}r(\gamma - 1)M_e^2 \tag{8.81}$$

where r is the recovery factor $\simeq 0.89$ for air.

The three equations of integrated momentum, entrainment and lag [cf. equations (7.27), (8.37) and (8.43)] can be written

$$\frac{d\theta}{dx} = \tfrac{1}{2}c_{fe} - (H + 2 - M_e^2)\frac{\theta}{u_e}\frac{du_e}{dx} \tag{8.82}$$

$$\theta\frac{d\bar{H}}{dx} = \frac{d\bar{H}}{dH_1}\left\{C_e - H_1\left[\tfrac{1}{2}c_{fe} - (H + 1)\frac{\theta}{u_e}\frac{du_e}{dx}\right]\right\} \tag{8.83}$$

and

$$\theta\frac{dC_E}{dx} = \frac{C_E(C_E + 0.02) + 0.2667c_{f0}}{(C_E + 0.01)(H + H_1)}[f(C_E, c_{f0})_{eq} - \lambda f(C_E, c_{f0})] + \left(\frac{\theta}{u_e}\frac{du_e}{dx}\right)_{eq} - \frac{\theta}{u_e}\frac{du_e}{dx}\left[1 + 0.075M_e^2\frac{1 + 0.2M_e^2}{1 + 0.1M_e^2}\right] \tag{8.84}$$

where

$$f(C_E, c_{f0}) = 2.8[(0.024C_E + 1.2C_E^2 + 0.32c_{f0})(1 + 0.1M_e^2)]^{1/2} \tag{8.85}$$

These equations can be solved numerically given the external flow conditions. The various dependent variables are obtained as follows.

For c_{f0}, the Gaudet–Winter (zero heat transfer) correlations [see Section 6.11 and equation (8.45)] are used, viz.

$$F_c c_{f0} = \{0.01013/[\log_{10}(F_R R_\theta) - 1.02]\} - 0.00075 \tag{8.86}$$

where $F_c = (1 + 0.2M_e^2)^{1/2}$, $F_R = 1 + 0.056M_e^2$

Then [cf. equation (8.51)]

$$1 - 1/\bar{H}_0 = 6.55[\tfrac{1}{2}c_{f0}(1 + 0.04M_e^2)]^{1/2} \tag{8.87}$$

and [cf. equation (8.50)]

$$(c_{fe}/c_{f0} + 0.5)(\bar{H}/\bar{H}_0 - 0.4) = 0.9 \tag{8.88}$$

For H_1 [cf. equations (8.46A) and (8.46B)]

$$H_1 = 3.15 + [1.72/(\bar{H} - 1)] - 0.01(\bar{H} - 1)^2 \tag{8.89A}$$

or

$$H_1 = 2 + 1.5[1.12/(\bar{H} - 1)]^{1.093} + 0.5[(\bar{H} - 1)/1.12]^{1.093} \quad \text{for } 1.3 < H < 4 \tag{8.89B}$$

For the wake the solution is continued for each half wake, with $c_{fe} = c_{f0} = 0$, and $\lambda = 0.5$, as before.

8.7 Concluding remarks

We have noted that the kinds of integral prediction methods that we have so far described are valid for attached boundary layers not too close to separation on surfaces of small curvature. The lag entrainment methods offer for such cases an accuracy as good as the better differential methods (see Chapter 9). Such cases are often categorised as cases of *weak interaction* between the boundary layer and the external flow, i.e. the latter is closely approximated by the inviscid flow past the surface and is determined with adequate accuracy by allowing for the effective displacement of the surface by the calculated displacement thickness which is small in such cases. However, in cases where the surface curvature is large or the flow streamlines are highly curved, as in a region of flow approaching or downstream of separation or in a region of interaction between a shock wave and a boundary layer, the presence of the boundary layer has a considerable effect on the external flow. The interaction with the external flow is then referred to as *strong*. Such cases are associated with significant pressure changes across the boundary layer because of strong pressure gradients normal to the surface, and the terms in the momentum equation involving normal eddy stresses can also be significant and need to be allowed for. In addition, the effects of flow curvature on the dynamic development of the turbulence and therefore on the eddy stresses have to be taken into account. The inclusion of these so-called secondary terms has attracted much attention in recent years, particularly with the development of wing sections with high rear loading at transonic speeds coupled with the commercial incentives for increasing accuracy of prediction.

The design process requires methods for predicting the whole flow, both the external and the boundary layer flows, and their interaction, since the surface pressure distribution is not known *a priori*. The solution of the all-embracing Navier–Stokes equations offers a possible way of meeting these requirements and it is one that is increasingly pursued as computing facilities become increasingly powerful and less costly per operation. However, the cost and complexity of this approach are still very formidable, and it must be noted that the equations are used in a time-averaged form for turbulent flows and, except for the simulation of low Reynolds number flows, the empirical modelling of the small scale turbulence is still a feature and source of uncertainty. For everyday use in the processes of design, development and research a technique of simultaneously solving the external inviscid flow equations and the boundary layer equations and their coupling has much to commend it and is widely adopted. For such a procedure the use of integral methods for the boundary layers and wake has proved both cost effective and accurate enough, particularly if the secondary terms are included where necessary. For fuller details of such techniques the reader is referred to the survey of Lock and Williams.[6.1]

Appendix – Note on the pitot traverse method of measuring boundary layer drag

It will be evident from Sections 8.3 and 8.5.3 that we can relate the momentum thickness measured over any section of the wake of an aerofoil to that far downstream and hence infer the boundary layer drag from such measurements. Indeed, equation (8.17) applies by simply replacing suffix c by, say, M where M denotes the measurement section chosen; the latter is taken to cut the wake normal to the local mean flow direction. We then have

$$C_{\mathrm{D}} = 2\theta_{\mathrm{M}}(u_{\mathrm{e}}/u_{\infty})^{(H_{\mathrm{M}}+5)/2} \tag{8.90}$$

The freedom to choose the measurement station is of practical advantage and convenience. For example, we need not go far downstream in the wake to find a measurement plane where the static pressure is effectively constant across the wake section, perhaps no more than $0.1c$ to $0.2c$ aft of the trailing edge. It is then relatively easy to deduce the local velocity from pitot pressure measurements, made with a traversing pitot tube or a rake of tubes across the wake, combined with one or two fixed static pressure tubes.

An alternative and even simpler approach than that leading to equation

(8.90) was developed by Melvill Jones *et al.* in the mid-1930s, and this approach has been widely adopted and is still generally used.[8.20] He assumed that the flow in the wake downstream of the measuring station was in streamlines on each of which the total pressure remained constant at the value determined at the measuring station. This is far from true in detail but it leads to a surprisingly reliable formula for the drag coefficient.

Consider incompressible flow. It we denote the total pressure by P then for an element δz_M of the wake and its corresponding element far downstream δz_∞ (defined by the same neighbouring pair of streamlines)

$$P_M(z) = (p + \tfrac{1}{2}\rho u^2)_M = (p + \tfrac{1}{2}\rho u^2)_\infty \qquad (8.91)$$

Hence $$u_M^2/u_\infty^2 = \frac{(P - p)_M}{\tfrac{1}{2}\rho u_\infty^2} = \frac{[(P - p_\infty)_M - (p - p_\infty)]_M}{\tfrac{1}{2}\rho u_\infty^2}$$

$$= g - c_p, \quad \text{say} \qquad (8.92)$$

where $g = (P - p)/\tfrac{1}{2}\rho u_\infty^2 \qquad c_p = (p - p_\infty)/\tfrac{1}{2}\rho u_\infty^2$

at the measuring station. Far downstream the velocity in the wake at δz_∞ is $(u)_\infty$, say, so that

$$(\rho u \delta z)_M = \rho(u)_\infty \delta z_\infty$$

and from equation (8.91)

$$(u)_\infty^2/u_\infty^2 = g$$

Hence $$\theta_\infty = \int_{W(M)} \frac{u}{u_\infty}\left[1 - \frac{(u)_\infty}{u_\infty}\right] dz_\infty$$

$$= \int_{W(M)} \frac{u_M}{u_\infty}\left[1 - \frac{(u)_\infty}{u_\infty}\right] dz_M$$

$$= \int_{W(M)} (g - c_p)^{1/2}\,(1 - g^{1/2})\,dz$$

and $$C_D = \frac{2}{c}\int_{W(M)} (g - c_p)^{1/2}\,(1 - g^{1/2})\frac{dz}{c} \qquad (8.93)$$

Thus, C_D can be determined from the measurements in the traverse plane M. Although the derivations of the alternative formulae (8.90) and (8.93) differ in the assumptions involved, they have been found in practice to lead to answers that are in close agreement[8.21] the simpler method of (8.93) is therefore to be preferred. When the pressure is constant across the wake at the measuring station it lends itself to even greater simplication.[8.22]

The method has been readily extended to axi-symmetric and to compressible flow. For further details the reader can consult references 8.21 and 8.23–8.25.

References

8.1 Lock, R.C. and Williams, B.R. (1987) 'Viscous–inviscid interactions in external aerodynamics'. *Prog. Aeron. Sc.*, **24**, 2, p. 51.
8.2 Young, A.D. (1939) 'The calculation of the total and skin friction drags of bodies of revolution at zero incidence'. ARC RM 1874.
8.3 Maskell, E.C. (1953) 'Approximate calculation of the turbulent boundary layer in two dimensional incompressible flow'. ARC Rep. 14564.
8.4 Buri, A. (1931) 'Eine Berechnungsgrundlage für die turbulente Grenzschicht bei beschleunigter und verzögeter Grundströmung'. Dissertation E.T.H., Zurich, 652.
8.5 Spence, D.A. (1956) 'The development of turbulent boundary layers'. *JAS*, **23**, p. 3.
8.6 Thompson, B.G.J. (1964) 'A critical review of existing methods of calculating the turbulent boundary layer'. ARC RM 3447.
8.7 Head, M.R. (1958) 'Entrainment in the turbulent boundary layer'. ARC RM 3152.
8.8 Newman, B.G. (1951) 'Some contributions to the study of the turbulent boundary layer near separation'. Dep. Supply Rep. ACA 53.
8.9 Schubauer, G.B. and Klebanoff, P.S. (1950) 'Investigation of separation of a turbulent boundary layer'. NACA Rep. 1030.
8.10 Kline, S.J. *et al.* (1969) Proc. AFOSR – IFP Stanford Conf. *Computation of Turbulent Boundary Layers*. Vol. I & II., Standford Univ. Press.
8.11 Kline, S.J. *et al.* (1981) Proc. 1980–81 AFOSR–HTTM Standford Conf. 'Complex Turbulent Flows', Thermosciences Div., Dept. of Mech. Eng., Stanford Univ.
8.12 Head, M.R. and Patel, V.C. (1969) 'Improved entrainment method for calculating turbulent boundary layer development'. *ARC RM 3643*.
8.13 Green, J.E. (1972) 'Application of Head's entrainment method to the prediction of turbulent boundary layers and wakes in compressible flow'. RAE Tech. Rep. 72079.
8.14 Green, J.E., Weeks, D.J. and Brooman, J.W.F. (1973) 'Prediction of turbulent boundary layers and wakes in compressible flow'. ARC RM 3791.
8.15 Bradshaw, P., Ferris, D.H. and Atwell, N.P. (1967) 'Calculation of boundary layer development using the turbulent energy equation'. *JFM*, **28**, p. 593.
8.16 Lock, R.C. (1985) 'Prediction of the drag of wings at subsonic speeds by viscous–inviscid interaction techniques'. AGARD–R–723, paper 10.
8.17 Truckenbrodt, E. (1973) 'Neuere Erkentnisse über die Berechnung von Strömungsgrenzschichten mittels einfacher Quadraturformeln', Pt. I. & II. *Ing. Arch.*, **43**, p. 9; ibid. **43**, p. 136 (1974).
8.18 Spence, D.A. (1961) 'The growth of compressible turbulent boundary layers on isothermal and adiabatic walls'. ARC RM 3191.
8.19 Luxton, R.E. and Young, A.D. (1965) 'Boundary layer drag of bi-convex wing sections with heat transfer at supersonic speeds'. ARC RM 3393.
8.20 Cambridge Univ. Aeronautical Lab. (1936) 'Measurement of profile drag by the pitot traverse method'. ARC RM 1688.
8.21 Young, A.D. (1940) 'Note on momentum methods of measuring profile drag at high speeds'. ARC RM 1963.

8.22 Young, A.D. (1938) 'Note on a method of measuring profile drag by means of an integrating comb'. ARC RM 2257.
8.23 Lock, C.N.H., Hilton, W.F. and Goldstein, S. (1940) 'Determination of profile drag at high speeds by a pitot traverse method'. ARC RM 1971.
8.24 Beavan, J.A. (1943) 'Note on methods at the NPL and in the USA for calculating the profile drag at high speeds from pitot traverse measurements'. ARC RM 2102.
8.25 Heaslet, M.A. (1945) 'Theoretical investigation of methods for computing drag from wake surveys at high subsonic speeds'. NACA ACR 5C21.

Chapter 9

Turbulence Modelling and Differential Predictive Methods

9.1 Introductory remarks

In Section 7.3 we saw that the time averaged equations of motion with the approximations of boundary layer theory [equation (7.6)] involve the Reynolds stresses $-\rho\overline{u'_\alpha u'_\beta}$ as well as other correlations of the form $\overline{\rho' u'_\alpha}$ and $\overline{\rho' u'_\alpha u'_\beta}$. The term $\overline{\rho' u'_\alpha}$ can be formally eliminated by using the mass-weighted velocity concept due to Favre[9.1] as in equations (7.11) and (7.12). For the terms of the form $\overline{\rho' u'_\alpha u'_\beta}$, it can be argued from the equation of state that

$$p'/p - \rho'/\rho = T'/T$$

and if we assume that p'/p is small compared with the other terms inside a boundary layer then

$$\rho'/\rho = -T'/T$$

In conditions of constant total energy, T_H, with T'_H assumed negligible, it follows that

$$T'/T \simeq O[(\gamma - 1)u'M^2/u]$$

if we take u in the direction of the local mean velocity. Hence the term $\overline{\rho' u'_\alpha u'_\beta}$ can be regarded as small compared with $\rho\overline{u'_\alpha u'_\beta}$ unless M is large. In practice, it seems that for attached boundary layers the term $\overline{\rho' u'_\alpha u'_\beta}$ can be neglected for $M < 5$, approx. Morkovin[9.2] has inferred that density fluctuations for Mach numbers in this range (i.e. other than hypersonic flows) do not affect the turbulence structure for attached boundary layers. Hence, concepts of the turbulence field deduced from low speed flow results can be readily generalised, provided that the variations of mean density and its gradients are properly accounted for. This is known as *Morkovin's hypothesis*. Similar arguments apply to the eddy transport terms that figure in the time average energy equation (7.13).

It follows that to solve the equations of motion and energy we must in some way model (i.e. approximate to) the Reynolds stresses and other eddy transport terms, or model the equations from which they can be determined, so as to relate them to the mean flow. For the foreseeable future such modelling processes must necessarily be empirically based and tailored for particular flow situations so that we must expect their ranges of validity to be limited.

In recent years a number of such modelling processes have been formulated. They can be broadly classified as follows:

(1) Simple applications of the Boussinesq eddy viscosity concept combined with the Prandtl mixing length concept (see Section 6.4) – the so-called zero equation models.
(2) The eddy viscosity concept is expressed as the product of a characteristic velocity and length derived from modelled, or simplified forms of transport equations for appropriate turbulence quantities, usually the turbulence kinetic energy and dissipation – the so-called one or two equation models.
(3) The additional use of modelled forms of the Reynolds stress transport equations but without the eddy viscosity concept – the so-called Reynolds stress transport models. These forms may be either algebraic or differential.
(4) The time averaged Navier–Stokes equations are solved for the larger eddies but the smaller eddies are empirically modelled using a sub-grid scale hypothesis. This is known as the *large-eddy simulation* process.

The auxiliary equations used in the above methods are themselves derived from the Navier–Stokes equations. In principle, it should be possible to solve the non-steady Navier–Stokes equations outright, given a large and powerful enough computer, and so avoid the approximations of a modelling process. However, in practice, the need to use grid sizes small enough to compute the dissipative eddies with adequate detail and accuracy renders such a procedure for flow problems at the Reynolds numbers of aeronautical interest out of the question in terms of the computing power likely to be available for many years to come.

Scales of representative length and velocity are either explicitly or implicitly formulated in all modelling processes. However, it is as well to appreciate the very wide range of length scales that can be associated with turbulent boundary layers. At one end of the scale we have the large energy containing eddies that have a typical length scale of the order of the boundary layer thickness. At the other end of the scale we have the small eddies that are largely responsible for the energy dissipation by viscous action. For these small eddies the characteristic Reynolds number based on a typical velocity and length must be of the order of unity, since the viscous and inertia forces acting on them must be of the same order.

Hence, the characteristic length of the dissipative eddies, l_d, must be such that $\Delta u' l_d/\nu = O(1)$, where $\Delta u'$ is a typical turbulent velocity change over the length l_d. The local rate of dissipation of energy per unit mass is of the order $\nu(\Delta u'/l_d)^2$ and assuming local equilibrium this is of the same order as the rate at which energy is created by the Reynolds stress τ working on the mean velocity gradient and so is of the order of $(\tau_w/\rho)(u_e/\delta)$. Equating these magnitudes and using the simple power law relations of Section 6.6 we infer that

$$l_d/\delta = O(0.003R_x^{11/5})^{-1/4}$$

Thus at $R_x = 10^6$, $l_d \simeq \delta/500$, and at $R_x = 10^8$, $l_d \simeq \delta/6000$. It will therefore be evident that any representation of the turbulent eddies by a single length scale involves a considerable simplification of the physics, and its validity will depend on the particular part of the turbulence spectrum that is important for the modelling process used.

It what follows we shall briefly describe a few typical modelling methods that are currently used for engineering purposes. For a more detailed survey the reader may consult such references as 9.3–9.5. The computational details in the numerical solutions of the equations involved are not dealt with here, they merit a separate book, reference 9.5 provides a good introduction.

9.2 Eddy viscosity/mixing length models (zero equation): two dimensional flow

The basic equations of continuity, motion and temperature for incompressible flow are (see Sections 7.2 and 7.3)

$$\partial u/\partial x + \partial w/\partial z = 0 \tag{9.1}$$

$$\rho \frac{Du}{Dt} = \rho u_e \frac{\partial u_e}{\partial x} + \frac{\partial}{\partial z}\left(\mu \frac{\partial u}{\partial z} - \rho\overline{u'w'}\right) \tag{9.2}$$

$$\rho c_p \frac{DT}{Dt} = \frac{\partial}{\partial z}\left(k \frac{\partial T}{\partial z} - c_p\rho\overline{w'T'}\right) \tag{9.3}$$

We note that for low Mach numbers the terms in Dp/Dt and $\tau\, \partial u/\partial z$ of equation (7.13) become negligible compared with the terms involving the temperature.[9.6]

The Boussinesq postulate of an eddy viscosity ν_t (see Section 6.4) leads to the expression

$$-\overline{u'w'} = \nu_t\, \partial u/\partial z \tag{9.4}$$

Likewise we can postulate an eddy temperature diffusivity k_t such that

$$-\overline{T'w'} = (k_t/\rho c_p)\partial T/\partial z \qquad (9.5)$$

If we accept these relations our problem then is to determine the quantities ν_t and k_t, which unlike their molecular counterparts (ν and k) are not properties of the fluid but are functions of the flow field including the turbulence.

The simplest approach is to use Prandtl's mixing length concept, so that in the inner region of the boundary layer we write [see equation (6.17)]

$$\nu_t = l^2|\partial u/\partial z| \qquad (9.6)$$

where for l we can either write

$$l = Kz$$

where K is Karman's constant (0.4–0.41), or, if we wish to include the laminar sub-layer, we can make use of Van Driest's formula [equation (6.20)]

$$l = Kz[1 - \exp(-z^+/A_0)] \qquad (9.7)$$

with $A_0 \simeq 26$. Outside the inner region we can use the empirical relation, equation (6.26)

$$\nu_t = 0.0168 u_e \delta^* \gamma_0 \qquad (9.8)$$

Here we may note that Cebeci and Smith[9.7] adopted this relation (but initially with γ_0 replaced by unity) for values of $R_\theta > 5000$, and an empirical additional factor involving a function of R_θ for $R_\theta < 5000$. Their method also included a factor for modelling the transition region between laminar and turbulent flow. The junction between the inner and outer regions of the boundary layer is determined by where the two distributions of ν_t result in the same value.

An alternative approach is to combine equation (2.6) with equation (6.27), viz.

$$l/\delta = 0.085 \tanh(Kz/0.085\delta) \qquad (9.9)$$

A related method is that of Baldwin and Lomax[9.8] which uses mean vorticity, ω, in place of the mean velocity gradient in equation (9.4) and expresses the eddy viscosity in the outer region in terms of the maximum of $F(z) = z|\omega|$, so that

$$\nu_t = 1.6 \times 0.0168 F(z)_{max} \cdot \gamma_0 \qquad (9.10)$$

They adopted another empirical expression for ν_t for separated flow.

To determine k_t it is usual to relate it to ν_t by means of the turbulent Prandtl number

$$\sigma_t = \nu_t/k_t \qquad (9.11)$$

and to assume a constant value for σ_t of 0.9–1.0 for air. It must be noted, however, that the experimental evidence supporting this assumption is meagre and more relevant data are needed.

Surface or flow curvature introduces an effect, that can be surprisingly large, on the dynamics of the turbulent boundary layer by modifying the stability of the flow.[9.9] There is a centrifugal force due to such curvature and if a fluid particle is slightly perturbed from the level at which the force is in balance with that due to the pressure gradient the two forces will no longer balance. The difference may then be such as to restore the particle to its original level, in which case the curvature is stabilising, or the particle may move further away from its initial level, in which case the curvature is destabilising. This is much the same effect that induces Taylor–Görtler vortices to form in a laminar boundary layer developing over a concave surface, see Section 5.5, and experimenters have sometimes observed similar vortical structures in turbulent boundary layers on concave surfaces. Hence, we find that the turbulence intensity increases for an attached turbulent boundary layer on a concave surface relative to that on a flat surface and conversely the intensity decreases when the surface is convex.

As a parameter of the effect of curvature we can use the ratio:

$$\frac{\text{additional strain rate due to curvature}}{\text{basic strain rate}} = \frac{\partial w}{\partial x}\Big/\frac{\partial u}{\partial z}$$

But $\partial w/\partial x = -u/r$, where r is the streamline radius of curvature and is positive if the streamline is convex to the external flow, and so

$$\partial w/\partial x \simeq -u/r_0$$

where r_0 is the surface radius of curvature. Hence the parameter can be written

$$p_c = -u/[r_0(\partial u/\partial z)] \tag{9.12}$$

It is then argued that the effect of this extra strain rate is to change the mixing length from l to l_0, say, where

$$l_0/l = 1 + \beta \,.\, p_c \tag{9.13}$$

where β is an empirical 'constant', dependent somewhat on the type of shear flow considered. For a boundary layer β is found to be about 10, so that the direct effect of the extra strain rate is strongly magnified by the curvature. We see that for a convex wall, $l_0/l < 1.0$ and ν_t (and hence the eddy stress) is reduced, whilst for a concave wall $l_0/l > 1.0$ and ν_t is increased as is the turbulence intensity.

The extensions of these methods to compressible flow are fairly straightforward, the direct compressibility effects on the mean density and temperature are allowed for but otherwise there is general acceptance of Morkovin's hypothesis. The basic equations that require to be solved are now (7.11), (7.12), (7.13) and (7.14). The methods assume that equations (9.4), (9.5), (9.7) and (9.8) are unchanged but for ν_t we must understand μ_t/ρ, where ρ is the local mean density, likewise $u_\tau = (\tau_w/\rho)^{1/2}$. It will be seen that in equation (9.7)

$$\frac{z^+}{A_0} = \frac{zu_\tau}{\nu A_0} = \frac{z}{\nu A_0}\left(\frac{\tau_w}{\rho_w}\frac{\rho_w}{\rho}\right)^{1/2} \tag{9.14}$$

Cebeci *et al.*[9.10] introduced a factor N in the numerator on the RHS of this equation, where N is a complex function of pressure gradient and wall suction velocity in cases of transpiration through the wall. Equation (9.11) is also assumed to hold for compressible flow and again the turbulent Prandtl number σ_t is usually taken for air in the range 0.9 to 1.0, although experimental results show variations of σ_t across the boundary layer and significant departures from unity have been measured.

The Baldwin–Lomax method referred to above was developed for compressible flow, with separation in mind as a possible outcome of shock-wave boundary layer interaction. For further details of these and other eddy viscosity methods see reference 9.3.

It must be emphasised that the simple correlation, at the heart of such methods, between the turbulent shear stress components and the local rates of strain of the mean flow, via the medium of a locally determined mixing length, leaves out of account effects of the history of the motion and are therefore strictly applicable only to boundary layers in near equilibrium. Indeed, as we shall see (Section 9.4) the correlation is equivalent to assuming that the production and dissipation terms in the transport equation for the turbulent kinetic energy are the only dominant ones and are in balance, whilst the diffusion terms are negligible. Hence an equilibrium condition is implied. It is also implied that at a point where the velocity gradient $\partial u/\partial z$ is zero the shear stress must also be zero. A number of experiments (e.g. wall jets) have demonstrated that this is not necessarily the case. Nevertheless, such methods have been found to be fairly successful in predicting the development of attached boundary layers provided separation conditions are not imminent, the pressure gradients are moderate and are not changing rapidly with streamwise distance.

9.3 The turbulence transport equations

9.3.1 Introduction

We have remarked that we can make use of the transport equations for selected turbulence quantities in the modelling process and to that extent give weight to the history of the turbulence as a factor additional to the local mean flow conditions.

The specific turbulence quantities whose transport equations are most often modelled are the eddy stresses, the turbulence kinetic energy and the turbulence dissipation.

We shall from now on adopt the commonly used symbol k to denote the mean turbulence kinetic energy per unit mass, viz. $\frac{1}{2}[\overline{u'^2} + \overline{v'^2} + \overline{w'^2}]$, instead of $\overline{e'}$. This is because there is no further risk of confusion with other usages of k. However, we shall retain e' to denote the instantaneous value of the turbulence kinetic energy per unit mass.

9.3.2 The eddy stress transport equations

For the instantaneous velocity component $U_i = u_i + u_i'$, say, the Navier–Stokes equation (2.7) can be written, in the absence of body forces,

$$\rho\frac{DU_i}{Dt} = -\frac{\partial}{\partial x_i}(p + p') + \mu\,\nabla^2 U_i + \frac{\mu}{3}\frac{\partial}{\partial x_i}\left(\frac{\partial U_\gamma}{\partial x_\gamma}\right)$$
$$-\frac{2}{3}\frac{\partial\mu}{\partial x_i}\frac{\partial U_\gamma}{\partial x_\gamma} + \frac{\partial\mu}{\partial x_\gamma}\left(\frac{\partial U_i}{\partial x_\gamma} + \frac{\partial U_\gamma}{\partial x_i}\right) \tag{9.15}$$

If we multipy this equation by u_j' and similarly multiply the equation for U_j by u_i' and add the two and then take the mass weighted time mean we obtain the transport equation for $\overline{u_i' u_j'}$. For incompressible flow this takes the form

$$\frac{D}{Dt}(\overline{u_i' u_j'}) = \overset{\text{A}}{-\left(\overline{u_i' u_\gamma'}\frac{\partial u_j}{\partial x_\gamma} + \overline{u_j' u_\gamma'}\frac{\partial u_i}{\partial x_\gamma}\right)}$$
$$\overset{\text{B}}{-\,2\nu\,\overline{\frac{\partial u_i'}{\partial x_\gamma}\frac{\partial u_j'}{\partial x_\gamma}}} + \overset{\text{C}}{\overline{\frac{p'}{\rho}\left(\frac{\partial u_i'}{\partial x_j} + \frac{\partial u_j'}{\partial x_i}\right)}}$$
$$\overset{\text{D}}{-\frac{\partial}{\partial x_\gamma}\left[\overline{u_i' u_j' u'_\gamma} + \delta_{j\gamma}\frac{\overline{u_i' p'}}{\rho} + \delta_{i\gamma}\frac{\overline{u_j' p'}}{\rho} - \nu\frac{\partial}{\partial x_\gamma}(\overline{u_i' u_j'})\right]} \tag{9.16}$$

The terms labelled A represent the production of the stress by action of the stresses on the mean velocity gradients, B represents the dissipation by viscous action, C is the fluctuating pressure–strain contribution, whilst D represents the diffusion effects due to turbulent and molecular motions.

In two dimensions (x, z) the equation for the stress/density $(-\overline{u'w'})$ in a boundary layer reduces to

$$\frac{\mathrm{D}}{\mathrm{D}t}(-\overline{u'w'}) = \overline{w'^2}\frac{\partial u}{\partial z} - \overline{\frac{p'}{\rho}\left(\frac{\partial u'}{\partial z} + \frac{\partial w'}{\partial x}\right)} + \frac{\partial}{\partial z}\left(\frac{\overline{p'u'}}{\rho} + \overline{u'w'^2}\right) - \nu[\overline{u'\nabla^2 w'} + \overline{w'\nabla^2 u'}] \qquad (9.17)$$

From (9.16) the corresponding transport equation for the averaged turbulence kinetic energy per unit mass

$$k = \tfrac{1}{2}(\overline{u'^2} + \overline{v'^2} + \overline{w'^2})$$

is obtained by putting i = j and summing for each of the velocity components. We obtain for incompressible flow

$$\frac{\mathrm{D}k}{\mathrm{D}t} = -\overline{u_i' u_j'}\frac{\partial u_i}{\partial x_j} - \frac{\partial}{\partial x_j}\left[\overline{u_i'\left(\tfrac{1}{2}u_j' u_i' + \frac{p'}{\rho}\right)}\right] - \epsilon \qquad (9.18)$$

where $\epsilon \simeq \nu\,\overline{(\partial u_i'/\partial x_j)^2}$.

We can again identify the first term on the RHS as representing the production rate of kinetic energy due to the rate of work done by the Reynolds stresses on the mean velocity gradients, the second group are diffusion contributions due to the turbulence components plus the fluctuating pressure and the third term ϵ is the dissipation rate of the turbulence kinetic energy by the action of viscous forces.

Since ϵ is the rate of conversion of the turbulent kinetic energy into heat we can expect it to be of the form (typical turbulence velocity)^2/(typical turbulence time scale). For two dimensional flow we can take the typical turbulence velocity as $|(\overline{u'w'})|^{1/2}$ and the typical time scale as $L/|(\overline{u'w'})|^{1/2}$, where L is an eddy length scale. Thus we can write

$$\epsilon = |(\overline{u'w'})|^{3/2}/L \qquad (9.19)$$

and we can regard this equation as a definition of the length L.

Where the generation and dissipation of the turbulence kinetic energy are roughly in balance and the diffusion terms are negligible, as for an equilibrium boundary layer or in the inner region of a boundary layer, then we have from equations (9.18) and (9.19)

$$|\overline{u'w'}|\frac{\partial u}{\partial z} = |\overline{u'w'}|^{3/2}/L$$

and hence $|\overline{u'w'}| = L^2(\partial u/\partial z)^2$

or $\nu_t = L^2(\partial u/\partial z)$ (9.20)

Thus, in this case we have recovered the Prandtl mixing length formula, so that for such flows we can identify L with the mixing length l of equation (9.6). However, in general the two lengths are not identical.

This discussion serves to reinforce the point that the mixing length concept is most readily justified for boundary layers in near equilibrium when the generation and dissipation terms for the turbulence kinetic energy are roughly in balance.

An alternative to equation (9.19) is obtained if $(k)^{1/2}$ is used as the representative velocity so that we write

$$\epsilon = (k)^{3/2}/L_\epsilon \qquad (9.21)$$

where L_ϵ is another turbulence length scale.

Another transport equation that is frequently modelled is that for ϵ, mainly because in combination with the other two transport equations it can be regarded as a modelling procedure for a length scale, in view of equations (9.20) and (9.21). This transport equation is also derived from the Navier–Stokes equation for the instantaneous velocity component U_i by a process that is fairly straightforward but tedious. The equation contains a number of terms that are unknown and they do not readily lead to a rational modelling procedure. Hence, sweeping empirical assumptions based on relatively inadequate data have to be made to recast the equation in the form of an acceptable model. For this reason the original equation is not presented here (it is to be found in reference 9.5). However, certain groups of terms can be labelled, as with the other two transport equations, as convective and diffusive transport, generation and viscous destruction terms. The modelling of the equation for ϵ will be further discussed in Section 9.7.

9.4 Single equation model using the eddy viscosity concept

Since ν_t has the dimensions velocity × length we can write

$$\nu_t = c_\mu\, k^{1/2}\, l_1, \quad \text{say} \qquad (9.22)$$

where k can be derived from a suitably modelled form of equation (9.18), l_1 is a length usually taken as the mixing length for attached boundary layers and is derived from an algebraic equation such as (9.7), and c_μ is an empirical constant ($\simeq$ 0.09).

In modelling the equation for k the main problem is the modelling of the diffusion terms as they are not directly measurable. The procedure generally adopted is to write

$$\overline{-u_i'\left(\frac{u_i' u_j'}{2} + \frac{p'}{\rho}\right)} = \frac{\nu_t}{\sigma_e}\frac{\partial k}{\partial x_j} \qquad (9.23)$$

in accordance with the practice of regarding diffusion terms as gradient driven, and σ_e is an empirical constant. The dissipation term is also assumed to be replaceable by [cf. equation (9.21)]

$$\sigma_d \, k^{3/2}/l \tag{9.24}$$

where σ_d is an empirical constant. The modelled equation for k is then

$$\frac{Dk}{Dt} = \nu_t \left(\frac{\partial u_i}{\partial x_j} + \frac{\partial u_j}{\partial x_i}\right)\frac{\partial u_i}{\partial x_j} + \frac{\partial}{\partial x_i}\left(\frac{\nu_t}{\sigma_e}\frac{\partial k}{\partial x_i}\right) - \frac{\sigma_d \, k^{3/2}}{l} \tag{9.25}$$

production diffusion dissipation

In the production term we have made use of the Boussinesq relation

$$-\overline{u_i' u_j'} = \nu_t \left(\frac{\partial u_i}{\partial x_j} + \frac{\partial u_j}{\partial x_i}\right)$$

This method is not applicable in the viscous sub-layer close to the wall, where the turbulence Reynolds number $k^2/\nu\epsilon = k^{1/2} L_\epsilon/\nu$ is small, and if the method is to be extended to the wall the growing importance of the viscous stresses as the wall is approached has to be allowed for in the equation. It is usual to assume the law of the wall there. For further details of 'Low Reynolds number flows' see the contribution by Launder in reference 9.3.

9.5 The Bradshaw method[8.15]

This is a single equation method but it does not involve the eddy viscosity concept. It now has a classical status amongst predictive methods involving turbulence modelling as it was the first to attract widespread attention to the use of an auxiliary transport equation for a turbulence quantity. It is still widely used with good success for attached boundary layers. An essential feature is the use of an experimental result that for such boundary layers the ratio of the shear stress/density in two dimensional flow, $-\overline{u'w'}$, to the mean turbulence kinetic energy per unit mass, k, is nearly constant $\simeq 0.3$. This result was used by Bradshaw *et al.* to express the turbulence kinetic energy equation (9.18) as an equation for the eddy stress but with the diffusion term and the length L [equation (9.19)] modelled as proportional to simple functions of z/δ.

Thus, in two dimensions equation (9.18) can be written

$$Dk/Dt = -\overline{u'w'}\frac{\partial u}{\partial z} - \frac{\partial}{\partial z}\left[\overline{w'\left(e' + \frac{p'}{\rho}\right)}\right] - \epsilon \tag{9.26}$$

Bradshaw *et al.* then assumed

$$\overline{w'\left(e' + \frac{p'}{\rho}\right)} = \frac{\tau}{\rho}\left(\frac{\tau_{max}}{\rho}\right)^{1/2} G(z/\delta) \tag{9.27}$$

where G is an empirically determined function. The use of τ_{max} as a scaling factor on G is suggested by the argument that diffusion is largely determined by the larger eddies whose intensity is in turn related to τ_{max}. Then with the aid of equation (9.19) and using the above result $\tau/\rho k = 2a_1 = 0.3$, say, we can rewrite equation (9.26) as an equation for τ in the form

$$\frac{D}{Dt}\left(\frac{\tau}{2a_1\rho}\right) - \tau\frac{\partial u}{\partial z} + \frac{\partial}{\partial z}\left[G\tau\left(\frac{\tau_{max}}{\rho}\right)^{1/2}\right] + \left(\frac{\tau}{\rho}\right)^{1/2}/L = 0 \qquad (9.28)$$

Bradshaw *et al.* then determined empirically a function $f_1(z/\delta)$ such that

$$L/\delta \simeq f_1(z/d) \qquad (9.29)$$

The log law velocity distribution was assumed to apply close to the wall to provide the inner boundary conditions for the controlling equations which were solved numerically.

9.6 The $k - \epsilon$ method (a two equation method)[9.11]

In this method modelled forms for both the k and ϵ transport equations are used along with the assumption of an eddy viscosity given by

$$\nu_t = c_\mu\, k^2/\epsilon \qquad (9.30)$$

The modelled form of the k equation is usually of the form of equation (9.25) but with ϵ left as the last term on the RHS to be determined from its modelled equation which is assumed similar in form to that for k, viz.

$$\frac{D\epsilon}{Dt} = c_{\epsilon 1}\,\nu_t\,\frac{\epsilon}{k}\left(\frac{\partial u_i}{\partial x_j} + \frac{\partial u_j}{\partial x_i}\right)\frac{\partial u_i}{\partial x_j} + \frac{\partial}{\partial x_i}\left(\frac{\nu_t}{\sigma_\epsilon}\frac{\partial \epsilon}{\partial x_i}\right) - \frac{c_{\epsilon 2}\,.\,\epsilon^2}{k} \qquad (9.31)$$

where $c_{\epsilon 1}$, $c_{\epsilon 2}$, σ_ϵ are constants determined empirically. It will be appreciated that there is little *a priori* justification for this equation beyond the general expectation that any transport equation should have production, diffusion and destruction terms. These modelled equations plus equation (9.30) provide the closure relations required for solving the mean flow boundary layer equations.

Various authors have arrived at values of the empirical constants that differ a little amongst themselves but they range as follows:

c_μ	$c_{\epsilon 1}$	$c_{\epsilon 2}$	σ_e	σ_ϵ
0.09	1.45–1.5	1.9–2.0	1.0	1.3

For cases where there is heat transfer the eddy heat transfer coefficient k_t is related to ν_t as before [see equation (9.11)] by

$$k_t = \nu_t/\sigma_t$$

where σ_t is the turbulent Prandtl number which for air is often assumed in the range 0.9–1.0.

This method is widely used and has been found to work well for attached boundary layers and for certain shear flows of simple geometry and small curvature, where the implied assumption is justifiable that the turbulence energy production and dissipation terms nearly balance. However, for more complex situations the above empirical constants need modifying, in ways that depend on the circumstances, to achieve acceptable agreement with experimental results. The limitations on the validity of the method derive from the eddy viscosity hypothesis and the inadequacies of the modelled dissipation equation. For a discussion of recent developments to improve the method see the contribution of Rodi in reference 9.3.

9.7 Algebraic stress models

To avoid the assumption of an eddy viscosity in the $k - \epsilon$ method with its implication of near equilibrium flow one can instead make use of the modelled eddy stress transport equations. To simplify this approach, Rodi[9.12] suggested the following assumptions:

(1) the convective terms in the equation for $\overline{u_i' u_j'} = (\overline{u_i' u_j'}/k) \times$ the convection terms in the modelled k equation.
(2) the diffusion terms in the equation for $\overline{u_i' u_j'} = (\overline{u_i' u_j'}/k) \times$ the diffusion terms in the modelled k equation.

Thus, the terms in the stress transport equation that contained gradients of the stress are replaced by terms that leave the equation an algebraic one in $\overline{u_i' u_j'}$ and to that extent is simpler to handle in combination with the equations for k and ϵ. A similar procedure can be adopted for scalar quantities, e.g. temperature, for problems involving heat transfer. Such procedures involve additional constants to be determined empirically.

Insofar as each eddy stress in a three dimensional boundary layer has its own transport equation this approach avoids the implication of the basic $k-\epsilon$ method that the eddy viscosity is isotropic, there is plenty of evidence in three dimensional flows that it is not.

9.8 Concluding remarks

It appears that for attached boundary layers in two dimensions the available methods are applicable with an accuracy that is acceptable for many engineering purposes, i.e. with an error in boundary layer drag coefficient that is within about $\pm$ 5%. The simpler methods, particularly the integral methods, then have the advantage of relatively low computational cost. For more complex flows, e.g. those involving limited regions of separation, rapidly changing external flow conditions or high surface curvature, interacting boundary layers and wakes, or three dimensional boundary layers, the more complex methods are on the whole more accurate but not always accurate enough. The accuracy is closely dependent on the available experimental data on which the assumed relations and constants are based, and such data are too sparse as yet for any method to be consistently and generally reliable. It will be evident that the approximations made in the modelling processes are often sweeping and in most cases the ranges of validity are not well established. The Stanford Conference of 1981–82[8.11] pointed to certain conclusions that still apply and are worth recalling. They are:

(1) All methods work well – sometimes, but no existing method is universally capable of predicting the full range of turbulent flows to engineering accuracy.
(2) For the present, efforts should be directed to relating the 'constants' involved to the particular types of flow under consideration – this is referred to as 'zonal modelling'. A body of experience has to be built up as to which methods and 'constants' are best suited to the type of problem in hand. Complex flows may require different methods for different zones.
(3) The dissipation equation is at present the weakest feature of the models that use it.
(4) Algebraic stress models proved not to be significantly better than eddy viscosity models. However, Launder[9.14] has shown that with recent improvements in computing techniques these models show up to advantage, particularly when applied to internal swirling flows.
(5) Much more experimental data are needed, both to improve our physical understanding, and so hopefully lead to better models, and to provide stringent test cases against which to assess prediction methods.
(6) Future effort should be directed to the development of Reynolds stress transport models free of the eddy viscosity concept.
(7) In the longer term, the large eddy simulation process should provide a more reliable if costly method to be mainly used to obtain results as 'bench marks' against which to test simpler methods.

It is as well to note that the computational process itself can lead to significant errors, since it involves finite mesh sizes, discretisation and truncations which are ready sources of numerical perturbations. These errors are not always adequately assessed or even understood.

It must likewise be emphasised that experimental results are also subject to errors, and experiments need very careful planning and assessment to ensure that their accuracy is such that they can be reliably used to validate results obtained by computation. At the best it seems that an experimental accuracy for drag of the order of one drag count (i.e. 0.0001 in the drag coefficient) is achievable.[9.13]

References

9.1 Favre, A. (1965) 'Equations des gaz turbulents combustibles'. *J. Phys. Atmos*, **4**, p. 361.

9.2 Morkovin, M.V. (1960) 'The structure of supersonic turbulent boundary layers'. AGARD Wind Tunnel and Model Testing Panel.

9.3 Introduction to the modelling of turbulence. VKI Lecture Series 1987–06. (1987)

9.4 Special Course on modern theoretical and experimental approaches to turbulent flow structure and its modelling. AGARD Rep. 755. (1987)

9.5 Bradshaw, P., Cebeci, T. and Whitelaw, J. (1981) *Engineering Calculation Methods for Turbulent Flow*. Acad. Press.

9.6 Squire, H.B. (1953) *Modern Developments in Fluid Dynamics. High Speed Flow* (ed. L. Howarth). OUP, Vol. II, Ch. XIV 'Heat transfer'.

9.7 Cebeci, T. and Smith, A.M.O. (1973) *Analysis of turbulent boundary layers*. Acad. Press, NY.

9.8 Baldwin, B.S. and Lomax, H. (1978) 'Thin layer approximations and algebraic models for separated flows'. AIAA Paper 78–257.

9.9 Bradshaw, P. (1973) 'Effects of streamline curvature in turbulent flow'. AGARDograph 169.

9.10 Cebeci, T., Smith, A.M.O. and Mosinskis, G.J. (1970) 'Calculation of compressible adiabatic turbulent boundary layers'. *AIAA J*, **8**, p. 1974.

9.11 Jones, W.P. and Launder, B.E. (1972) 'The prediction of laminarisation with a two equation model of turbulence'. *Int. J. Heat Mass Transfer*, **15**, p. 301.

9.12 Rodi, W. (1976) 'A new algebraic relation for calculating the Reynolds stresses'. *ZAMM*, **56**, p. 219.

9.13 Carter, E.C. and Pallister, K.C. (1987) 'Development of testing techniques in a large transonic wind tunnel to achieve a required drag accuracy and flow standards for modern civil transports'. AGARDograph 429, paper 11.

9.14 Launder, B.E. (1989) 'Fluid dynamics of three-dimensional turbulent shear flows and transition'. AGARD CP438, paper 26.

Chapter 10

Some Complex Problems for Further Study

10.1 Introduction

It will be appreciated that the material dealt with in the foregoing chapters has covered what might be described as the fundamentals of the subject, but aircraft designers often have to deal with problems of much greater complexity. For example, if they are concerned with a high performance civil transport aircraft of moderate sweep and high aspect ratio the main operating conditions determining the aerodynamic design will be the cruise at high subsonic Mach numbers and modest lift coefficients and the landing and take-off conditions at relatively low speeds but high incidences with high lift devices (leading and trailing edge flaps) extended. For these conditions they need to be able to predict the aerodynamic characteristics, particularly the lift, drag and pitching moment coefficients with an accuracy that is becoming very stringent because of commercial and competitive pressures. Ideally, they would like to be able predict these characteristics with the required accuracy solely on the basis of computation for which boundary layer theory must play a key part. However, the theory is still far from the stage where the accuracy requirements can be confidently met once the flow ceases to be classifiable as quasi two dimensional and attached. Sweep, tip effects, junctions, shock–boundary layer interactions, high incidences, incipient or developed regions of flow separation, streamwise vortical flows, interactions of wing wakes and the boundary layers on flaps and of slat wakes and wing boundary layers all introduce features that are in the main not yet predictable with the required accuracy.

Hence, for a long time to come the designer must be guided by a judicious mixture of experimental results as well as computation together with empiricisms based on past experience. In any case, the computational methods must involve three dimensional boundary layer theory if the problem areas are to be tackled with any pretence of an adequately representative flow model.

The designer of a military aircraft is concerned with even more complex three dimensional effects since the aspect ratio is relatively low, and a very wide range of operating conditions, in terms of incidence and speed, has to be considered. Some forms of separation combined with vortical flows may well be design features.

In consequence, there has been increasing attention paid in recent years to the development of three dimensional boundary layer theory, but in the main these developments are fairly simple extensions of two dimensional boundary. layer theory and include the assumptions and empiricisms of the latter. There is little extra experimental evidence culled from three dimensional flows to support the use of these assumptions; on the contrary, there is evidence to show that three dimensional boundary layers can behave in ways not predictable on the basis of two dimensional experience. Nevertheless, some progress has been made. Transition from laminar to turbulent flow is also more complex in three dimensions than in two, wing sweep, for instance, introduces destabilising factors that can play a major role in provoking transition.

The need to include separated flows has focussed attention on the use of so-called 'inverse' methods of boundary layer prediction, i.e. methods in which the external velocity (or pressure) distribution is not presumed known but some overall boundary layer characteristic, such as the displacement thickness or skin friction distribution is prescribed and the external velocity distribution is deduced. In two dimensional flow, at a point of separation the boundary layer equations when formulated for the usual direct method of solution become singular and it is impossible to continue a solution through the separation point. However, with the inverse form of the equations the singularity does not arise at the separation point, and provided the region of separation is not extensive and is fairly shallow the calculation can proceed to the point of reattachment. It must be emphasised that separation in three dimensional flow is a more complex phenomenon than in two dimensions and is not simply defined in terms of a local criterion.

The need to reduce drag to improve performance and to reduce fuel consumption has in recent years led to a revival of interest in the use of suction through the surface on which a boundary layer is developing to delay transition, as was discussed in Section 5.10, and in addition it has led to investigations of possible methods for reducing the skin friction in turbulent boundary layers by 'manipulating' the turbulence structure. These latter methods have been inspired by the growing knowledge of transient but coherent structures in turbulent shear flows revealed by recent advances in flow visualisation techniques and digital sampling of instantaneous flow data. This work has yet to be applied to operational aircraft but certain developments look promising and their operational use may not be far off.

In what follows we shall briefly summarise some of the salient features of the above topics but with little mathematical detail. The reader is advised to study the relevant references should he or she wish to pursue these topics further.

10.2 Three dimensional boundary layers, general[10.1]

In three dimensions we use a coordinate system that conforms to the surface, and this involves a network of two families of lines, not necessarily orthogonal, mapped over the surface plus the normals to the surface. Hence, curvatures of the coordinates are introduced with associated centrifugal and Coriolis terms in the equations as well as metrics of the system.

However, in the simple case of a developable surface (one that can be obtained by rolling up a plane flexible sheet) with small surface curvature, cartesian networks can be conveniently used. Then, if x, y are the co-ordinates of the surface network and z is normal to the surface, with the same arguments as were developed in Section 2.7 together with the boundary layer assumptions, the mean equations of motion for incompressible flow are [cf. equation (2.32)]

$$\left.\begin{aligned} \rho \mathrm{D}u/\mathrm{D}t &= -\partial p/\partial x + \partial \tau_{xz}/\partial z \\ \rho \mathrm{D}v/\mathrm{D}t &= -\partial p/\partial y + \partial \tau_{yz}/\partial z \end{aligned}\right\} \qquad (10.1)$$

where in laminar flow

$$\tau_{xz} = \mu\, \partial u/\partial z, \qquad \tau_{yz} = \mu\, \partial v/\partial z$$

and in turbulent flow

$$\tau_{xz} = \mu\, \partial u/\partial z - \rho\overline{u'w'}, \qquad \tau_{yz} = \mu\, \partial v/\partial z - \rho\overline{v'w'} \qquad (10.2)$$

The resultant shear stress $\boldsymbol{\tau} = [\tau_{xz}, \tau_{yz}]$.

The boundary layer assumptions require that $w \ll u$ or v, and $\partial p/\partial z = 0$. They are readily satisfied for simple attached boundary layers on nearly flat surfaces but require critical examination and, if necessary, revision in regions such as wing–body junctions, wing tips, on shapes of large surface curvature or in flows approaching separation. As already noted, the methods of predicting attached two dimensional boundary layers, both integral and differential, can in principle be extended to include the additional equation of three dimensional boundary layers, given the requisite boundary conditions.

For more general surfaces a popular system of orthogonal coordinates involves the projections on the surface of a family of external streamlines just outside the boundary layer and an orthogonal family of lines on the surface. The latter are projections of equi-potential lines if the external flow is irrotational. Such a system is usually referred to as *streamline coordinates*.

In general, if there is a well defined undisturbed stream direction, x, the flow that develops in the transverse orthogonal direction, y, is referred to as *secondary flow* and its presence implies the development of streamwise vorticity, a feature that, of course, is not present in two dimensional flow. With streamline coordinates the secondary flow is usually called *cross-flow*. We note that then the cross-flow velocity, v, is zero at the surface and also at the outer edge of the boundary layer where additionally $\partial v/\partial z = 0$ [see Fig. 10.1(b)]. Cross-flows are readily generated by the lateral pressure gradients associated with lateral curvature of the external streamlines, where these gradients are balanced by the centrifugal pressure gradients. Since these gradients are independent of z, it follows that in the boundary layer where the velocity is less than in the external flow the flow curvature is greater than externally and hence a cross-flow tends to develop towards the external flow centre of curvature [Fig. 10.1(a)]. The final balance there is achieved by a combination of centrifugal, viscous and convection effects.

If the lateral pressure gradient changes sign at some stage along an external streamline the corresponding cross-flow changes tend to lag behind because of the viscous and convection effects. We then tend to get cross-over or S-shaped profiles for the cross-flow for some streamwise distance after the pressure gradient sign has changed [see Fig. 10.1(c)].

When the cross-flow velocity is small, i.e. $v \ll u$, it can be shown that for laminar flow the streamwise flow equations are uncoupled from the cross-flow and are formally those for the boundary layer on a body of revolution at zero incidence, which can be solved by the methods for two dimensional flow.

For the laminar boundary layer on a developable surface in incompressible flow a number of similar solutions have been obtained[10.2] paralleling the Falkner–Skan solutions of two dimensional flow (see Section 3.4). These are useful as test cases for checking approximate prediction methods but in general they involve rotational external flows.

For integral methods using streamline coordinates assumptions must be made about the forms of the cross-flow velocity distributions as functions of z. In the case of the laminar boundary layer these, like those of the streamwise component (see Chapter 4), are usually in the form of polynomials or transcendental functions involving a few non-dimensional parameters to be determined from the solutions of the integral equations.[10.3] In the case of the turbulent boundary layer various simple, empirical forms have been used, e.g. Mager's profile:[10.4, 10.5]

$$v/u = \tan \beta_0 (1 - z/\delta)^2 \qquad (10.3)$$

where β_0 is the angle between the local x direction and the local limiting surface streamline (the direction of the resultant shear stress at the

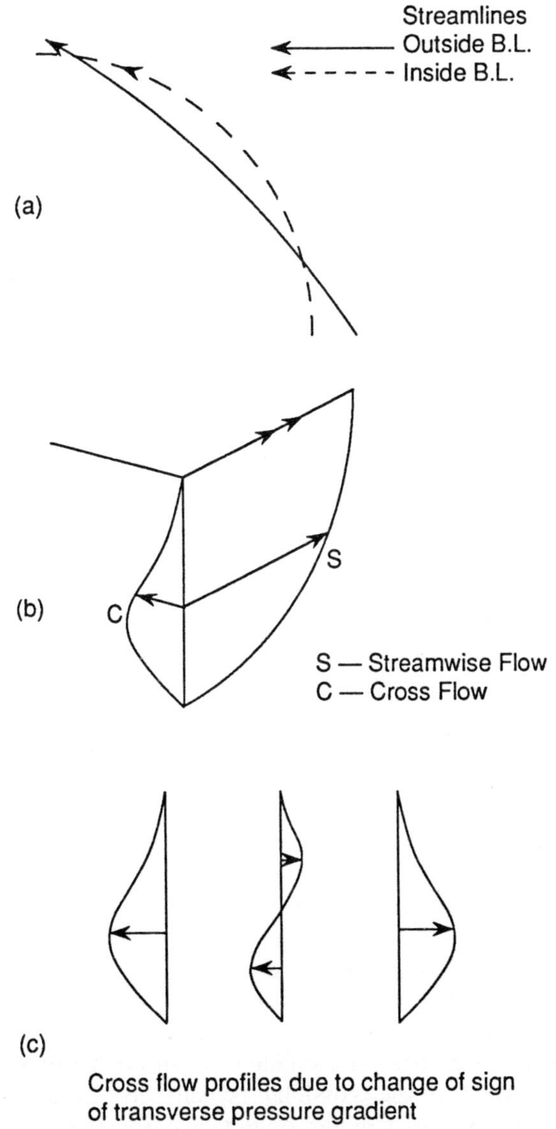

Fig. 10.1 Cross-flows in three dimensional boundary layers.

surface). It will be evident that such a profile form does not include cross-over profiles and so cannot accurately represent cases where the external pressure gradients change sign.

A more satisfactory if less simple approach is that of Cousteix[10.6] who made use of extensions of self similar (or equilibrium) boundary layers (cf. Section 6.5). Another approach is that of Cross[10.7] who has developed a modification of Cole's combination of the law of the wall with his law of the wake (cf. Section 6.5). Cross assumed that the law of the wall component lined up with the limiting streamline but the wake component direction changed with distance from the surface in a manner that he determined empirically.

For differential methods involving the eddy viscosity concept it has been usual to assume that the eddy viscosity is isotropic so that

$$\tau_{xz}/\tau_{yz} = (\partial u/\partial z)/(\partial v/\partial z)$$

This is equivalent to assuming that the shear stress and velocity gradient vectors coincide in direction. However, experimental evidence shows that these vectors do not coincide, particularly where the rate of change of flow conditions with streamwise distance is large. Rotta[10.8] has plausibly introduced a simple discriminating factor T in the shear stress/rate of strain relations which resulted in non-isotropy, but it has not been found to consistently improve the agreement with experiment nor is it found to be constant across a boundary layer.

For useful discussions of the main types of calculation method for three dimensional turbulent boundary layers that have been developed and a comparison of their predictions with experiment the reader is referred to references 10.9–10.11. Not surprisingly, the level of agreement is generally less satisfactory than for two dimensional flow and it is clear that the turbulence models used need to be considerably improved. Here it is as well to note certain special features of three dimensional boundary layers that future modelling processes and prediction methods must reflect.

We have already noted that cross-flows or secondary flows and the associated streamwise vorticity are physical features which render three dimensional boundary layers different from two dimensional boundary layers. We have seen that cross-flows can be generated by lateral pressure gradients associated with flow curvature (skew-induced): however, they can also be generated by turbulent eddy stresses (e.g. in the flow in non-circular sectioned pipes or ducts) although such sources of streamwise vorticity are often small compared with lateral pressure gradients. A factor that has perhaps not been sufficiently stressed or explored as yet is the possible effect the associated lateral rates of strain and curvature may have on the dynamics of the turbulence. Experiments have shown significant changes in turbulence intensity and related characteristics in the presence of lateral curvature.[10.12] It seems plausible that a lateral pressure field in combination with mean rates of strain may stabilise or destabilise the turbulent flow in much the same way as longitudinal curvature has been shown to do. If this is the case then it implies that turbulence models should include other rates of strain additional to $\partial u/\partial z$ and $\partial v/\partial z$, such as $\partial u/\partial y$ and $\partial v/\partial y$.

A related feature of three dimensional boundary layers that also needs emphasising is that initial conditions at leading edges, etc., vary in general with spanwise distance and these variations can lead to the initiation of streamwise vorticity. Given the right conditions for the rapid growth of this component, such as instability due to flow curvature, strong vortices can develop from seemingly small beginnings. These vortices can have major effects on the downstream aerodynamic loading distribution and the consequent aerodynamic characteristics of the aircraft. Such effects are indeed associated with leading edge notches and strakes and can be deliberately introduced as design features to be exploited to improve aerodynamic performance. However, it is known that the development of leading edge vortices can sometimes be triggered by minute surface changes and our knowledge of this triggering process, and of how it can be controlled, is still very inadequate.

Finally, we have already noted that separation of three dimensional boundary layers is a much more complex phenomenon than for two dimensional boundary layers. For the latter it is associated for steady flow with the simple local criterion of zero surface shear stress or $\partial u/\partial z = 0$. In three dimensions[10.13, 10.14] the vanishing of both shear stress components at the surface takes the form of:

(1) nodes, with the limiting streamlines (or friction lines) converging (separation) or diverging (reattachment) to or from the node, but with one streamline to which the other streamlines are tangential [see Fig. 10.2(a)];
(2) saddle points, each of which has two characteristic streamlines through it such that the streamlines on either side of one of the characteristic streamlines merge with it (separation) or emerge tangentially from it (attachment), except that the other characteristic streamline crosses the first [see Fig. 10.2(b)];
(3) spiral foci or nodes, where the streamlines follow spiral forms emerging (attachment) or converging into (separation) the node [Fig. 10.2(c)].

Two nodal points of attachment will have a saddle point of separation between them and the characteristic line tangential to the streamlines of the latter can be regarded as a separation line from which a separation (vortex) sheet will spring. Spiral points of separation generally result in a vortical flow emerging from the surface.

Separated vortical flows in three dimensions are not necessarily undesirable, they can be well defined and steady engendering helpful lift and at an acceptable cost in drag.

10.3 Transition in three dimensional boundary layers

The prediction of transition in three dimensional boundary layers is even more difficult and less certain than in two dimensions. All the physical

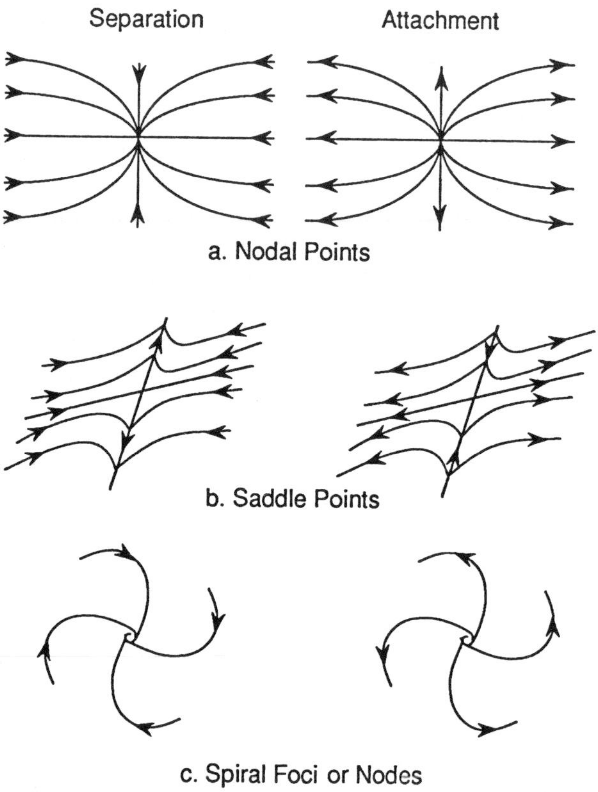

Fig. 10.2 Sketches illustrating different types of limiting streamlines near singularities (zero shear stress).

features and unresolved problems of the latter as discussed in Chapter 5 are present but in addition there are other factors more particularly associated with wing sweep that have to be taken into account.

We have seen that cross-flow generally arises in a three dimensional boundary layer and the cross-flow velocity profile has at least one point of inflection. Therefore, even if the streamwise profile is uninflected there will be a range of planes normal to the surface in which the component velocity profile, resolved from both the cross-flow and streamwise profiles, will be inflected. Theory then demonstrates that in such planes the flow is unstable to a corresponding range of small disturbances in the form of Tollmien–Schlichting waves above a related range of critical Reynolds numbers, as in two dimensional flow. Hence, the presence of cross-flow ensures that at all but relatively low Reynolds numbers there is a family of directions for which disturbances can grow and can therefore generate turbulence. The turbulence is then convected and spreads downstream.

Thus, cross-flow can be a ready source of transition to turbulence and this tendency is known as *cross-flow instability*.

This instability can therefore be expected wherever there are regions of high lateral curvature of the external streamlines such as just aft of the leading edge of a swept wing. At the leading edge (see Fig. 10.3) the external chordwise velocity component is zero and then rapidly increases to a peak with distance downstream. The external spanwise component, however, changes relatively little in that region and so the local streamlines are therefore in the spanwise direction immediately at the leading edge and then turn rapidly towards the chordwise direction over a relatively short distance aft of the leading edge. This is therefore a region in which we can expect a strong cross-flow to develop. The lateral flow curvature will depend on the wing section and the sweep angle, it tends to increase with reduction of the nose radius of curvature and with increase of the wing sweep angle. A useful criterion for cross-flow instability is the cross-flow Reynolds number χ based on the local maximum cross-flow velocity and the boundary layer thickness, it seems that the critical value for χ varies with the form parameter H of the streamwise flow but lies in a range from about 250 to 350.[10.15] For other criteria see Arnal.[10.16]

Another feature of swept wings is what is known as *leading edge contamination*. The boundary layer thickness at the leading edge is finite (see Sections 3.4 and 4.2.2) and increases with the nose radius of curvature. In the presence of the spanwise flow it can be rendered turbulent by finite disturbances from further inboard due to excrescences of some kind there or from the wing–fuselage junction where the fuselage boundary layer can be expected to be turbulent. Once the flow somewhere along the

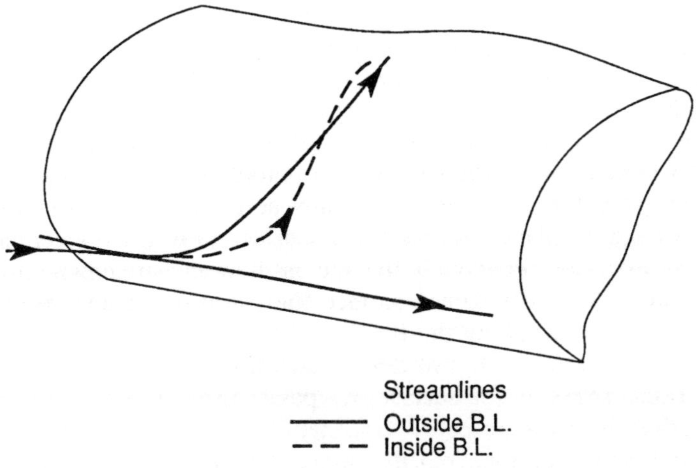

Fig. 10.3 Flow near leading edge of swept wing.

leading edge is turbulent it can engender turbulent flow further outboard as well as in the downstream boundary layer developing from that region of the leading edge – hence the term leading edge contamination. A criterion that has been postulated is that there is a certain critical value of the Reynolds number, $\tilde{R}$, based on the external flow spanwise velocity component and the spanwise momentum thickness of the boundary layer on the leading edge. For values of $\tilde{R}$ greater than the critical it is postulated that leading edge contamination can occur provided the disturbance source is strong enough. It has been suggested on the basis of experiments that the critical value of $\tilde{R}$ is about 250. In assessing the transition position in a three dimensional boundary layer one must therefore take account of at least three factors, the Tollmien–Schlichting waves acting on the streamwise flow as in two dimensions, cross-flow instability and leading edge contamination.

We have already noted that in recent years there has been a revival of interest in the development of wings with extensive regions of laminar flow in order to reduce the overall drag. It is fortunate that surface suction of the boundary layer (see Section 5.10), in reducing boundary layer thickness and in removing inflections from the velocity profile, is also effective in reducing or eliminating both cross-flow instability and leading edge contamination. It appears that the amount of suction required for a typical transport aircraft is small enough not to result in a serious performance penalty.[10.17]

10.4 Inverse methods

We have seen that separation of the boundary layer is associated with the convection of vorticity away from the surface, so that unless the separated flow region is very limited and shallow, the normal velocity component w can there become of the same order as u. Then $\partial w/\partial z$ near the surface can become large and so by continuity either or both $\partial u/\partial x$ and $\partial v/\partial y$ can become large and the boundary layer assumptions become inapplicable. One is not, therefore, surprised that the boundary layer equations exhibit singular behaviour there. We may also note that the boundary layer equations are parabolic, so that the flow at a point is determined by upstream or local conditions only, but in a separated region with fluid moving upstream close to the surface the downstream conditions must play a part in determining the flow.

We recall that in steady two dimensional flow separation occurs where the frictional stress on the surface, τ_w, passes through zero from positive to negative. In a classic paper Goldstein[10.18] showed in steady two dimensional flow that the velocity component u, the stress τ_w and the displacement thickness in a laminar boundary layer varied as $(x_s - x)^{1/2}$

and w varied as $1/(x_s - x)^{1/2}$, where x_s denotes the separation point, so that the boundary layer equations as normally formulated could not be integrated through and beyond that point. However, it seems from later work by Stewartson[10.19] that there are certain conditions involving the external velocity (or pressure) distribution under which the singularity at the separation point need not arise. It has been postulated that in practice such conditions usually hold.

In any case, the approach to separation by computation using finite differences often runs into considerable difficulties because of the rapid increase in sensitivity to errors which reflects the fact that the equations are not well conditioned there. The so-called direct viscous–inviscid coupling method involves an iterative process to reconcile the inviscid pressure distribution past the surface plus displacement thickness (or equivalent transpiration velocity) with the corresponding calculated boundary layer displacement thickness. It is often impossible to take this process beyond the separation point.

However, an alternative method that appears to work well, at least for shallow separated flows of limited extent, is the so called *inverse method*.[10.20–10.22] Here, the distribution of an overall characteristic quantity of the boundary layer is postulated, such as the displacement thickness or skin friction, and the pressure (or external velocity) distribution is inferred from the numerical solution of the resulting equations. This process, like the direct method, is usually iterative. Inverse methods appear to work well where the departures from the boundary layer assumptions are not too severe and the sensitivity to downstream conditions is not strong. It is usual to combine direct methods for the attached flow regions with inverse methods to deal with the separated flows up to reattachment. Both integral and differential methods have been developed in inverse forms.

It should be noted that the inverse method is the natural one for dealing with the internal flows in ducts and pipes. In such flows boundary layers develop along the walls but the central flow that defines the outer boundary conditions for the boundary layers can only be determined from the integrated continuity equation ensuring constant mass flow rate along the duct or pipe.

10.5 Some concepts for reducing skin friction in the turbulent boundary layer

10.5.1 Coherent structures in turbulent shear flow

We have noted that the representation of the turbulence in a boundary layer as a stochastic process to be treated statistically is useful but somewhat simplistic since it ignores potentially important phenomena. Modern experimental techniques of flow visualisation and methods of conditional sampling, (i.e. registering the instantaneous incidence and magnitude of quantities, e.g. the velocity components, when their values are greater than a set threshold) have demonstrated the existence of coherent structures in the boundary layer.[10.23–10.27] These are transient but last long enough to be recognised as recurring flow forms. These forms are slow moving vortical streaks very close to the surface, bursts or sweeps of rapidly moving jets of fluid a little further out in the boundary layer and large eddies in the outer parts of the boundary layer. The streaks are often in opposite handed pairs and tend to lift at their downstream end into the log law region where they appear to join to form hairpin vortices, there they are stretched and become more intense. The bursts may well be due to the transient secondary instabilities associated with inflectional velocity profiles induced by the lifting vortices. They are accompanied by locally high values of the Reynolds shear stress, and after they have been convected downstream they may be followed by a relatively long period of calm and low Reynolds shear stress, so that the maximum stress at a point may be orders of magnitude greater than the mean value. The large eddies are coagulations of vortical elements and their size is of the same order as the mean boundary layer thickness.

We are still far from understanding or explaining the nature of the interactions of these coherent structures and how they complement each other but it is believed that by changing one of them the others will be modified because of their interdependence. It follows that by 'manipulating' one of them it should be possible to modify the mean shear stress and in particular the skin friction at the surface.

These ideas have received considerable attention in recent years and have stimulated some interesting research, particularly in the US, and two developments have shown significant progress and may well find application on operational aircraft in the next decade or so.[10.28] One of these developments aims to confine and order the streaks in the sub-layer by means of fine streamwise grooves or *riblets*. The other is directed at the large eddies – small aerofoils in the form of stiff ribbons are mounted in the boundary layer more or less parallel to the surface and normal to the mean flow. They are called *LEBUs – large eddy break-up devices*. The aim is to modify the large eddy structures by seemingly 'decapitating' them, although the relatively small scale vortices shed into the wakes of the ribbons must be an important factor in the resulting effects. We will say a little about both these developments.

10.5.2 Riblets

A variety of geometries of riblets have been tested and it appears that near optimum results are obtained with riblets of simple triangular section with $h^+ = hu_\tau/\nu \simeq 10-15$, and $s/h \simeq 1-2$, where h is the height of a riblet and s is the distance between adjacent riblets. In practice this means that the grooves are barely perceptible to the touch and of the order of 0.05 mm in depth at full scale Reynolds numbers. Fig. 10.4 illustrates the order of skin friction reduction that has been measured as a function of $s^+ = su_\tau/\nu$, the latter parameter offers the best collapse of the experimental data.

It will be seen that reductions of the order of 8% result from the optimum arrangement of riblets, but for s^+ greater than about 30 the riblets increase the skin friction drag. With increase of the Reynolds number the optimum height decreases and one must expect some decrease of the drag reduction.

It has been argued that these results may be simply explained as due to the fact that the slow moving part of a boundary layer will be thickened over a groove centre line and the frictional stress will be correspondingly reduced whilst the opposite effects will arise over a groove apex. These effects are non-linear and the former effects dominate, notwithstanding the actual increase in surface area due to the grooving. Gaudet[10.29] has examined data obtained at the RAE and has found that the riblet results are similar to those of small sand roughnesses insofar as the plots of u/u_τ against $z^+ = zu_\tau/\nu$ are parallel to and displaced from the corresponding

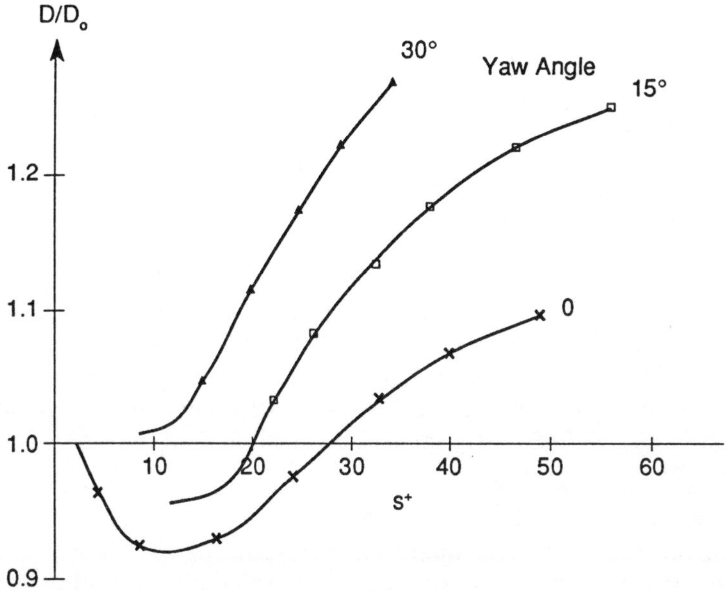

Fig. 10.4 Some typical results for effect of V-groove riblets on drag of flat plate. D = drag with riblets, D_0 = drag of smooth plate.

smooth surface plot (see Section 6.12.2). The significant point is that the displacement can be up or down depending on whether the riblets cause a reduction or an increase of the skin friction, respectively. Gaudet inferred that the optimum overall skin friction drag reduction could be about 7% at a Reynolds number of about 10^7 (in terms of free stream velocity and wing chord) and about 5% at a Reynolds number of 10^9.

The available experimental data indicate that the sensitivity of the effects to yaw of the grooves relative to the local free stream direction is fairly small, it needs a yaw angle of about 30° to nullify the drag reduction (see Fig. 10.4). The possible reductions in frictional drag with riblets are important enough to justify further work on the engineering and operational problems of achieving and maintaining such surfaces.

10.5.3 Large eddy break up devices (LEBUs)

Here too various arrangements have been tested but a typical arrangement is sketched in Fig. 10.5(a). It will be seen to be a tandem arrangement of two flat stiff ribbons of rectangular section that approximates to an aerofoil section, each ribbon being of the order of 1–2 boundary layer thicknesses (δ) in chord, $4-8\delta$ apart and set at a height of $\delta/2-3\delta/4$ above the surface.

Typical results for the skin friction distribution measured downstream of the device and with $h/\delta \simeq 0.6$, where h is the height of the ribbons above the surface, are shown in Fig. 10.5(b). It will be seen that the device results in a significant reduction of skin friction which extends for well over 100δ downstream and is most marked at about 30δ. The latter position varies with h/δ. The overall reduction in skin friction drag can be of the order 5–10%. The wake from the device evidently spreads downstream as it mixes with the boundary layer slowing the latter and this may contribute to the reduction of the skin friction. Smoke visualisation experiments show a marked smoothing of the flow in the outer parts of the boundary layer with the upper parts of the large billowing eddies replaced by more steamline flow. There is evidence that the drag reduction as a percentage of the basic smooth surface drag increases with Reynolds number.

However, we must take account of the drag increment due to the drag of the ribbons and their supports and as yet no device has been developed and tested which shows a net reduction of drag. These complete drag tests have been done in wind tunnels at relatively low Reynolds numbers where the ribbons themselves may have a high drag. It is to be expected that at full scale Reynolds numbers, corresponding to a large transport aircraft, the drag of the ribbons will be considerably reduced and hence there can be a significant net gain. There is also room for improvement by suitably shaping the cross-section of the ribbons by chamfering the trailing edge and rounding the leading edge.

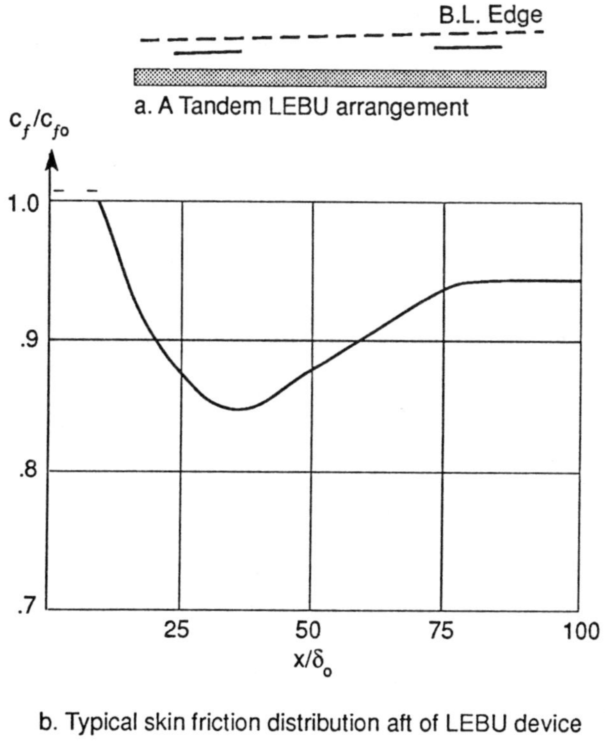

Fig. 10.5 Typical LEBU arrangement and results. c_{f0} is turbulent skin friction coefficient without LEBU.

There are clearly problems involved in the operational use of LEBUs, they must be made strong and stiff enough to withstand the loads on them and avoid serious aeroelastic effects. One can conclude that such devices may well find application for production aircraft in due course but riblets are likely to be applied earlier.

References

10.1 Cooke, J.C. and Hall, M.G. (1962) 'Boundary layers in three dimensions'. *Prog. Aero. Sc.*, **2**, p. 222.

10.2 Hansen, A.G. (1964) *Similarity Analysis of Boundary Value Problems in Engineering*. Prentice Hall, NJ.

10.3 Smith, P.D. and Young, A.D. (1969) 'Approximate solutions of the three dimensional laminar boundary layer momentum integral equations'. ARC CP 1064.

10.4 Mager, A. (1952) 'Generalisation of the boundary layer momentum integral equations to three dimensional flows including those of a rotating system'. NACA Rep. 1067.
10.5 Smith, P.D. (1972) 'An integral prediction method for three dimensional compressible turbulent boundary layers'. ARC RM 3739.
10.6 Cousteix, J. (1974) 'Analyse théorique et moyens de prévisions de la couche limite turbulente tri-dimensionalle'. ONERA Publ. 157.
10.7 Cross, A.G.T. (1983) 'Three dimensional turbulent boundary layer calculations presented at the 1982 Eurovisc Workshop'. B. Ae. Brough Note YAD 5020.
10.8 Rotta J.C. (1977) 'A family of turbulence models for three dimensional thin shear layers'. *Proc. 1st Symp. Turbulent Shear Flows*. **1**, p. 1027. Pennsylvania State Univ.
10.9 Humphreys, D.A. and Lindhout, J.P.F. (1988) 'Calculation methods for three dimensional turbulent boundary layers'. *Prog. Aero. Sc.*, **25**, 3, p. 107.
10.10 Van den Berg, B., Humphreys, D.A., Krause, E. and Lindhout, J.P.F. (1985) 'Comparison of three dimensional turbulent boundary layer calculations with experiment'. NLR MP 85023 U.
10.11 'Computation of Three Dimensional Boundary Layers including Separation'. AGARD Rep. 741 (1986).
10.12 Van den Berg, B., Elsenaar, A.E., Lindhout, J.P.F. and Wesseling, P. (1982) 'Measurements in an incompressible three dimensional turbulent boundary'. *JFM*, **70**, p. 431.
10.13 Lighthill, M.J. (1963) *Laminar Boundary Layers* (ed. Rosenhead). OUP, Ch. II.
10.14 Peake, D.J. and Tobak, M. (1980) 'Three dimensional interactions and vortical flows with emphasis on high speeds'. AGARDograph 252.
10.15 Poll, D.I.A. (1984) 'Transition description and prediction in three dimensional flows'. AGARD Rep. 709, paper 5.
10.16 Arnal, D. (1981) 'Three dimensional boundary layers, laminar-turbulent transition'. AGARD Rep. 741, paper 4.
10.17 Reshotko, E. (1984) 'Laminar flow control and viscous simulation'. AGARD Rep. 709, paper 8.
10.18 Goldstein, S. (1948) 'On laminar boundary layer flow near a position of separation'. *Qu. J. Mech.*, **1**, p. 43.
10.19 Stewartson, K. (1958) 'On Goldsteins' theory of laminar separation'. *Qu. J. Mech.*, **11**, p. 399.
10.20 Catherall, D. and Mangler, K. (1966) 'The integration of the two dimensional laminar boundary layer equations past the point of vanishing skin friction'. *JFM*, **26**, p. 163.
10.21 Horton, H.P. (1975) 'Numerical investigation of regular laminar boundary layer separation'. AGARD CP 168, paper 7.
10.22 Horton, H.P. (1981) 'Comparisons between inverse boundary layer calculations and detailed measurements in laminar separated flows'. *Aero. Qu.*, Aug, p. 168.
10.23 Kline, S.J., Reynolds, W.C., Schraub, F.A. and Rundstadtler P.W. (1967) 'The structure of turbulence'. *JFM*, **30**, p. 741.

10.24 Roshko, A. (1976) 'Structure of turbulent shear flows; a new look'. *AIAA J*, **14**, p. 10.
10.25 Falco, R.E. (1977) 'Coherent motions in the outer regions of turbulent boundary layers'. *Phy. of Fluids*, **20**, p. 124.
10.26 Head, M.R. and Bandyopadhyay, P. (1983) 'New aspects of turbulent boundary layer structure'. *JFM*, **107**, p. 297.
10.27 Fiedler, H.E. (1988) 'Coherent structures in turbulent flows'. *Prog. Aero. Sc.*, **25**, 3, p. 231.
10.28 'Improvement of aerodynamic performance through boundary layer control and high lift devices'. AGARD CP 365 (in particular papers 11, 17, 19) (1984).
10.29 Gaudet, L. (1987) 'An assessment of the drag reduction properties of riblets and the penalties of off design conditions'. RAE Tech Memo Aero 2113.

Index

Bold indicates main entries, subsequent pages should also be referred to.